Phthalocyanines

Properties and Applications

Phthalocyanines

Properties and

Applications

Edited by

C. C. Leznoff and A. B. P. Lever

o 3671847

CHEMISTRY

C. C. Leznoff
Department of Chemistry
York University
North York, Ontario M3J 1P3
CANADA

A. B. P. Lever
Department of Chemistry
York University
North York, Ontario M3J 1P3
CANADA

Library of Congress Cataloging-in-Publication Data

Phthalocyanines: properties and applications/edited by C. C. Leznoff and A. B. P. Lever.

   p.   cm.
   Bibliography: p.
   Includes index.
   ISBN 0-89573-753-1
   1. Phthalocyanines. I. Leznoff, C. C. (Clifford C.) II. Lever, A. B. P. (Alfred Beverley
Philip)
QD441.P37   1989
547.8′6—dc20   89-16518
                                                          CIP

British Library Cataloguing in Publication Data

Phthalocyanines: properties and applications.

   1. Phthalocyanines
   I. Leznoff, C. C. II. Lever, A. B. P. (Alfred Beverley Philip)
547.8′6

ISBN 3-527-26955-X W. Germany

© 1989 VCH Publishers, Inc.

Printed in the United States of America

ISBN 0-89573-753-1 VCH Publishers
ISBN 3-527-26955-X VCH Verlagsgesellschaft

Printing History:
10 9 8 7 6 5 4 3 2 1

Published jointly by:

VCH Publishers, Inc.
220 East 23rd Street
Suite 909
New York, NY 10010

VCH Verlagsgesellschaft mbH
P.O. Box 10 11 61
D-6940 Weinheim
Federal Republic of Germany

VCH Publishers (UK) Ltd.
8 Wellington Court
Cambridge CB1 1HW
United Kingdom

QD 441
P 371
1989

CHEM

Contents

Contributors

William R. Barger, Chemistry Division, Naval Research Laboratory, Washington, D.C. 20375, U.S.A.

Ehud Ben-Hur, Department of Radiobiology, Nuclear Research Center-Negev, Beer Sheva, Israel

G. Ferraudi, Radiation Laboratory, University of Notre Dame, Notre Dame, Indiana 46556, U.S.A.

C. C. Leznoff, Department of Chemistry, York University, North York, Ontario M3J 1P3, Canada

Tebello Nyokong, Department of Chemistry, National University of Lesotho, Roma, South Africa

Ionel Rosenthal, Department of Food Science, Agricultural Research Organization, The Volcani Center, Bet Dagan 50250, Israel

Arthur W. Snow, Chemistry Division, Naval Research Laboratory, Washington, D. C. 20375, U.S.A.

Martin J. Stillman, Department of Chemistry, University of Western Ontario, London, Ontario N6A 5B7, Canada

Dieter Wöhrle, Institut für Organische und Makromolekulare Chemie, Universität Bremen, Bremen, Federal Republic of Germany

Series Preface

Since their synthesis early this century, phthalocyanines have established themselves as blue and green dyestuffs par excellence. They are an important industrial commodity (output 45,000 tons in 1987) used primarily in inks (especially ballpoint pens), coloring for plastics and metal surfaces, and dyestuffs for jeans and other clothing. More recently their use as the photoconducting agent in photocopying machines heralds a resurgence of interest in these species. In the coming decade, their commercial utility is expected to have significant ramifications. Thus future potential uses of metal phthalocyanines, currently under study, include (I) sensing elements in chemical sensors, (II) electrochromic display devices, (III) photodynamic reagents for cancer therapy and other medical applications, (IV) applications to optical computer read/write discs, and related information storage systems, (V) catalysts for control of sulfur effluents, (VI) electrocatalysis for fuel cell applications, (VII) photovoltaic cell elements for energy generation, (VIII) laser dyes, (IX) new red-sensitive photocopying applications, (X) liquid crystal color display applications, and (XI) molecular metals and conducting polymers.

In recent years there has been a growth in the number of laboratories exploring the fundamental academic aspects of phthalocyanine chemistry. Interest has been focused, inter alia, on the synthesis of new types of soluble and unsymmetrical phthalocyanines, on the development of new approaches to the synthesis of polynuclear, bridged, and polymeric species, on their electronic structure and redox properties, and their electro- and photocatalytic reactivity.

It is timely therefore to consolidate these academic and applied aspects of phthalocyanine chemistry into a series of monographs that will present our current knowledge of this fascinating area of chemistry. In this series of monographs, we plan to bring together for the first time detailed critical coverage of the whole field of phthalocyanine chemistry in a sequence of chapters that should prove stimulating for industrial, medical, and academic researchers.

C. C. Leznoff
A. B. P. Lever
March 1989

Preface

This is the first volume of a planned series that will cover all areas of phthalocyanine chemistry at the frontier of the science. This first volume begins with a detailed critical survey of the synthetic methods used to prepare and characterize metal-free phthalocyanines (Leznoff). Emphasis is concentrated on the design and synthetic strategies of new systems and their purification and characterization. This chapter describes most known characterized metal-free phthalocyanines including patents and poorly accessible references, and provides the entrée to begin studies in the field. This is followed by a survey of polymer bound and polymeric phthalocyanines (Wöhrle) outlining the variety of such species. The chapter is subdivided by the type of polymer and deals with synthesis, characterization, and utility. Chapter 3 (Stillman and Nyokong) deals with the mammoth task of presenting and analyzing the electronic and magnetic circular dichroism spectra of metal-free and metallophthalocyanines in their standard Pc(−2) oxidation state. Since color is the single most important attribute to phthalocyanine chemistry, this chapter will prove invaluable to all researchers in the field.

Ferraudi (Chapter 4) discusses the photochemical and physical properties of the metallophthalocyanines in homogeneous solution. This is the first in a series of several chapters that will explore the electronic properties of these molecules. Chapters 5 and 6 present more applied aspects of phthalocyanine chemistry. Snow and Barger (Chapter 5) deal with the exciting new applications of phthalocyanines to the field of chemical sensors with special emphasis on interdigital electrodes, Langmuir–Blodgett films, and surface acoustic wave devices. This volume closes with a review (Rosenthal and Ben-Hur) that concentrates on the potential uses of metallophthalocyanine species in photodynamic cancer therapy, but includes other uses such as phthalocyanines as staining media.

C. C. Leznoff
A. B. P. Lever
March 1989

1

Syntheses of

Metal-Free

Substituted

Phthalocyanines

C. C. Leznoff

A. INTRODUCTION

The first synthesis of a phthalocyanine was recorded in 1907 when Braun and Tcherniac heated o-cyanobenzamide at a high temperature [1]. The structure of this metal-free, unsubstituted phthalocyanine was determined only about a quarter of a century later by the comprehensive researches of Linstead [2–6] and the X-ray diffraction analyses of Robertson [7–9], while examining both metal-free phthalocyanine (1) and metallophthalocyanines (2) (Fig. 1).

The early studies on phthalocyanine (1) itself and particularly metalloph-thalocyanines (2) have been amply reviewed [10–12]. Although specialized aspects of phthalocyanine chemistry have been more recently summarized [13–17] and even collected in texts [18,19], no full description of the synthesis of metal-free, substituted phthalocyanines is currently available. In this chapter we will describe the common methods of preparing metal-free, substituted phthalocyanines and list in a series of tables most of the known metal-free substituted phthalocyanines and some of their analogs. Most importantly, we will indicate the level of *characterization and purification* described for each compound. Some of the early literature, particularly the patent literature, describes substituted metallophthalocyanines on which little purification was attempted and the level of characterization was limited to mention of the deep purple or dark blue color of the phthalocyanine produced. Although satisfactory for the purpose of making dyes, a color criterion is insufficient characterization for the modern chemist. Thus, in general, each of the compounds listed in Tables 1–7 in this chapter have at least a satisfactory elemental analysis or exhibit a parent ion in its mass spectrum or preferably both, especially for unsymmetrically substituted phthalocyanines for which elemental analysis alone can be unreliable. Other spectroscopic evidence such as the ultraviolet-visible (UV), nuclear magnetic resonance (NMR), and infrared spectra (IR) give one ancillary proof of structure and level of purification.

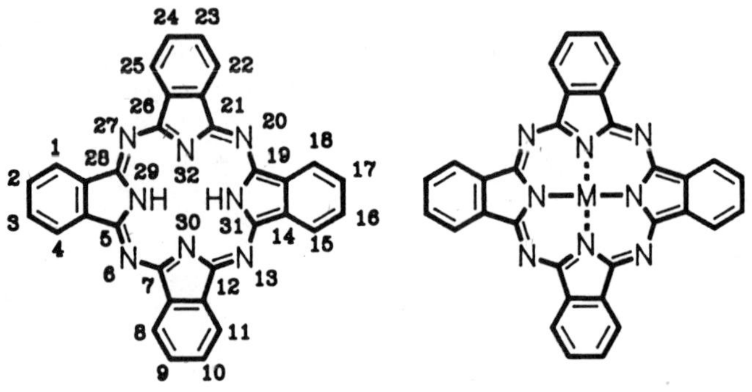

Figure 1 Metal-free phthalocyanine (1) and metallophthalocyanines (2).

B. METHODS OF PREPARATION

The very first synthesis of phthalocyanine (1) involved the reaction of *o*-cyanobenzamide (3) in refluxing ethanol [1] from which a blue 1 was recovered in low yield (Scheme I, Method IA). Linstead's group [2] confirmed this result and showed that yields up to 40% of 1 could be obtained if magnesium or antimony metal or magnesium salts such as its oxide and carbonate are mixed with 3 and heated over 230°C and the resultant metallophthalocyanine is demetallated with cold concentrated H_2SO_4 (Scheme I, Method IB). The use of substituted analogs of 3 to make substituted phthalocyanines has not been common.

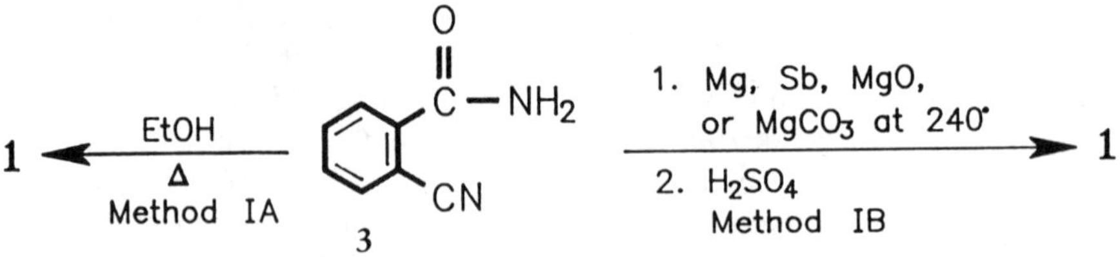

Scheme I

Although phthalonitrile (4) was not easily prepared in early studies [3], it was shown that 4 on treatment with sodium or lithium *n*-pentoxide in *n*-pentanol or other alcohols at 135–140°C (Method IIA) gave disodium phthalocyanine (2), which could be directly demetallated to phthalocyanine (1) with conc. H_2SO_4 [3,20] (Scheme II). In a variation of Method IIA 4 was treated with ammonia gas in 2-*N*,*N*-dimethylaminoethanol to give 1 in 90% yield without the necessity of acid treatment [21]. As substituted phthalonitriles are now readily prepared and since phthalocyanines are stable to strongly basic conditions [2], it is possible to prepare a wide variety of substituted phthalocyanines by this method. For example, 4-phenoxyphthalonitrile (5) and 4-thiophenoxyphthalonitrile (6) give 2,9,16,23-tetraphenoxyphthalocyanine (7) and 2,9,16,23-tetrathiophenoxyphthalocyanine (8) as mixtures of isomers (see Section E) in 39 and 25% yield, respectively [22]. In other variations of Method IIA, a solution of phthalonitrile (4) under standard conditions was preirradiated with UV light prior to heating giving phthalocyanine (1) in higher yields [22a], while substitution of 1,8-diazabicyclo[5.4.0]undec-7-ene (DBU) or 1,5-diazabicyclo[4.3.0]non-5-ene (DBN) as bases for alkoxides gave 1 in good yield [22b]. Phthalonitrile can be fused with magnesium or sodium metal above 200°C to give metallophthalocyanines [3,23], from which phthalocyanine (1) can be liberated by treatment with conc. H_2SO_4 (Method IIB, Scheme II). Using hydroquinone, tetrahydropyridine, or 4,4'-dihydroxybiphenyl as co-reactants, the substituted phthalonitriles 5 and 6 were fused at 180°C in a sealed tube to afford 7 and 8 in 81 and 43% yield, respectively [24,25] (Method IIC, Scheme II). Thus, although Methods IIA–C all use phthalonitriles as starting materials for phthalocyanine formation these are very different reactions and hence do not necessarily occur through similar mechanisms or identical intermediates (see Section D).

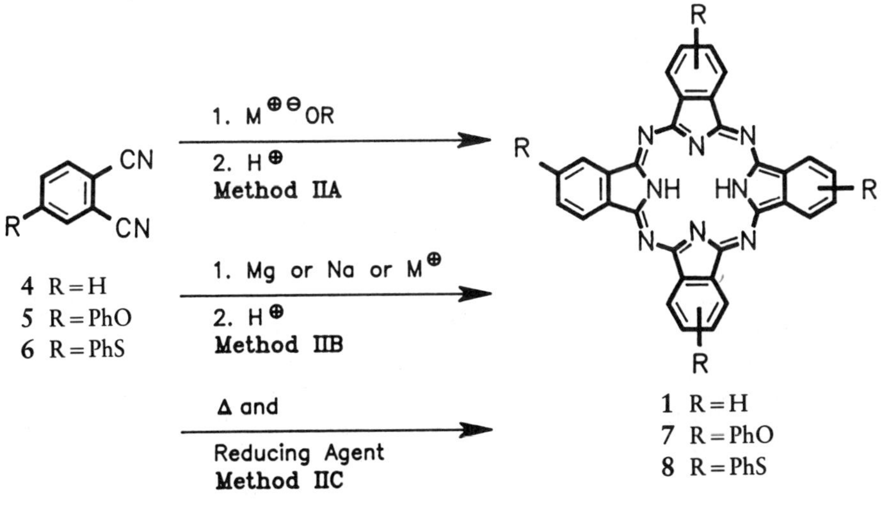

1. $M^{\oplus\ominus}OR$

2. $H^{\oplus}$
Method IIA

1. Mg or Na or $M^{\oplus}$

2. $H^{\oplus}$
Method IIB

Δ and

Reducing Agent
Method IIC

4 R=H
5 R=PhO
6 R=PhS

1 R=H
7 R=PhO
8 R=PhS

Scheme II

Phthalonitrile (4) can readily be converted into 1,3-diiminoisoindoline (9) by bubbling gaseous ammonia into a solution of 4 in sodium methoxide in methanol at room temperature [21], although refluxing methanol [26] or methanol-dioxane [27] may be used for insoluble phthalonitriles. Elvidge and Linstead [5] had shown that 1,3-diiminoisoindoline (9) on treatment with $NiCl_2$ in hot formamide gave the nickel metallophthalocyanine (2) in 96%

yield. When **9** was heated in the presence of a hydrogen donor such as succinonitrile or boiling tetralin, phthalocyanine (**1**) was produced in 34 and 45% yield, respectively (cf. Method IIC) (Scheme III, Method IIIA). It was later shown that **9** can be converted to **1** in 85% yield by simply refluxing **9** in 2-*N*,*N*-dimethylaminoethanol [21] (Scheme III, Method IIIB). Octasubstituted phthalocyanines, for example, have been readily formed by Method IIIB from 5,6-bis(ethoxymethyl)-1,3-diiminoisoindoline (**10**) or 5,6-bis(phenoxymethyl)-1,3-diiminoisoindoline (**11**) to give 2,3,9,10,16,17,23,24-octa(ethoxymethyl)-phthalocyanine (**12**) or 2,3,9,10,16,17,23,24-octa(phenoxymethyl)phthalocyanine (**13**) in 80% yield [28] (Scheme III).

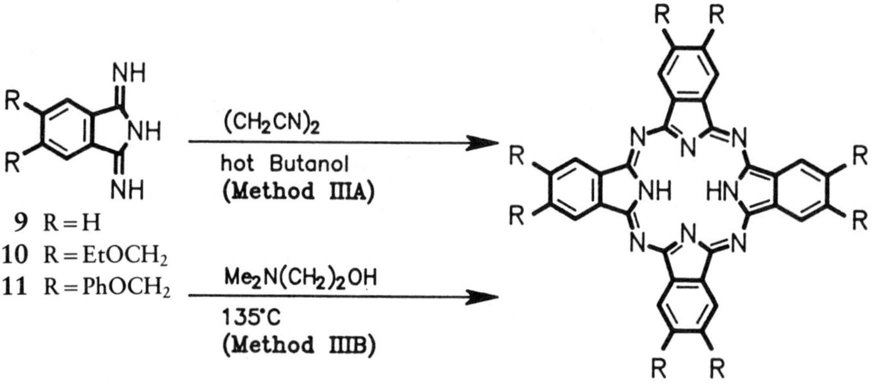

9 R=H
10 R=$EtOCH_2$
11 R=$PhOCH_2$

1 R=H
12 R=$EtOCH_2$
13 R=$PhOCH_2$

Scheme III

In the above examples cited, all the substituents on the phthalocyanine ring are identical. If one wanted to prepare phthalocyanines containing different substituents in some of the benzo rings, one could use a mixed condensation of two different substituted-1,3-diiminoisoindolines according to Method IIIB, but statistical mixtures of compounds, likely inseparable by chromatographic methods, would be bound to occur. If one of the substituted diiminioisoindolines were attached to an insoluble polymer and reacted with a large excess of a second diiminoisoindoline in solution, then an unsymmetrical substituted phthalocyanine could be formed exclusively on the polymer and subsequently liberated from the polymer as shown in Scheme IVa (Method IV). Thus, a polymer-bound trityloxyalkoxy-1,3-diiminoisoindoline (**14**) condensed with 5-isopropoxy-1,3-diiminoisoindoline (**15**) in a *mixed* diiminoisoindoline condensation in 2-*N*,*N*-dimethylaminoethanol to give, after Soxhlet extraction to remove 2,9,16,23-tetraisopropoxyphthalocyanine (**16**) from polymer-bound phthalocyanine (**17**) and mild acid treatment, the

Scheme IVa

unsymmetrical tetrasubstituted 2-(6′-hydroxyhexoxy)-9,16,23-triisopropoxy-phthalocyanine (18a) in 24% yield [29,30]. A 2-(4′-hydroxybutoxy)-9,16,23-triisopropoxyphthalocyanine (18b) was similarly made [29,30] (Method IVA). It was possible to vary this synthesis in such a way that the one different substituent was itself derived from a bisdiiminoisoindoline. In this way binuclear and multinuclear phthalocyanines can be formed. For example, treatment of the bis-1,3-diiminoisoindoline 19 with a large excess of 5-neopentoxy-1,3-diiminoisoindoline (20) gave 2,9,16,23-tetraneopentoxyph-thalocyanine (21) and the metal-free binuclear phthalocyanine 22 containing a 5-atom bridge [26,31] (Scheme IVb, Method IVB) in a 10% yield.

In method IVa, the polymer-bound unsymmetrically tetrasubstituted phthalocyanine 17 was separated from the symmetrically substituted phthalo-cyanine 16 by filtration of the polymer-bound phthalocyanine. In Method IVb, the difference between the molecular weight of the binuclear phthalo-cyanine 22 and the mononuclear phthalocyanine 21 allowed separation by chromatographic methods but conceptually both are *mixed* diiminoisoindoline condensations with special features allowing separation of the pure com-pound. We reserve the word *mixed* condensations to indicate two differently substituted precursors bearing the same functional groups, while *crossed* condensations refer to two differently substituted precursors having different functional groups.

In a closely related approach, a 20% macroporous divinylbenzene-styrene copolymer containing a 4-benzyloxyphenylphthalonitrile (23) reacted with an excess of 4-phenoxyphthalonitrile (5) in the presence of zinc acetate to give 2,9,16,23-tetraphenoxyphthalocyaninato zinc(II) (24) and polymer-bound metallated, unsymmetrical, tetrasubstituted phthalocyanine (25). Sox-hlet extraction of 25 removed all of 24 and acid cleavage of 25 gave pure 2-(4′-hydroxyphenyl)-9,16,23-triphenoxyphthalocyaninato zinc(II) (26) in 22% yield (Scheme V, Method V). As Methods IVA, B are mixed *diiminoisoindo-line* condensations this method is a mixed *phthalonitrile* method that exhibited in this example only some demetallation of 26 on acid cleavage but 26 was only characterized as its zinc derivative [32].

In a variation of Methods IV and V, 4-neopentoxyphthalonitrile (27) is condensed with the bisdiiminoisoindoline 28 in a *crossed* condensation in refluxing 2-N,N-dimethylaminoethanol to give a mixture of the monomeric phthalocyanine 21, the binuclear phthalocyanine 29 linked at the benzo rings by a single oxygen atom and a trinuclear phthalocyanine 30, in which the central ring is depicted as a 2,16-disubstituted isomer but is likely a mixture with a 2,9-disubstituted isomer [33] (Scheme VI, Method VI). In this procedure, trinuclear 30 is preferentially formed compared to dimer 29 as self-condensation of the diiminoisoindolines 28 is likely somewhat more favored than either the crossed condensation of 28 with 27 or the self-condensation of 27. Thus, one self-condensation of 28 followed by two crossed condensations with 27 gives trinuclear 30. A related approach toward a mononuclear phthalocyanine analog has been recorded [34,35].

Scheme IVb

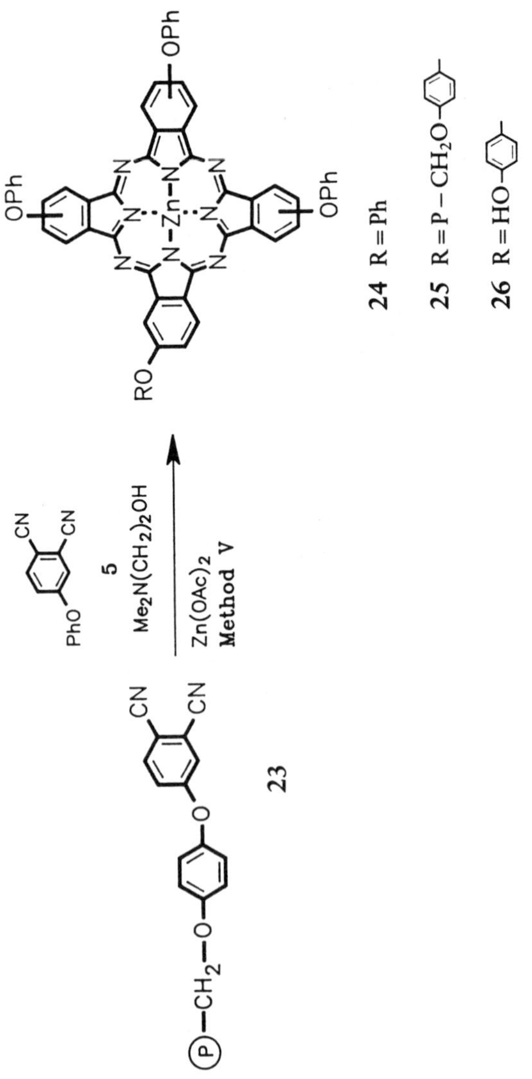

Scheme V

24 R = Ph

25 R = P–CH₂O–⬡

26 R = HO–⬡

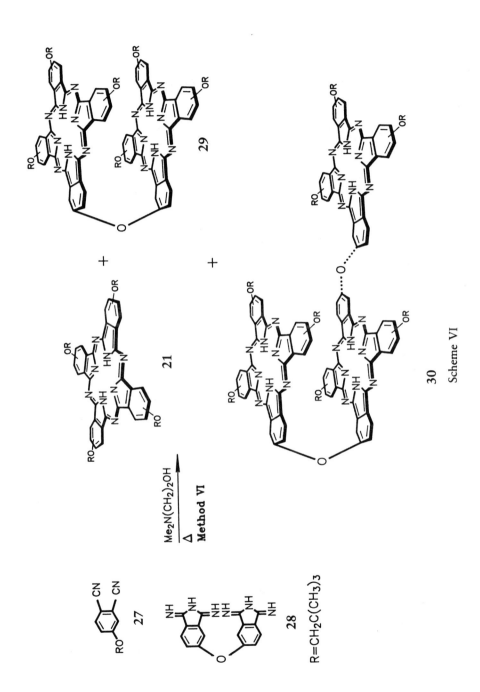

Scheme VI

Another interesting variation of a crossed condensation involving a diiminoisoindoline leads to one of the rare bonafide syntheses of a pure disubstituted phthalocyanine. When 5-phenyl-1,3-diiminoisoindoline (31) is treated with 1,3,3-trichloroisoindolenine (32) [36] at room temperature (rt) in the presence of an acid acceptor such as triethylamine and a reducing agent such as hydroquinone, almost pure 2,16- and 2,17-diphenylphthalocyanine (33) are produced in 7% yield [37]. Phthalocyanine 33 is characterized by both elemental analysis and mass spectroscopy and hence the determination of the structures of 33 is reliable. This observation has been confirmed in our own laboratory and only traces of trisubstituted phthalocyanine were found on examination of the phthalocyanine product by mass spectroscopy. In this synthesis, self-condensation of 31 at room temperature does not occur and self-condensation of 32 is unlikely so that only phthalocyanine 33 is produced. Interestingly no 2,9-diphenylphthalocyanine is produced by this method. Another patent [38] had also reported the use of 32 in making disubstituted phthalocyanines but their reaction mixture was *heated* under different conditions. Repetition of their conditions in our laboratory gave mixtures of un-, mono-, di-, tri-, and tetrasubstituted phthalocyanines as determined by mass spectroscopy (Scheme VII).

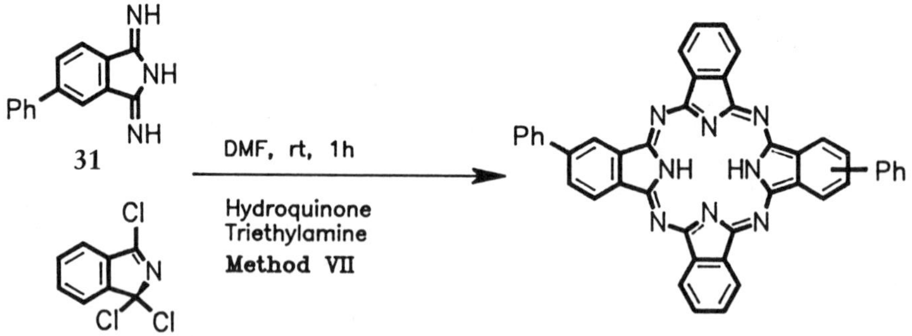

Scheme VII

In another example of a crossed condensation reaction involving a diiminoisoindoline, 1,3-diiminoisoindoline (9) reacted with 5-neopentoxy-1H-isoindole-1,3(2H)dithione (34) in a 1:1 or 15:1 ratio, in 2-N,N-dimethyl-aminoethanol for 24 h at 80–90°C to give in 10% yield mixtures of phthalocyanine (1), 2-neopentoxyphthalocyanine (35), 2,16- and 2,9-di-neopentoxyphthalocyanine (36 and 37), 2,16,23-trineopentoxyphthalo-cyanine (38) (as a mixture of isomers), and a trace of 2,9,16,23-tetraneopen-toxyphthalocyanine (21) (as a mixture of isomers) [39] (Scheme VIII, Method VIII).

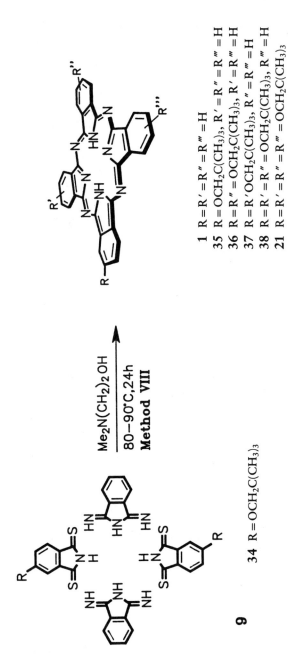

1 R = R' = R'' = R''' = H
35 R = OCH₂C(CH₃)₃, R' = R'' = R''' = H
36 R = R'' = OCH₂C(CH₃)₃, R' = R''' = H
37 R = R'OCH₂C(CH₃)₃, R'' = R''' = H
38 R = R' = R'' = OCH₂C(CH₃)₃, R''' = H
21 R = R' = R'' = R''' = OCH₂C(CH₃)₃

34 R = OCH₂C(CH₃)₃

$$\xrightarrow[\substack{80-90°C,24h \\ \textbf{Method VIII}}]{Me_2N(CH_2)_2OH}$$

Scheme VIII

Since neither **9** nor **34** self-condensed under the experimental conditions, the formation of substituted phthalocyanines other than the desired disubstituted **36** was surprising and led to some insights into possible mechanisms of phthalocyanine formation [39]. Exhaustive flash [40] and vacuum [41] liquid chromatography afforded small samples (<5 mg) of almost pure **35** and **36** ($+37$) with only a trace of **36** in **35** and **38** in **36** ($+37$) as determined by mass spectral analysis of each chromatographic fraction. In condensations leading to unsymmetrical phthalocyanines, phthalocyanines that are "separated" by column and even preparative thin layer chromatography and elute as separate spots can contain other phthalocyanine compounds, whose presence could most easily be confirmed by mass spectral analysis [39].

The use of phthalic anhydride (**39**) or related compounds such as phthalic acid, phthalimide, or phthalamide for the preparation of metallophthalocyanines such as **2** has been extensively reviewed [11–13,19] and *metallated* substituted phthalocyanines can be made by this method. In typical recent procedures [42–44] outlined in Scheme IXa, trimellitic anhydride (**40**) or 4-nitrophthalic anhydride (**41**), dissolved in nitrobenzene at 170–190°C with urea, $CoCl_2$, and often a catalyst such as ammonium molybdate, gave 2,9,16,23-tetracarboxyphthalocyaninato cobalt(II) (**42**) [43] or 2,9,16,23-tetranitrophthalocyaninato cobalt(II) (**43**) [44] in high yield (Method IXA). Depending on the metal used in this synthesis, it would be possible to prepare metal-free substituted phthalocyanines by this method followed by demetallation with strong acid as in Methods IIA and IIB, but this method has been rarely used. Weber and Busch [45] have used the sodium salt of 4-sulfophthalic acid instead of the anhydride essentially by Method IXA to produce tetrasodium (2,9,16,23-tetrasulfophthalocyaninato) cobalt(II) (see Table 1).

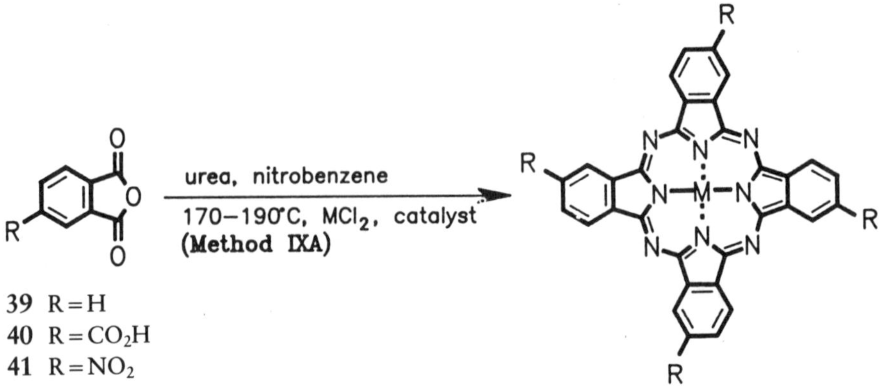

urea, nitrobenzene
———————————→
170–190°C, MCl$_2$, catalyst
(Method IXA)

39 R=H
40 R=CO$_2$H
41 R=NO$_2$

2 R=H
42 R=CO$_2$H
43 R=NO$_2$

Scheme IXa

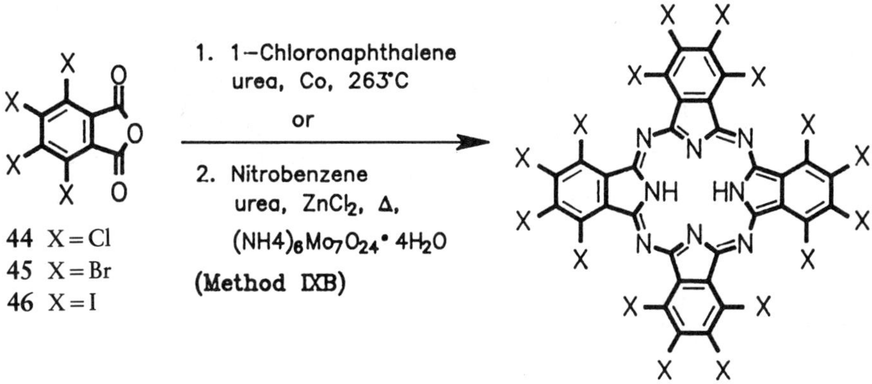

44 X = Cl
45 X = Br
46 X = I

1. 1−Chloronaphthalene
   urea, Co, 263°C

   or

2. Nitrobenzene
   urea, ZnCl₂, Δ,
   (NH4)₆Mo₇O₂₄ · 4H₂O

   (**Method IXB**)

47 X = Cl
48 X = Br
49 X = I

Scheme IXb

When perhalosubstituted phthalic anhydrides (**44–46**) were treated under conditions similar to Method IXA, the hexadecachlorophthalocyanine (**47**) [46], hexadecabromophthalocyanine (**48**) [46], and hexadecaiodophthalocyanine (**49**) [47] were obtained in up to 80% yield [47] (Scheme IXb, Method IXB).

It is important to note that although cobalt or zinc salts were used in these condensations, metal insertion did not occur and **47–49** were isolated directly as metal-free phthalocyanines. One can regard this particular method as a substituent-directed metal-free phthalocyanine synthesis.

Very recently [48] it was shown that phthalocyanine formation can occur as low as −20°C. Thus, 1-imino-3-methylthio-6-neopentoxyisoindolenine (**50**) or 1-imino-3-methylthio-5-neopentoxyisoindolenine (**51**) underwent self-condensation at *room temperature* in 2-N,N-dimethylaminoethanol to give in 5–18% yield 2,9,16,23-tetraneopentoxyphthalocyanine (**21**) as a *nonstatistical* mixture of isomers. In the presence of zinc acetate, however, at −15 to −20°C, **50** or **51** gave almost pure 2,9,16,23-tetraneopentoxyphthalocyaninato zinc(II) (**52**) as a single isomer in 5–11% yield (Scheme X, Method X).

It is envisioned that imino displacement of the methylthio group in **50** and **51** is more rapid than imino displacement of another imino group and, hence, exclusive consecutive cyclic displacement of four methylthio groups yields one exclusive isomer of a tetrasubstituted phthalocyanine and the reaction is rapid enough to proceed at very low temperatures for phthalocyanine formation. Unfortunately, extensive by-product formation leads to low yields.

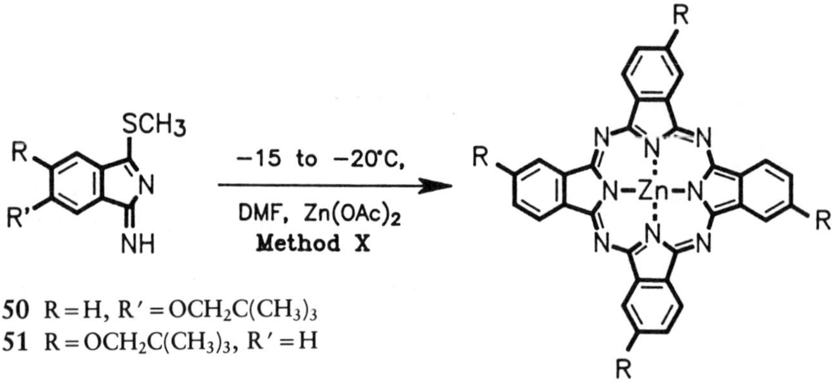

50 R = H, R' = OCH$_2$C(CH$_3$)$_3$
51 R = OCH$_2$C(CH$_3$)$_3$, R' = H

52 R = OCH$_2$C(CH$_3$)$_3$

Scheme X

A tetracyclohexenotetrazaporphyrin can be readily dehydrated by subli- mation at 300–320°C, heating with sulfur, boiling with palladium in chloronaphthalene [49], or treatment with 2,3-dichloro-5,6-dicyanobenzo- quinone (DDQ) [50] to give phthalocyanine (1) itself (Scheme XI, Method XI). Substituted phthalocyanines have not been generally made by this method.

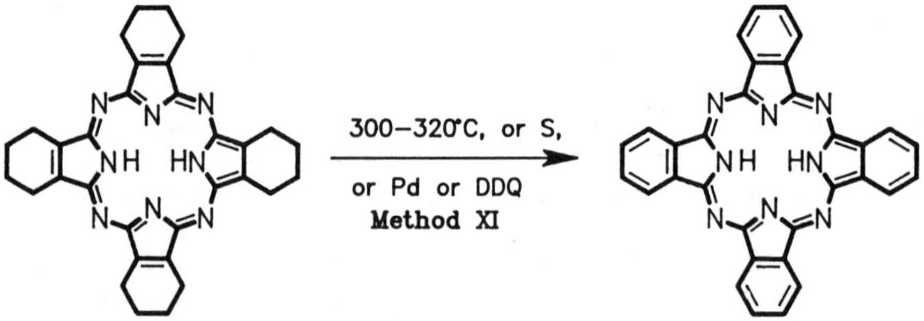

53   1

Scheme XI

Finally, substituted phthalocyanines can be prepared by converting a substituted or unsubstituted phthalocyanine, already prepared by Methods I– XI described above, into a substituted phthalocyanine containing different or new substituents. For example, metal-free 2,3,9,10,16,17,23,24-octa- cyanophthalocyanine (54) can be readily converted by treatment with base to the metal-free 2,3,9,10,16,17,23,24-octacarboxyphthalocyanine (55) in 64% yield [51]. Furthermore, conversion of 55 into its anhydride 56a with acetyl chloride and subsequent reaction of 56a with a series of amines, including ammonia, to give **56b** and 1- and 2-aminoanthraquinones, *n*-decylamine, and 5-*p*-aminophenyl-10,15,20-triphenylporphyrin led to a series of functionally changed phthalocyanines containing quinone or alkyl groups **57a, b, c** as substituents and an unusual mixed tetraporphyrin-monophthalocyanine (**58**) [52] (Scheme XIIa, Method XIIA). Thus, simply transforming existing functionalized phthalocyanines into interesting new substituted phthalo-cyanines by functional group transformation is readily achieved. One difficulty with this method is the fact that functional group transformation should approach 100% as separation of the one functionalized phthalocyanine (the starting material) from a new substituted phthalocyanine (the product) is often by no means facile.

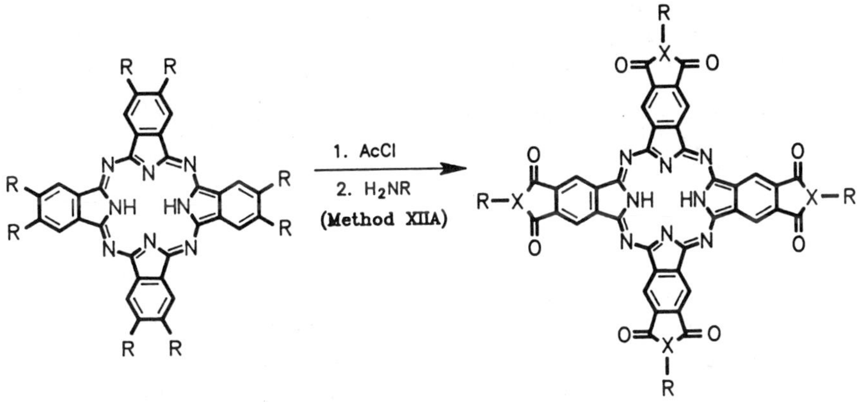

54 R = CN
55 R = CO$_2$H

56a X = O, R = nil
56b X = NH, R = nil

57a X = N, R =

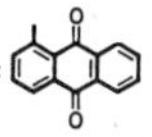

57b X = N, R =

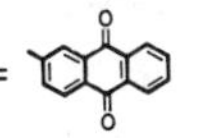

57c X = N, R = (CH$_2$)$_9$CH$_3$
58 X = N,
R =

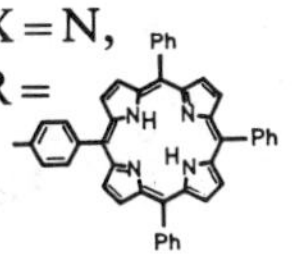

Scheme XIIa

As phthalocyanines are aromatic systems [14], the common reactions associated with aromatic chemistry such as nucleophilic aromatic substitution and, of course, electrophilic aromatic substitution can be applied. When 1,4,8,11,15,18,22,25-octachlorophthalocyanine (**59**) is treated with 4-methyl-phenylthiol in quinoline and potassium hydroxide at 180–210°C for 1–2 h, a low yield of 1,4,8,11,15,18,22,25-octa-(4-methylphenylthio)phthalocyanine (**60**) was obtained [53] after chromatography, but characterization was achieved only by elemental analysis and UV-VIS spectroscopy (Scheme XIIb, Method XIIB).

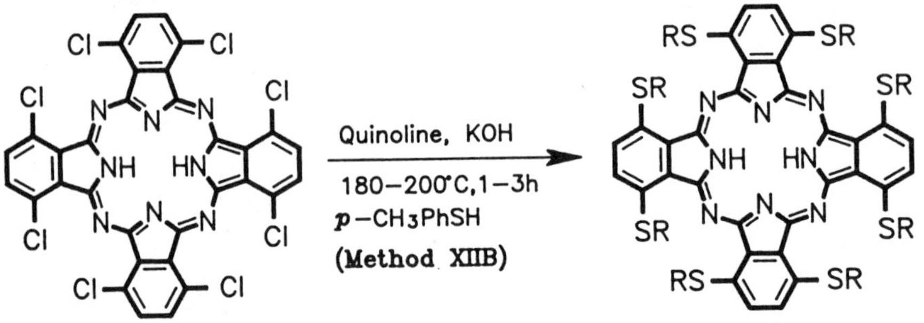

Quinoline, KOH

180–200°C, 1–3h

p–CH₃PhSH

(**Method XIIB**)

59

60 R = *p* – CH₃Ph

Scheme XIIb

Although electrophilic aromatic substitution reactions on phthalocyanine (**1**), or metallophthalocyanines (**2**) including halogenation [12,19,54], sulfonation [12,19], and nitration [55], have been known for some time, the products of such reactions are invariably mixtures of compounds. Although possibly suitable as dyes, these mixtures are not unique compounds and, as such, will not be included in this review. For example, mass spectroscopic analysis of "mononitrophthalocyaninato copper(II)" shows it to exist as a mixture of the unsubstituted, mono- and dinitrophthalocyanines, while analysis of a sample of aluminum chlorophthalocyanine exhibits a mixture of mono-, di-, tri-, tetra-, penta-, hexa-, hepta-, and octachlorinated compounds [56,57]. Thus, substituted phthalocyanines synthesized by electrophilic aromatic substitution reactions (Method XIIC), reported to give pure compounds, must be accompanied by mass spectral data providing proof of structure as elemental analysis alone of, for example, a monosubstituted phthalocyanine may well represent a statistical mixture of un-, mono-, and disubstituted phthalocyanines.

C. METHODS OF PURIFICATION

Phthalocyanine (**1**) and its metallated derivatives (**2**) have been readily purified by sublimation [6,23] and by dissolution in concentrated sulfuric acid followed by precipitation on ice [2]. These classical methods of purification of

1 and 2 are not widely used for purifying organic compounds and were applicable to 1 and 2 only because these compounds are, in general, extremely stable to heat (up to 550°C) and strong acid. In addition, the extreme insolubility of 1 and 2 made the more usual recrystallization and chromatographic methods inapplicable, although some metallated phthalocyanines that are somewhat soluble can be purified by extraction and recrystallization [23].

For *substituted* phthalocyanines, the increased molecular weights compared to 1 or 2, along with possible dipole interactions between substituent groups, means that sublimation is normally an unsuccessful method of purification. Thus, octasubstituted phthalocyanines 12 and 13 did not sublime [28], while 2,9,16,23-tetra-*t*-butylphthalocyanine (61) actually exhibited an enhanced tendency to sublime [58]. It is likely that the bulky *t*-butyl groups in 61 prevent aggregation compared to 1 and enhances the possibility of sublimation. Despite the ability of 61 to sublime, the authors purified the organic solvent soluble 61 by chromatography on alumina. Some substituted phthalocyanines exhibit enhanced rates of decomposition when dissolved in concentrated sulfuric acid even in the cold. For example, tetra-*t*-butylphthalocyanine (61) (Fig. 2) [58], 1,4,8,11,15,18,22,25-octamethoxyphthalocyanine (62) (Fig. 2), and the 2,3-naphthalocyanine (63) (Fig. 2) all decompose in conc. sulfuric acid. On the other hand 2,3,9,10,16,17,23,24-octaphenylphthalocyanine (64) (Fig. 2) undergoes sulfonation of the phenyl groups in conc. sulfuric acid [59] while the tetra-2,3-triphenylenoporphyrazinato copper(II) (65) (Fig. 2) [octabenzo-2,3-naphthalocyaninato copper(II)] was insoluble in conc. sulfuric acid [59]. Therefore, dissolution of substituted phthalocyanines in conc. sulfuric acid and reprecipitation can only rarely be used as a purification method of substituted phthalocyanines.

For highly insoluble phthalocyanines such as 63 and 65 that cannot be purified by sublimation or dissolution in conc. sulfuric acid and precipitation, only simple washing procedures using water and organic solvents can be used. Unlike 63 and 65, most substituted phthalocyanines can be purified by methods normally applied to organic compounds in general, such as extraction with an organic solvent and recrystallization [23] or alumina [58], silica gel [26,30], or gel permeation [27] chromatography.

i. Purification Methods

For reference to Tables 1–7 listed in Sections E–J, we will summarize the methods of purification of substituted phthalocyanines under nine headings a–i. Although the published methods of purification may not conform exactly to a–i, they can usually be easily accommodated within the general methods outlined in a–i.

Substituted phthalocyanines can be thus purified by

a. dissolution in conc. sulfuric acid, followed by precipitation in cold water or ice;

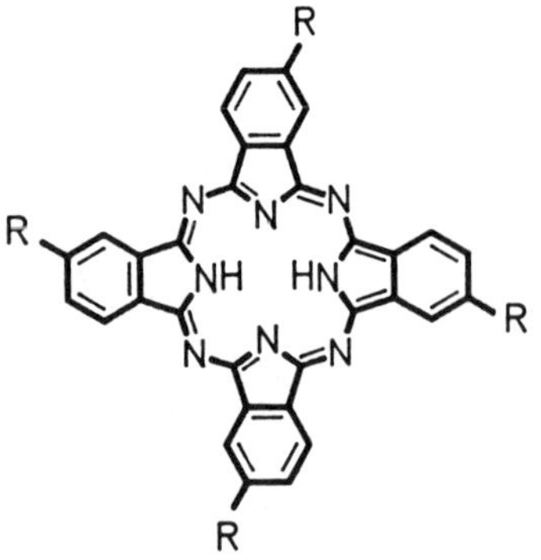

61 R = C(CH₃)₃
66 R = OH

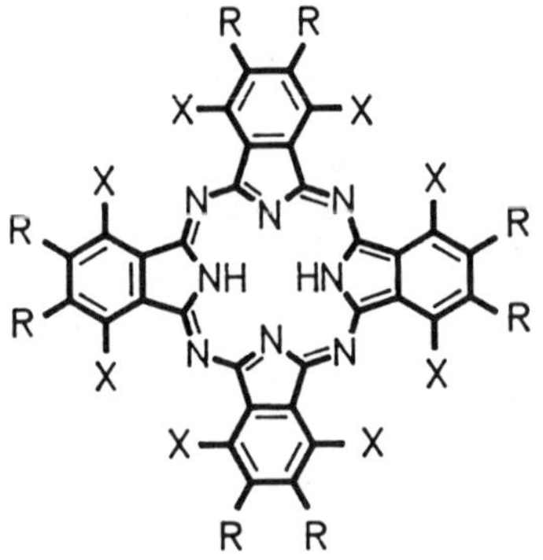

62 R = H, X = OCH₃
64 R = Ph, X = H

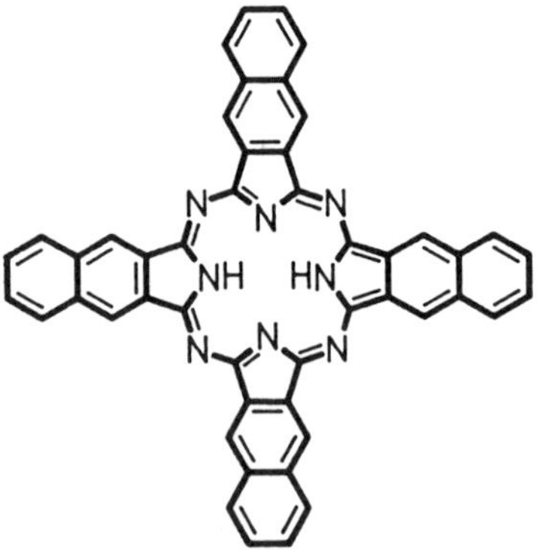

63

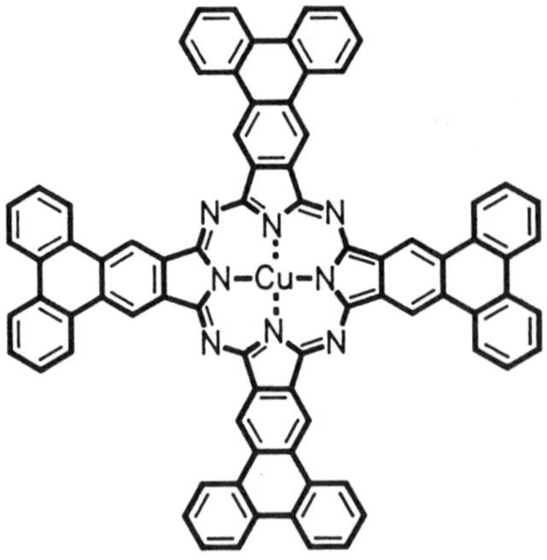

65

Figure 2 Some substituted phthalocyanines that cannot be readily purified in cold concentrated sulfuric acid.

b. dissolution in conc. hydrochloric acid, followed by precipitation in aqueous base, for amino-substituted phthalocyanines [60];

c. column chromatography on alumina and solvent evaporation or recrystallization;

d. column chromatography on silica gel using normal, flash [30,40], or vacuum [30,41] methods followed by solvent evaporation or recrystallization;

e. gel permeation chromatography [27,61];

f. washing insoluble substituted phthalocyanines with a variety of solvents to remove impurities leaving a purified residue;

g. extracting soluble substituted phthalocyanines from insoluble impurities with a variety of solvents and evaporation of the solvent or recrystallization of the extracted substituted phthalocyanine;

h. sublimation methods; and

i. other methods, including thin-layer chromatography (TLC) and high-performance liquid chromatography (HPLC).

A few comments should be mentioned concerning problems associated with purification of substituted phthalocyanines by purification methods a–i. The problems associated with methods a and h were discussed above. For method b, the main difficulty is that unwanted amino impurities could be both solubilized and reprecipitated by the given method. Chromatographic methods c and d can provide excellent separation of soluble substituted phthalocyanines but a word of caution is advised. As all phthalocyanines exhibit strong aggregation effects [27,62,63], it often happens that bands, eluting from a column or spots on TLC that supposedly represent a pure substituted phthalocyanine, can incorporate unsubstituted phthalocyanine 1 or other phthalocyanines and thus a pure band on column chromatography or a single spot on TLC is by itself an insufficient criterion of purity and should be accompanied by, particularly, mass spectral and other spectroscopic data [39]. Gel permeation chromatography (Method e) can separate molecules according to size [27]. By this method binuclear phthalocyanines, existing in extended conformations, can be separated from mononuclear phthalocyanines, but not those binuclear phthalocyanines that exist in folded conformations [26]. In addition, molecules separated by gel permeation chromatography must be further purified on silica or alumina columns to remove very minor impurities picked up from the 1% cross-linked divinylbenzene-styrene gel permeation chromatography column itself [27]. Very insoluble substituted phthalocyanines can be separated from soluble impurities by washing with solvents (method f), but this method leaves other insoluble impurities behind. It has been found in our laboratory that if any solvent, even dimethylformamide, quinoline, dimethyl sulfoxide, methanol, or other unusual solvents can solubilize the phthalocyanine to some extent, a rapid filtration through silica gel or alumina may remove polymeric and even more insoluble impurities. This method was applied to the recent synthesis and purification of 2,9,16,23-tetrahydroxyphthalocyanine (66) [64]. On the other hand, relying solely on solvent extraction (method g) to isolate soluble, substituted phthalocyanines can give mixtures of phthalocyanines or phthalocyanines containing impurities. Therefore, method g is best coupled with chromatographic methods c–e. Preparative TLC (method i) can be used to separate small quantities of phthalocyanines, but recoveries from TLC are often low and aggregation phenomena discussed above may still give incomplete separations. Our own attempts at using high-performance liquid chromatography on silica or alumina to separate substituted phthalocyanines have not been successful.

D. MECHANISMS OF PHTHALOCYANINE FORMATION

As shown in Section B, phthalocyanines can be prepared by many different routes, and although some of these synthetic routes may proceed through common intermediates, it is not necessary that all routes proceed through a common mechanism. The mechanisms of imidine condensations involving imidines such as 9 for their own sake and in relation to the use of 9 in phthalocyanine formation have been reviewed [34] and hence will not be recapitulated here. The reaction of a diamine with 9 and the isolation of the stable 2:1 adduct 67 provided some evidence of the plausibility of the course of the amino–imidine and hence imidine–imidine condensations [34,65] (Fig. 3) and its relevancy to phthalocyanine formation via 9.

In the recent synthesis of substituted phthalocyanines via crossed condensations of the 1,3-diiminoisoindoline 9 and a substituted dithioimide (Scheme VIII), typical acyclic intermediates such as dimeric 68 were postulated as necessary to give the product distribution of substituted phthalocyanines observed (Fig. 4). Although such intermediates are reasonable they were not isolated [39].

In the preparation of phthalocyanines via Method IIA, Borodkin [66] isolated a sodium derivative of methoxyiminoisoindolenine (70) as an intermediate. Later Hurley et al. [67] isolated the nickel complexes 71 and 72 as intermediates leading to nickel phthalocyanines (Fig. 5) and more recently a dimeric lithium salt 73 was isolated and shown to condense to a tetranitrophthalocyanine [68] (Fig. 5).

In the room temperature formation of phthalocyanines as outlined in Schemes VII and VIII, it is likely that the initial attack of 31 on 32 (Scheme VII) or 9 on 34 (Scheme VIII) is faster than self-condensation of 31 or 9, respectively. Intermediates such as 68 or 69, however, are probably much more reactive that 9 itself and this makes the interception of these intermediates difficult. The low-temperature condensation of 50 and 51 to the

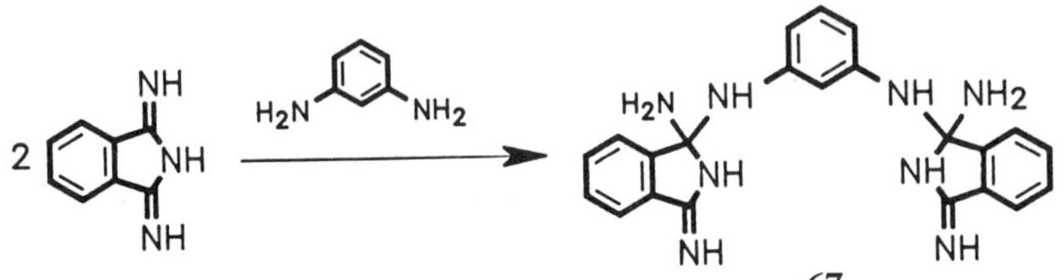

Figure 3 A stable adduct (67) of an imidine with a diamine as a model of an intermediate in phthalocyanine formation.

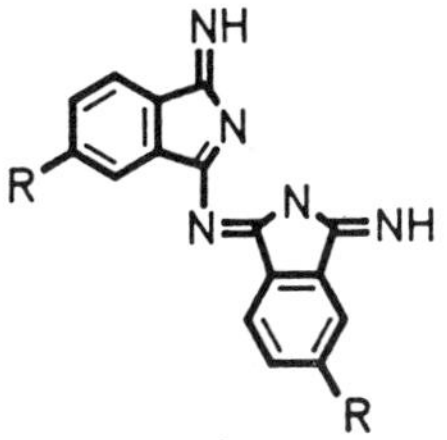

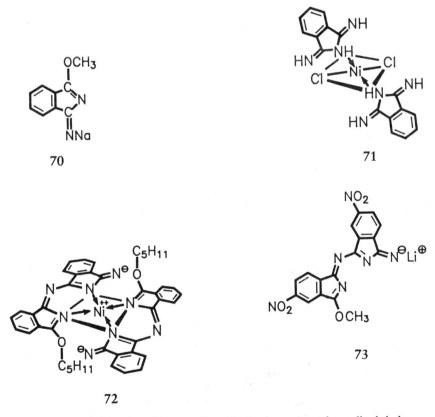

R=H or OCH₂C(CH₃)₃
68

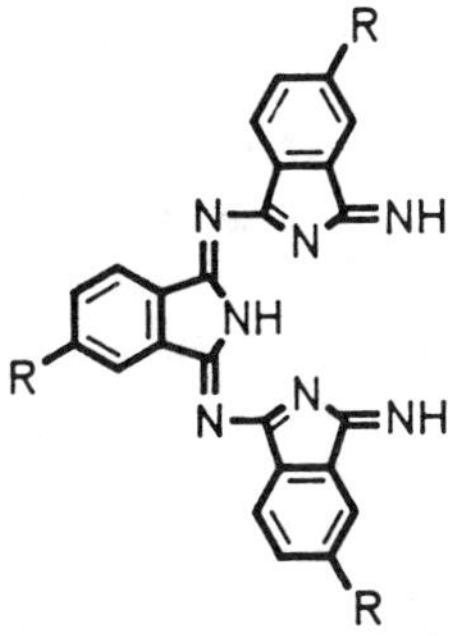

R=H or OCH₂C(CH₃)₃
69

Figure 4 Postulated dimeric and trimeric intermediates of **9** and **20** in phthalocyanine formation.

70

71

72

73

Figure 5 Some stable salts of intermediates in the formation of metallophthalocyanines by Method IIA.

symmetrical 2,9,16,23-tetraneopentoxyphthalocyaninato zinc(II) (**52**) as mostly a single isomer (Scheme X) is indicative that the imino group of **50** and **51** initiates attacks at the carbon having the methylthio group, perhaps using the zinc atom as a template. This type of attack is more rapid than self-condensation of **9**. At room temperature, however, the reactivity of intermediates such as **68** and **69**, that can be formed in the reaction outlined in Scheme X, is sufficiently high that they can intercept the course of reaction and give other isomers of **52**. On the other hand, the reaction of 4-*t*-butylphthalonitrile (**74**) with powdered zinc (Method IIB, Scheme II also Scheme XIII) leads to a different single isomer of a tetrasubstituted phthalocyanine compared to **52**, namely, 2,9,17,24-tetra-*t*-butylphthalocyaninato zinc(II) (**75**) [69]. The intermediacy of charged **76**, **77** and radical anion **78**, **79** species are postulated to explain the exclusive formation of **75** [69] (Scheme XIII). Thus, the mechanisms of phthalocyanine formation via Methods IIA or X and IIB are indicated to be quite different. It is clear that studies into the synthesis of substituted phthalocyanines can give insights into the mechanisms of phthalocyanine formation [39,48,68,69].

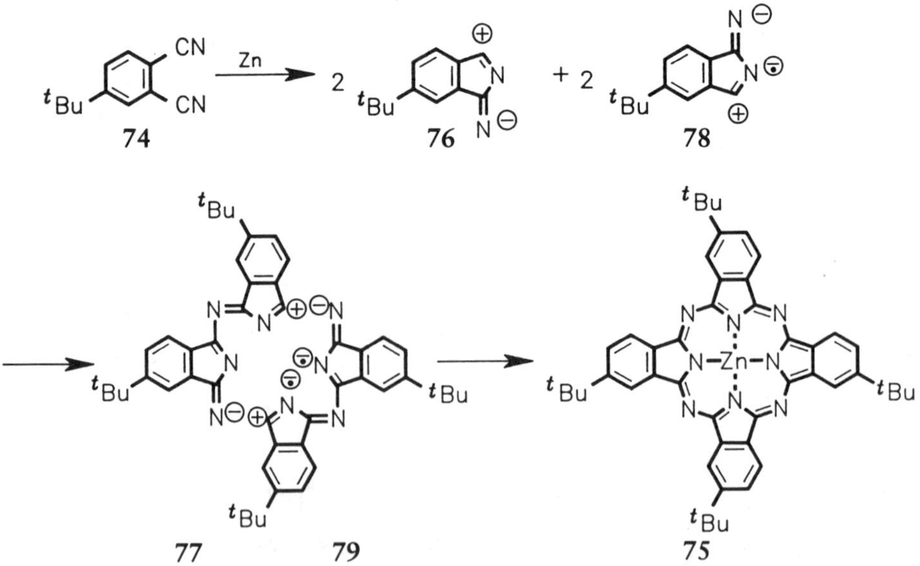

Scheme XIII

As Elvidge and Linstead have pointed out, condensation of four molecular proportions of **9** gave a hydrophthalocyanine that must be reduced to phthalocyanine **1** itself by an appropriate reductant [5,34] but at exactly what stage of phthalocyanine formation the reductant acts has not yet been clarified. Thus, although many clues to phthalocyanine formation by several synthetic methods give a broad, tantalizing picture of phthalocyanine formation, definitive studies remain to be accomplished.

E. TETRASUBSTITUTED PHTHALOCYANINES

In Sections E–J, we shall describe most of the metal-free substituted phthalocyanines that have been prepared and adequately characterized since the early reviews of Lever [11] and Moser and Thomas [12] and, hence, will cover the period from 1965 until the present. The compounds listed in Tables 1–7 have been characterized by elemental analysis and visible-ultraviolet spectroscopy as a minimum, and for unusually substituted phthalocyanines such as metal-free phthalocyanines containing different substituents or multinuclear phthalocyanines, a mass spectroscopic analysis was essential. Some substituted phthalocyanines are mentioned in the literature as having been prepared, but their characterization is only mentioned "en passant" or as a private communication and these are not included in the tables. Some substituted phthalocyanines have been made in their metallated forms but not in their metal-free forms. We have included some of the more interesting metallated substituted phthalocyanines in Section J as examples, but this list for metallated phthalocyanines is far from complete as this chapter is devoted to the metal-free derivatives. Finally, we are sure that some well-characterized, bonafide metal-free substituted phthalocyanines have been inadvertently omitted from Tables 1–7 and we apologize for any oversights incurred.

The lack of source material that can be rapidly scanned for metal-free substituted phthalocyanines, indicating the method of preparation, purification procedure, level of characterization, yield, metallated derivatives, and their references, has prompted the production of this chapter. The data given in Tables 1–7 should partially alleviate this need.

i. 1,8,15,22-Identically Substituted Phthalocyanines

Some metal-free phthalocyanines 80–84 identically substituted at the 1, 8, 15, and 22 positions of the phthalocyanine ring are outlined in Fig. 6. As shown in Table 1, the yields of 80–84, all by Method IIA, range from 25 to 69% and, hence, at least for ordinary sized substituents, the formation of phthalocyanines with substituents at these potentially sterically demanding positions have been realized. On the other hand, condensation of 3-(perfluoro-2-butyl)phthalonitrile did not give a phthalocyanine [69a]. It should be noted that 80–84 represent a mixture of compounds composed of the given structure and their 1,8,15,25-, 1,11,15,25-, and 1,11,18,22-isomers (see Section E,iv).

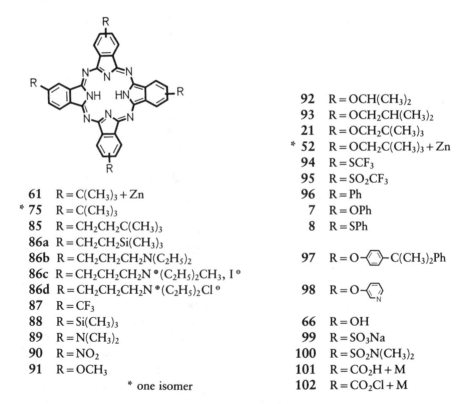

80 R = CF$_3$
81 R = CF$_3$S
82 R = CH$_3$O
83 R = PhO
84 R = PhS

Figure 6 Some metal-free 1,8,15,22-tetrasubstituted phthalocyanines.

ii. 2,9,16,23-Identically Substituted Phthalocyanines

A wide variety of metal-free phthalocyanines 7, 8, 21, 61, 75, and 85–100 substituted at the 2,9,16,23-positions of the phthalocyanine ring have been prepared as illustrated in Fig. 7. For these compounds, yields should not be affected by steric effects and yet they ranged from 11 to 70% (Table 1).

61   R = C(CH$_3$)$_3$ + Zn
* 75   R = C(CH$_3$)$_3$
85   R = CH$_2$CH$_2$C(CH$_3$)$_3$
86a R = CH$_2$CH$_2$Si(CH$_3$)$_3$
86b R = CH$_2$CH$_2$CH$_2$N(C$_2$H$_5$)$_2$
86c R = CH$_2$CH$_2$CH$_2$N$^{\oplus}$(C$_2$H$_5$)$_2$CH$_3$, I$^{\ominus}$
86d R = CH$_2$CH$_2$CH$_2$N$^{\oplus}$(C$_2$H$_5$)$_2$Cl$^{\ominus}$
87   R = CF$_3$
88   R = Si(CH$_3$)$_3$
89   R = N(CH$_3$)$_2$
90   R = NO$_2$
91   R = OCH$_3$

92   R = OCH(CH$_3$)$_2$
93   R = OCH$_2$CH(CH$_3$)$_2$
21   R = OCH$_2$C(CH$_3$)$_3$
* 52   R = OCH$_2$C(CH$_3$)$_3$ + Zn
94   R = SCF$_3$
95   R = SO$_2$CF$_3$
96   R = Ph
7   R = OPh
8   R = SPh

97   R = O—⟨ ⟩—C(CH$_3$)$_2$Ph

98   R = O—⟨N⟩

66   R = OH
99   R = SO$_3$Na
100 R = SO$_2$N(CH$_3$)$_2$
101 R = CO$_2$H + M
102 R = CO$_2$Cl + M

* one isomer

Figure 7 Some metal-free 2,9,16,23-tetrasubstituted phthalocyanines.

Only the zinc derivative of the tetrabutyl phthalocyanine **61** and a zinc derivative of tetraneopentoxyphthalocyanine **52** have been prepared as one isomer only. All other compounds listed consist of a mixture of the 2,9,16,23-, 2,9,16,24-, 2,10,16,24-, and 2,9,17,24-isomers (see Section E,iv). Tetrasubstituted phthalocyanine tetraacid **101** and acid chloride **102** have been reported as their metallated derivatives (Table 1).

iii. 2,9,16,23-Nonidentically Substituted Phthalocyanines

All tetrasubstituted phthalocyanines so far synthesized contain one substituent in each of the benzo rings and no phthalocyanines such as 1,2,3,4- or 1,2,3,8-tetrasubstituted phthalocyanines have been made. Three sufficiently characterized phthalocyanines (**18a**, **18b**, **26**) (Scheme IVa, VI) containing *three* identical and *one* different substituent have been prepared by solid phase methods. Because of the difficulty of separating nonidentically substituted phthalocyanines from *symmetrically* identically substituted phthalocyanines, it is essential that the mass spectroscopic analysis be accomplished on this class of compounds both to show the parent ion of the given compound and the absence of identically substituted *contaminants* (Table 1).

iv. The One Isomer Problem

It was recognized early in the history of phthalocyanines by Linstead and co-workers [70,71] that condensation of unsymmetrical *ortho* dinitriles such as naphthalene-1,2-dicarbonitrile [70] or pyridine-2,3-dicarbonitrile [71] gave a mixture of four isomeric phthalocyanines, but identification of the components of the mixture was not possible in that day. Much more recently Marcuccio et al. [72] examined the ^{1}H NMR and ^{13}C NMR spectra of the tetrasubstituted phthalocyanines **21**, **85**, and **86a** and showed that their NMR spectra were consistent with a statistically produced mixture of their 2,9,16,23-, 2,10,16,24-, 2,9,17,24-, and 2,9,16,24-isomers in the required ratio of 1:1:2:4, respectively, giving, in at least one example, eight equivalent NMR absorbance peaks for the eight possible environments of the CH_2Si group represented by the four isomers of **86a**.

There have been only a very few attempts to separate or exclusively synthesize one isomer of a tetrasubstituted phthalocyanine. Thus, Wöhrle et al. [73] prepared the metal-free and zinc derivatives of the tetramethyl quarternized derivatives of tetra-2,3-pyridinoporphyrazine (1,8,15,22-tetraazaphthalocyanine) (**103**) and (**104**), respectively, and analysis by ^{1}H NMR spectroscopy exhibited a nonstatistical distribution of the four possible isomers of **104**, namely **104**, and some of the 1,11,15,25- (**105**), 1,8,18,25- (**106**), and 1,8,15,25- (**107**) isomers (Fig. 8).

A statistical distribution of **104** to **107** would give **104–107** in a 1:1:2:4

Table 1 Metal-Free Tetrasubstituted Phthalocyanines and Some of Their Metallated Derivatives[a]

Comp[b]	Syn[c]	Pur[d]	Yield (%)	Characterization[e]					Metal[f]	Reference[g]
				E.A.	MS	VIS	NMR	O		
1,8,15,22-Substituted derivatives										
80	IIA	a	69	x		x			Cu	74
81	IIA	a	54	x		x			Cu	75
82	IIA	a	46	x		x			Zn	76
83	IIA	b	36	x		x			Cu,Zn,Co	22
84	IIA	b	25	x		x			Cu,Zn,Co	22
2,9,16,23-Substituted derivatives										
61	IIA	c	56	x		x			Mg,Zn,Al,Ni, Pd, Cu	58
61	IIIB	a	70	x	x	x	x	x	Si,Ge,Sn	77
75[h]	IIB	d	—	x[b]		x	x		Zn,Mg	69
85	IIIB	d	40	x	x	x	x	x	Co	72
86a	IIIB	d	28	x	x	x	x	x		72
86b	IIIB	c	28	x	x	x	x	x	Co,Zn	77a
86c	XIIA	f	89	x		x		x		77a
86d	XIIA	f	90	x	x	x	x	x	Cu,Co,Zn	77a
87	IIIB	f	67	x	x	x		x	Ge,Sn	78
88	IIA	f	69	x		x			Cu	79
88	IIIB	a	70	x	x	x	x	x	Si,Ge,Sn	77
89	IIA	b	22			x			Co,Cu,VO	60
90	IIIB	f	50					x		21
90	IIA	f	20	x						68
91	IIA	a	47	x		x				76
92	IIIB	d	38	x	x	x	x	x	Cu	29,30
93	IIIB	d	27	x	x	x	x	x		26
21	IIIB	d	45	x	x	x	x	x	Cu,Co,Zn	26

No.[a]	Synthesis[c]	Purification[d]	No.[b]	E.A.	MS	VIS	NMR	O	Metal	Ref.[f,g]
52[i]	X	d,i	11		x	x	x	x	Zn	48
94	IIA	a	65	x		x			Cu	75
95	IIA	a	55	x		x				75
96	IIA	f	42	x	x	x	x	x	Mg,Zn,Al,Sn,Pb,Nb,Mn,Fe,Co,Ni,Pd,VO	80
7	IIA,C	f,d	39	x	x	x	x	x	Cu,Co,Zn	22,24
8	IIA,C	f,d	29	x	x	x	x	x	Cu,Co,Zn	22,24
97	IIA,C	f	49	x	x	x	x	x	Cu,Ni,Pd,Pt,Mg,Co,Zn,Pb,Bi	81,82
98	IIIB	b	16	x	x	x	x		Zn	83
66	XIIA	d	53	x	x	x	x	x	Zn	64
99	XIIC	f	35	x	x	x	x		Co,Ni,Cu,Mn,Fe[j]	84,45
100	IIA	d	31	x	x	x			Cu,VO	85
101[k]	XIIA	f	80	x		x	x		Co,Fe	43
102[k]	XIIA	d	100	x	x	x	x		Co,Fe	43
2,9,16,23-Nonidentically substituted derivatives										
18a	IVA	d	24	x	x	x	x			29,30
18b	IVA	d	12	x	x	x	x			29,30
26[k]	V		22	x	x	x	x	x	Zn	32

[a] For numbering system of phthalocyanines see Fig. 1.

[b] Compound number (see figures for structures).

[c] Method of synthesis IA–XIIB as outlined in Schemes I–XII in Section B.

[d] Method of purification a–i as outlined in Section C.

[e] Methods of characterization where E.A. = elemental analysis, MS = mass spectrometry, VIS = UV-VIS spectroscopy, NMR = nuclear magnetic resonance spectroscopy, and O = other, usually representing infrared spectroscopy, but could include other methods.

[f] Refers only to metallated derivatives described in the references in which the metal-free phthalocyanine was prepared.

[g] Refers to references where the given metal-free phthalocyanine was first adequately described or whose synthesis was substantially improved.

[h] This is the 2,9,17,24-tetra-t-butylphthalocyaninato zinc(II) compound as one isomer.

[i] This entry is the 2,9,16,23-tetraneopentoxyphthalocyaninato zinc(II) compound as one isomer.

[j] The metallated phthalocyanines were made via Method IXA using sodium 4-sulfophthalic acid.

[k] These entries describe only the metallated phthalocyanines.

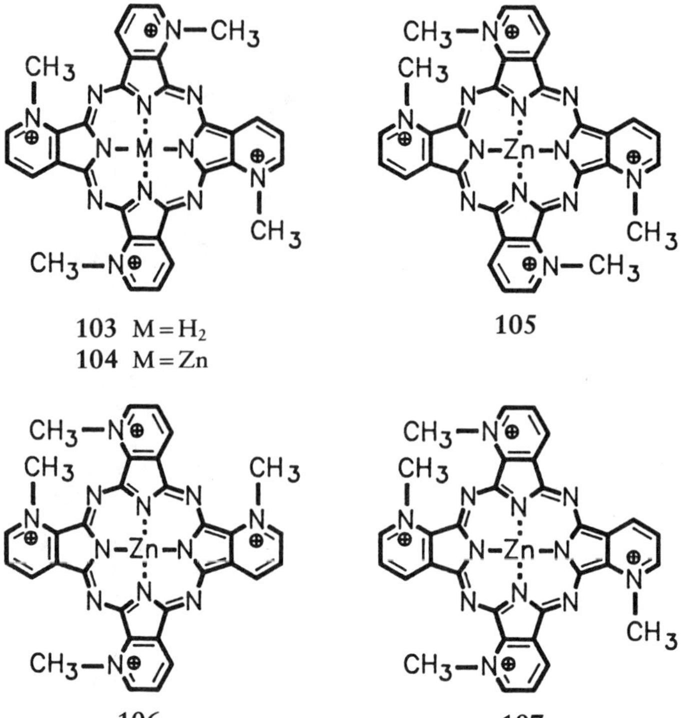

103 M = H$_2$
104 M = Zn

105

106

107

Figure 8 Four possible isomers of quaternized tetraazaphthalocyanines.

ratio, but since the number of ^{1}H NMR peaks for each of **104–107** is 1, 1, 2, and 4, respectively, then a statistical mixture of **104–107** would give eight absorptions for the methyl groups of equal intensity. Since only two or three ^{1}H NMR absorptions for the methyl groups were observed, the authors suggest that selective formation of isomer **104** occurs because this isomer will have the least steric interaction. It is an open question, however, whether phthalocyanine formation of the nonquaternized precursor of **104** was selective due to electronic effects or whether the alkylation step in forming **104** was selective due to steric effects or a combination of both.

Very few studies have been directed toward the exclusive formation of one isomer of tetrasubstituted phthalocyanines, but as shown in Scheme X (Method X), the 2,9,16,23-tetraneopentoxyphthalocyaninato zinc(II) (**52**) was produced in low yield by a one-isomer directed synthesis [48]. As shown in Scheme XIII using Method IIB, a 2,9,17,24-tetra-*t*-butylphthalocyaninato zinc(II) was also produced as pure isomer [69]. These recent directed syntheses show that the selected formation of a specific tetrasubstituted phthalocyanine as one isomer is possible. On the other hand, the great solubility of some tetrasubstituted phthalocyanines [26] results not only from the steric bulk of the substituents preventing aggregation, but also from the presence of the very similar isomers. Thus, a mixture of similar isomers may be practically more useful than a single, less soluble isomer.

F. OCTASUBSTITUTED PHTHALOCYANINES

Unlike tetrasubstituted phthalocyanines, most octasubstituted phthalocyanines are symmetrical compounds and contain two substituents in each of the benzo rings. As unique symmetrical compounds, octasubstituted phthalocyanines are generally less soluble in organic solvents than tetrasubstituted phthalocyanines.

i. 1,4,8,11,15,18,22,25-Identically and Nonidentically Substituted Phthalocyanines

Some metal-free 1,4,8,11,15,18,22,25-octasubstituted phthalocyanines 59, 60, 62, 108–130 are outlined in Fig. 9.

As shown in Table 2, the yields of these phthalocyanines with eight substituents at the sterically demanding 1,4,8,11,15,18,22,25 positions are low, but the phthalocyanines can still be formed even with fairly bulky substituents. Two unusual properties of the phthalocyanines mentioned in Table 2 are that (1) some of these compounds, e.g., 118–123, exhibit very high λ_{max} values (~ 800 nm) in the Q band region in the visible spectra [53,89] and (2) some (121–122) have melting points.

The methods of synthesis of the nonidentically substituted phthalocyanines 124–130 can lead to mixtures of the given products, starting materials, and substituted phthalocyanines containing increasing amounts of the minor substituent. Until mass spectroscopic or X-ray diffraction studies are done on these compounds, their structures must remain less than 100% confirmed. It should also be pointed out that compounds 127–130 are mistakenly identified in *Chemical Abstracts* as heptakis compounds, the chlorosubstituent being ignored.

ii. 2,3,9,10,16,17,23,24-Identically and Nonidentically Substituted Phthalocyanines

Some metal-free 2,3,9,10,16,17,23,24-octasubstituted phthalocyanines 12, 13, 54, 55, 56a, b, 57a, b, c, and 131–144 are shown in Fig. 10.

As shown in Table 3, the yields of these phthalocyanines can vary from 20 to 85%, noting that the near quantitative yields of 142 and 56a do not

Table 2 Metal-Free 1,4,8,11,15,18,22,25-Octasubstituted Phthalocyanines and Some of Their Metallated Derivatives[a]

Comp[b]	Syn[c]	Pur[d]	Yield (%)	Characterization[e]					Metal[f]	Reference[g]
				E.A.	MS	VIS	NMR	O		
Identically substituted derivatives										
108	IIA	—	18–25	x		x	x	x	Cu	86
109	IIA	—	18–25	x		x	x	x	Cu	86
110	IIA	—	18–25	x		x	x	x	Cu	86
111	IIA	—	18–25	x		x	x	x	Cu	86
112	IIA	—	18–25	x		x	x	x	Cu	86
113	IIA	—	18–25	x		x	x	x	Cu	86
114	IIA	—	18–25	x		x	x	x	Cu	86
115	IIA	a	53	x		x			Cu	74
62	IIA	b	63	x[b]			x		Cu,Ni,Co	87
116	IIA	—	18–25	x[b]		x	x	x[i]	Cu	86
117	IIA	—	18–25	x[b]		x	x	x[i]	Cu	86
60	XIIB	d	2	x[b]		x		x[i]	Cu	53
118	XIIB	d	—	x[b]		x		x[i]	Cu	53
119	XIIB	d	—	x[b]		x			Cu	53
120	XIIB	d	—	x[b]		x			Cu	53
121	XIIB	d	—	x[b]		x			Cu	53
122	XIIB	d	—	x[b]		x		x[i]	Cu	53
123	XIIB	d	—	x[b]		x		x[i]	Cu	53
59	IXB	—	—	x[b]		x			Cu	53
Nonidentically substituted derivatives[j]										
124	IIA	—	—			x				88
125	IIA	—	—			x				88
126	IIA	—	—							88
127	XIIB	d	—	x[b]		x		x[i]		53
128	XIIB	d	—	x[b]				x[i]		53
129	XIIB	d	—	x[b]		x		x[i]		53
130	XIIB	d	—	x[b]		x		x[i]		53

[a-g] These footnotes are the same as for Table 1.
[b] Elemental analysis is stated but not given.
[i] These phthalocyanines have melting points.
[j] These unsymmetrical phthalocyanines have no necessary mass spectral data.

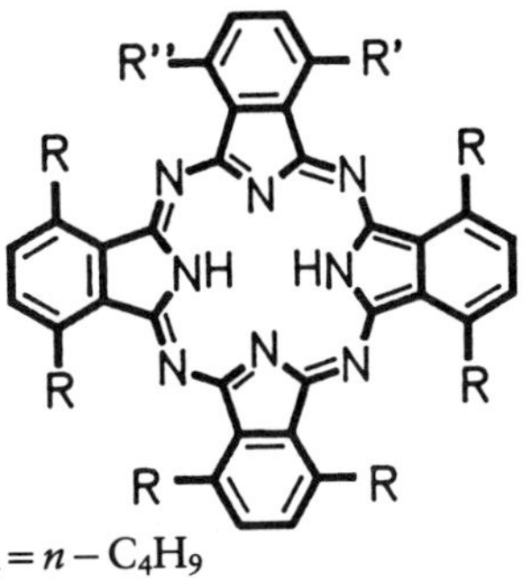

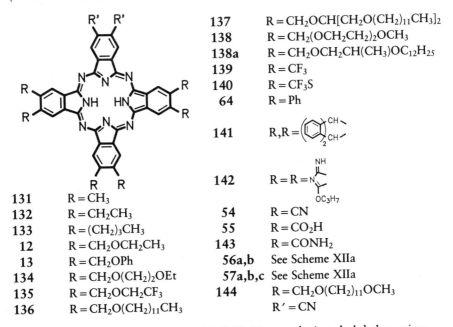

108 R = n – C$_4$H$_9$
109 R = n – C$_5$H$_{11}$
110 R = n – C$_6$H$_{13}$
111 R = n – C$_7$H$_{15}$
112 R = n – C$_8$H$_{17}$
113 R = n – C$_9$H$_{19}$
114 R = n – C$_{10}$H$_{21}$
115 R = CF$_3$
 62 R = OCH$_3$
116 R = OC$_5$H$_{11}$
117 R = OC$_8$H$_{17}$
 60 R = p – CH$_3$PhS

118 R = m – CH$_3$PhS
119 R = p – ^{t}BuPhS
120 R = p – CH$_3$OPhS
121 R = p – CH$_3$(CH$_2$)$_7$OPhS
122 R = p – CH$_3$(CH$_2$)$_7$OPhS
123 R = ^{i}PrS
 59 R = Cl
124 R = n – C$_8$H$_{17}$
    R$'$ = R$''$ = – (CH$_2$)$_3$CO$_2$H
125 R = n – C$_9$H$_{19}$
    R$'$ = R$''$ = – (CH$_2$)$_3$CO$_2$H
126 R = n – C$_{10}$H$_{21}$
    R$'$ = R$''$ = – (CH$_2$)$_3$CO$_2$H
127 R = R$'$ = p – ^{t}BuPhS
    R$''$ = Cl
128 R = R$'$ = p – (n – Nonyl)PhS
    R$''$ = Cl
129 R = R$'$ = p – (n + Dodecyl)PhS
    R$''$ = Cl
130 R = 3,4 – (CH$_3$)$_2$PhS
    R$'$ = R$''$ = Cl

Figure 9 Some metal-free 1,4,8,11,15,18,22,25-octasubstituted phthalocyanines (R = R$'$ = R$''$ unless otherwise indicated).

131 R = CH$_3$
132 R = CH$_2$CH$_3$
133 R = (CH$_2$)$_3$CH$_3$
 12 R = CH$_2$OCH$_2$CH$_3$
 13 R = CH$_2$OPh
134 R = CH$_2$O(CH$_2$)$_2$OEt
135 R = CH$_2$OCH$_2$CF$_3$
136 R = CH$_2$O(CH$_2$)$_{11}$CH$_3$

137   R = CH$_2$OCH[CH$_2$O(CH$_2$)$_{11}$CH$_3$]$_2$
138   R = CH$_2$(OCH$_2$CH$_2$)$_2$OCH$_3$
138a   R = CH$_2$OCH$_2$CH(CH$_3$)OC$_{12}$H$_{25}$
139   R = CF$_3$
140   R = CF$_3$S
 64   R = Ph

141   R,R = (image)

142   R = R = (image)

 54   R = CN
 55   R = CO$_2$H
143   R = CONH$_2$
56a,b See Scheme XIIa
57a,b,c See Scheme XIIa
144   R = CH$_2$O(CH$_2$)$_{11}$OCH$_3$
       R$'$ = CN

Figure 10 Some metal-free 2,3,9,10,16,17,23,24-octasubstituted phthalocyanines (R = R$'$ unless otherwise stated).

Table 3   Metal-Free 2,3,9,10,16,17,23,24- and 1,3,8,10,15,17,22,24-Octasubstituted Phthalocyanines and Some of Their Metallated Derivatives[a]

Comp[b]	Syn[c]	Pur[d]	Yield (%)	E.A.	MS	VIS	NMR	O	Metal[f]	Reference[g]
2,3,9,10,16,17,23,24-Substituted derivatives										
131	IIA	a	20	x		x		x	Ni,Cu[b]	90
132	IIIB	f	23	x	x	x				92
133	IIA	a	38	x		x		x	Ni	90
12	IIIB	a	85	x	x	x	x	x	Cu	28
13	IIIB	a	80	x	x	x		x	Cu	28
134	IIIB	a	75	x	x	x	x	x	Cu	28
135	IIIB	a	82	x	x	x		x	Cu	28
136	IIIB	f,g	40–60	x			x	x	Cu,Zn	93
137	IIIB	—	—					x		93
138	IIIB	d	40	x			x	x		93
139	IIIB	f	55	x	x	x		x	Ge,Sn	78
138a	IIIB	—	—	x			x	x		93a
140	IIA	a	75	x		x			Cu	75
64	IIA	f	27	x		x			Zn,Ga,VO,Cu, Ni,Pd	59
141	IXA[i], IIB	c	30	x		x			Cu,Zn,Al,VO,	94

Compound	Method	Note	Yield (%)					M	Ref.
142	XIIA	f	96	x	x		x	Cu,Co,Zn,Fe,Ni	51
54	XIIA,IIIB	f	49	x	x		x	Cu,Co[j], Ni[j], VO[j]; Fe[j]	51
55	XIIA	a,f	30	x	x				51,95
143	XIIA	f	79	x	x		x	Cu	95
56a	XIIA	h	99	x	x		x	Cu	51,95
56b	XIIA	f	77	x	x		x	Cu	95
57a	XIIA	g	38	x	x		x		52
57b	XIIA	g	77	x	x		x		52
57c	XIIA	g	61	x	x		x		52
144[k]	IIIB	d	20	x	x	x	x		97
1,3,8,10,15,17,22,24-Substituted derivatives[l]									
145	IIA	c	47	x	x			Zn	76
146	IIA	b	22	x	x			Cu,Co,Zn,VO	60
147	IXA	c	—	x	x			Mg,Cu,Zn,VO,Pd	98

a–g These footnotes are the same as for Table 1.

h The Cu derivative is described in ref. [91].

i Method IXA but using an o-dinitrile.

j These metal derivatives are described in refs. [42,96].

k The lack of mass spectral data for this nonidentically substituted phthalocyanine leaves the structure still tentative.

l These compounds all exist as a mixture of four isomers.

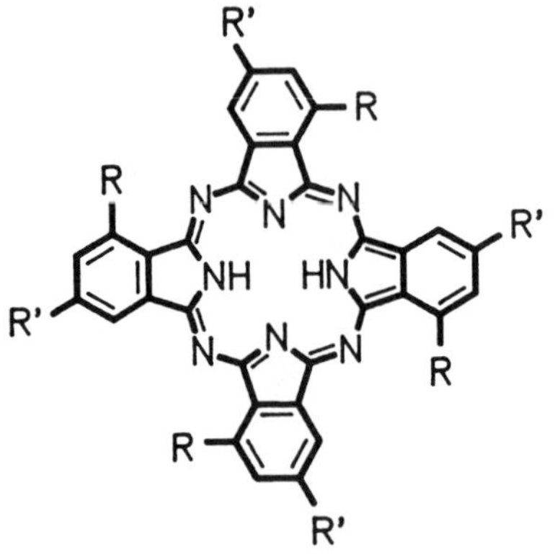

145 R = Br, R' = 'Bu
146 R = N(CH₃)₂, R' = 'Bu
147 R = NO₂, R' = 'Bu

Figure 11 Some metal-free 1,3,8,10,15,17,22,24-octasubstituted phthalocyanines.

involve phthalocyanine formation and are due to functional group interconversions of preformed phthalocyanines. Phthalocyanine **144** is the only nonidentically substituted compound in this class but, unfortunately, the method of synthesis could give mixtures of compounds that may not be separable even by chromatography. Mass spectroscopic examination of **144** could resolve this point.

iii. 1,3,8,10,15,17,22,24-Octasubstituted Phthalocyanines

The 1,3,8,10,15,17,22,24-octasubstituted phthalocyanines **145–147** (Table 3) that have been prepared (Fig. 11) have four like substituents at the 1,8,15,22-positions and a different like substituent at the 3,10,17,24-positions. These compounds exist as a mixture of the given structure and their 1,3,8,10,15,17,23,25-, 1,3,9,11,15,17,23,25-, and 1,3,9,11,16,18,22,24-isomers (see Section E,iv).

G. HEXADECASUBSTITUTED PHTHALOCYANINES

1,2,3,4,8,9,10,11,15,16,17,18,22,23,24,25-Hexadecasubstituted phthalocyanines (Fig. 12) can be prepared by nucleophilic aromatic substitution of hexadecahalosubstituted phthalocyanines or by condensation of the appropriate tetrasubstituted phthalonitrile (Table 4).

i. Identical Substituents

A series of identically substituted hexadecasubstituted phthalocyanines **47–49** and **148–151** have been prepared as shown in Fig. 12. Interestingly, the

Table 4 Metal-Free Hexadecasubstituted Phthalocyanines and Some of Their Metallated Derivatives[a]

Comp[b]	Syn[c]	Pur[d]	Yield (%)	E.A.[h]	MS	VIS	NMR	O[i]	Metal[f]	Reference[g]
Identical substituents										
47	IXB	—		x	x					46
48	IXB	—		x	x					46
49	IXB	a	80	x		x				47
148	IIA	d	—	x		x			Cu	53
149	IIA	d	—	x		x		x	Cu	53
150	IIA	d	—	x		x				53
151	XIIB	d	—	x		x			Cu,Zn,Pb	53
152	IIA	d	—			x	x			99
Nonidentical substituents[j]										
153	IIA	d	—	x		x		x		53
154	IIA	a	—	x		x		x		53
155	IIA	d	—	x		x		x		53
156	IIA	d	—	x		x		x	Cu	53
157	IIA	d	—	x		x		x	Cu	53
158	IIA	d	—	x		x		x	Cu	53
159	IIA	d	—	x		x		x	Cu	53

[a–g] These footnotes are the same as for Table 1.
[h] The elemental analysis for ref. [53] are only stated.
[i] These data refer to melting points.
[j] The lack of mass spectral data and the methods of synthesis leave the structures of these compounds as only tentative.

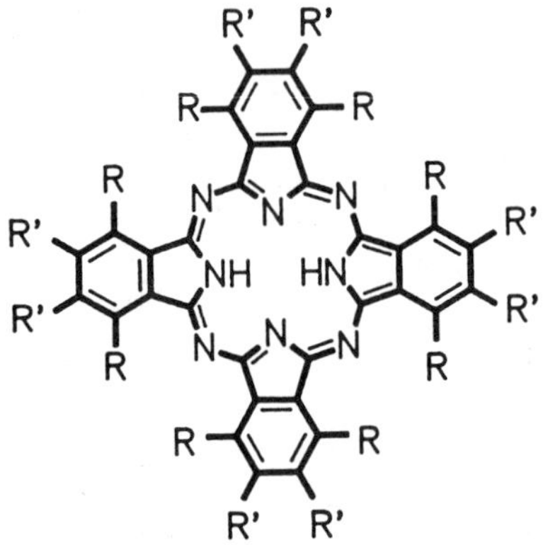

47	R = Cl
48	R = Br
49	R = I
148	R = PrS
149	R = CH$_3$(CH$_2$)$_7$S
150	R = PhS
151	R = p – CH$_3$PhS
152	R = OCH$_3$, R' = Cl

Figure 12 Some symmetrical metal-free hexadecasubstituted phthalocyanines (R = R' unless otherwise stated).

metal-free hexadecafluorophthalocyanine is still unknown although there have been attempts at its synthesis [46].

1,4,8,11,15,18,22,25-Octamethoxy-2,3,9,10,16,17,23,24-octa-chlorophthalocyanine (**152**) is one example of a symmetrical hexadecaphthalocyanine containing two different substituents. The data for these compounds are shown in Table 4.

ii. Nonidentical Substituents

A series of pentadecaalkylthio-monoisoamyloxyphthalocyanines **153–159** have been prepared as shown in Fig. 13 (Table 4) [53]. Although compounds **153–154** were purified by chromatography, the presence of other nonseparable phthalocyanines admixed with these compounds cannot be excluded because of the absence of mass spectral analysis. Unfortunately, the chemical analysis on many compounds in Table 4 were stated as having been accomplished in the patent but actual analytical results were not given. In addition, the absence of mass spectral data means that the structure for nonidentically substituted compounds remains tentative. As discussed in Section F,i the compounds in Table 4 have interesting UV-visible spectra and actually exhibit melting points.

H. MULTINUCLEAR PHTHALOCYANINES

A series of metal-free binuclear phthalocyanines (**22, 29, 160–172**) covalently linked through their benzo rings is described in Fig. 14. (All references and data on these compounds are shown in Table 5.) The two

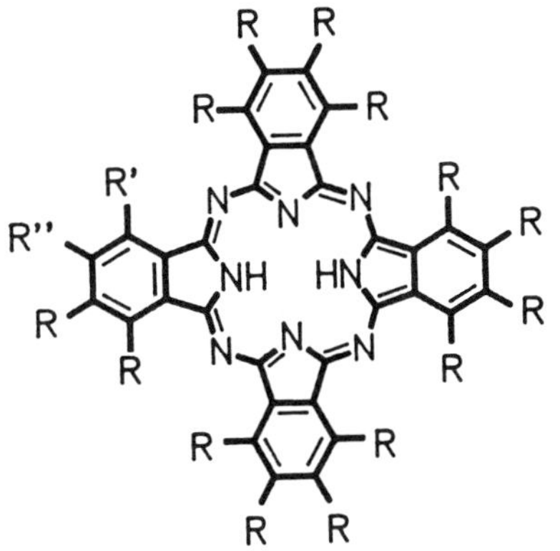

153 R = EtS
     R′ or R″ = isoamyloxy

154 R = ^{i}PrS
     R′ or R″ = isoamyloxy

155 R = n – BuS
     R′ or R″ = isoamyloxy

156 R = n – C$_5$H$_{11}$
     R′ or R″ = isoamyloxy

157 R = ⟨◯⟩–S

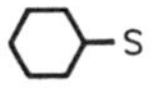

     R′ or R″ = isoamyloxy

158 R = CH$_3$CH$_2$(CH$_3$)CH
     R′ or R″ = isoamyloxy

159 R = PhCH$_2$
     R′ or R″ = isoamyloxy

Figure 13 Some nonidentically substituted hexadecasubstituted phthalocyanines (R = R′ if R″ = isoamyl, R = R″ if R′ = isoamyl).

phthalocyanine rings are bridged by 0 (**168**), 1 (**29**), 2 (**160–162, 171–172**), 3 (**169**), 4 (**163, 166, 167**), and 5 (**22, 164, 165, 170**) atoms. A planar binuclear phthalocyanine in which two phthalocyanine rings share a common benzo ring **173** actually contains less than two full phthalocyanine rings and hence has been referred to as a (– 1) linked binuclear phthalocyanine **173** (Fig. 15). A trinuclear **30** (Scheme VI) and a tetranuclear phthalocyanine **174** (Fig. 15) have been described. A mixed binuclear phthalocyanine–porphyrin **175** containing one phthalocyanine ring and one porphyrin ring linked by an oxygen atom has been made, as well as a mixed pentanuclear phthalocyanine–porphyrin **58** (Scheme XIIa) containing one phthalocyanine ring and four porphyrin moieties. Certain multinuclear phthalocyanines containing extended conformations such as **29, 171, 172, 173,** and **174** are readily purified by gel permeation chromatography while certain binuclear phthalocyanines that can readily attain cofacial conformations such as **22, 164,** and **169** cannot be purified by this method. All multinuclear phthalocyanines or mixed phthalocyanine–porphyrin compounds are sufficiently soluble in organic solvents due to the solubilizing effects of the bulky neopentoxy substituents or the porphyrinyl groups.

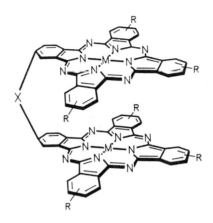

160   X = (CH₂)₂, R = CH₂CH₂C(CH₃)₃
161   X = (CH₂)₂, R = CH₂CH₂Si(CH₃)₃
162   X = (CH₂)₂, R = OCH₂C(CH₃)₃
163   X = (CH₂)₄, R = OCH₂C(CH₃)₃
29   X = O, R = OCH₂C(CH₃)₃
164   X = OCH₂C(CH₃)₂CH₂O, R = OCH₂C(CH₃)₃
22   X = OCH₂C(CH₃)(CH₂CH₃)CH₂O, R = OCH₂C(CH₃)₃
165   X = OCH₂C(CH₃)(CH₂OTr)CH₂O, R = OCH(CH₃)₂, M = Cu

166   X = [structure], R = OCH₂C(CH₃)₃

167   X = [structure], R = OCH₂C(CH₃)₃

168   X = NIL (Direct Linkage), R = OCH₂C(CH₃)₃

169   X = [structure], R = OCH₂C(CH₃)₃

170   X = [structure], R = OCH₂C(CH₃)₃

171   X = C≡C, R = OCH₂C(CH₃)₃
172a   X = CH=CH(cis), R = OCH₂C(CH₃)₃
172b   X = CH=CH(trans), R = OCH₂C(CH₃)₃

Figure 14 Metal-free binuclear phthalocyanines containing covalent bridges of 0–5 atoms through the benzo rings (M = H₂ unless otherwise stated).

Table 5 Metal-free Multinuclear Phthalocyanines and Some of Their Metallated Derivatives and Related Compounds[a]

Comp[b]	Syn[c]	Pur[d]	Yield (%)	Characterization[e]					Metal[f]	Reference[g]
				E.A.	MS	VIS	NMR	O		
160	IVB	d	8	x	x	x	x	x		72
161	IVB	d	7	x	x	x	x	x		72
162	IVB	d	10	x	x	x	x	x	Co	72
163	IVB	d	1.4	x	x	x	x	x		72
29	IVB,VI	d	11	x	x	x	x	x		33
164	IVB	d	17	x	x	x	x	x	Co	26
22	IVB	d	10	x	x	x	x	x	Cu,Co	26
165[i]	IIB	d	1	x	x	x		x	Cu,Zn,[b] Co	26
166	IVB	d,e	13	x	x	x	x	x	Cu	72
167	IVB	d,e	10	x	x	x	x	x	Co,Cu	72
168	IVB	d,e	33	x	x	x	x	x	Cu,Co	101
169	IVB	d	8.7	x	x	x	x	x	Cu,Co,Zn	101,102
170	IVB	d	12	x	x	x	x	x	Co,Zn	101
171	IVB	d,e	7	x	x	x	x	x	Co	103
172a[j]	IVB	d,e	22	x	x	x	x	x	Co,Zn	103
172b	IVB	d,e	5	x	x	x	x	x		103
173[k]	IVB	d,e	12[l]	x	x	x	x	x	Co,Cu,Zn	104
30	IVB,VI	d,e	6.6	x	x	x		x		33
174	IVB	d,e	12	x	x	x	x	x	Co,Zn[b]	27
58[m]	XIIA	g	73		x	x	x	x		52
175[m,n]	XIIB	d,i	—		x	x	x	x	Zn	105

[a–g] These footnotes are the same as for Table 1.
[b] The preparation of these zinc derivatives is described in ref. [100].
[i] This binuclear phthalocyanine was prepared as its copper derivative only.
[j] The cis isomer was completely characterized and separated from the trans isomer formed in 5% yield.
[k] This is called a (−1) bridged phthalocyanine as it contains less than two full phthalocyanine rings.
[l] Private communication.
[m] These entries are mixed phthalocyanines–porphyrin multinuclear macrocyclic compounds having one phthalocyanine ring.
[n] Obtained as the zinc derivative only.

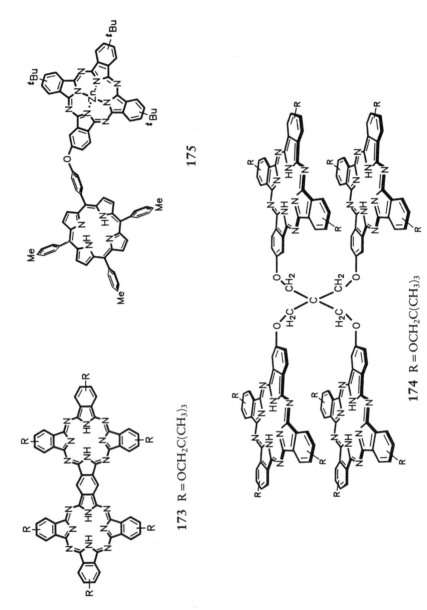

Figure 15 Other multinuclear phthalocyanines and related macrocyclic compounds containing a phthalocyanine ring (see also 30 and 58).

I. POLYAROMATIC AND HETEROCYCLIC AROMATIC PHTHALOCYANINE ANALOGS

Phthalocyanine-like compounds in which further benzo rings are added to the benzo rings of phthalocyanine, or in which heteroatoms replace carbon atoms of the benzo ring, cannot be called *phthalocyanines* because of the definition of what a phthalocyanine is. However, polyaromatic phthalocyanine analogs 63, 65, 177–183, 185 (Fig. 16) and heterocyclic phthalocyanine analogs 103, 184, 186–192 (Fig. 16) are so closely related to phthalocyanines that they can be regarded as substituted phthalocyanines. On the other hand, other analogs of phthalocyanines, such as the azaporphyrins or porphyrazines, have been adequately covered [17] and hence are not further reviewed here. The data and references for the above compounds are given in Table 6.

The unsubstituted metal-free naphthalene analogs 63 and 180, the triphenylene analog 65, and the metallated anthracene analog 181 are all highly insoluble compounds, but some of the substituted polyaromatic phthalocyanines analogs such as 176–179 and 182, 183 exhibit some solubility in organic solvents. Most of the heterocyclic aromatic compounds 184, 186–192 are quite insoluble except in acid solution in water. It should be noted, of course, that phthalocyanine analogs 103, 176, 177, 180, 187, 188, 191, and 192 all exist as a mixture of isomers (see Section E,iv).

J. OTHER PHTHALOCYANINES

A series of metal-free and some metallated, unusually substituted phthalocyanines 193–202 are described in Fig. 17 and Table 7. Although a wide variety of symmetrically substituted metal-free tetra- (Table 1), octa- (Tables 2 and 3), and hexadecasubstituted phthalocyanines (Table 4) have been described, no examples of a metal-free dodecasubstituted phthalocyanine, prepared from a trisubstituted phthalonitrile, could be found. On the other hand, an example of a *metallated* dodecasubstituted phthalocyanine (194) is shown in Fig. 17. There are also a large number of metallated substituted phthalocyanines known for which the metal-free parent compound has not been described. Phthalocyanine 195 is one example of this class of compounds, which are normally outside the scope of this chapter.

Simple monosubstituted phthalocyanines are extremely rare. Some examples of this type are abstracted in *Chemical Abstracts*, but careful reading of these references and patents invariably shows the entries refer to mixtures of compounds or poorly characterized compounds. Only 2-neopentoxyphthalo-

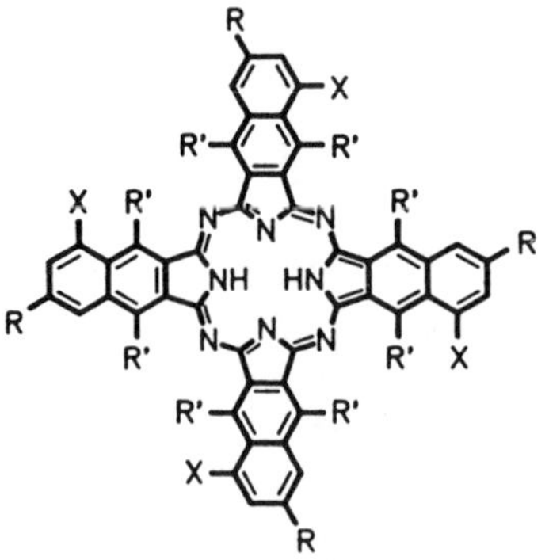

63 R = R′ = X = H
176 R = ′Bu, R′ = X = H
177 R = ′Bu, X = Br, R′ = H
178 R = X = H, R′ = O − n − Bu
179 R = X = H, R′ = O − n − C_5H_{11}

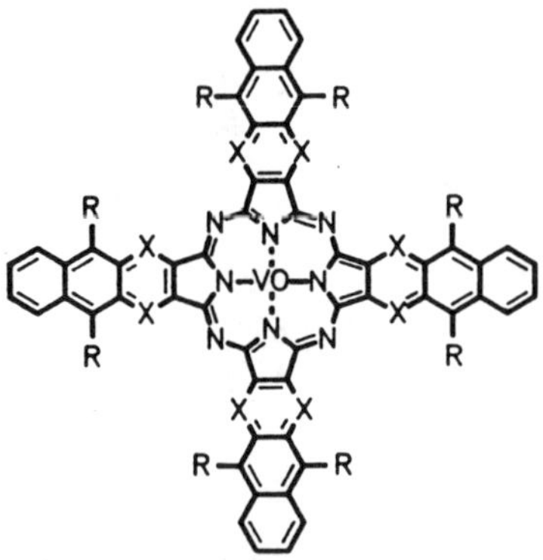

181 R = H, X = CH
182 R = Ph, X = CH
183 R = Br, X = CH
184 R = H, X = N
185 R = Oxo, X = CH

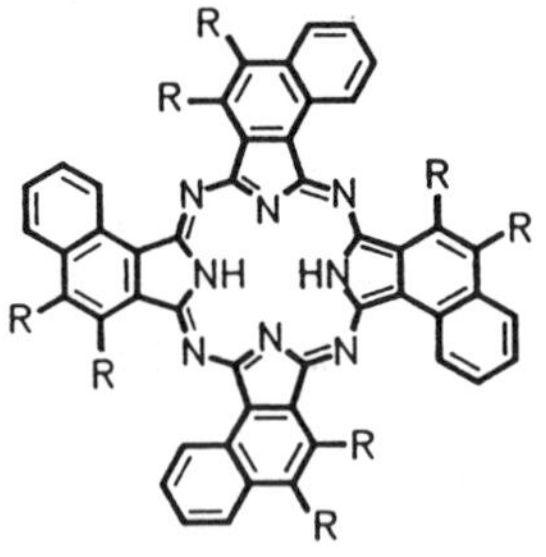

180a R = H
180b R,R = Benzo

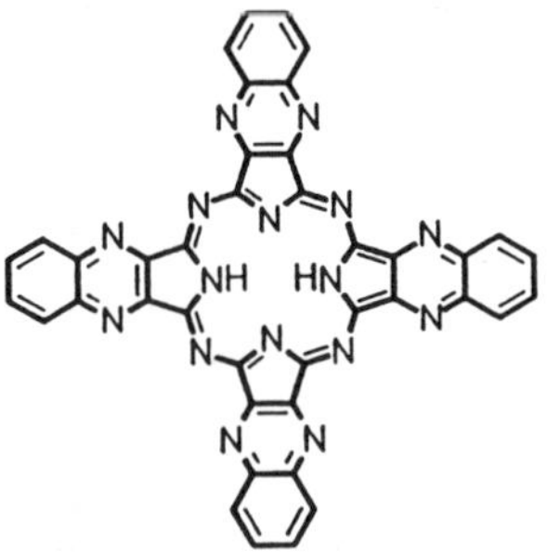

186

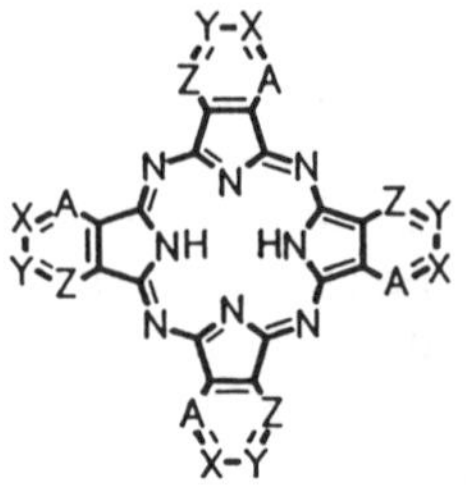

187 A = N, X = Y = Z = CH
188 X = N, A = Y = Z = CH
189 A = Z = N, X = Y = CH
190 X = Y = N, A = Z = CNH_2

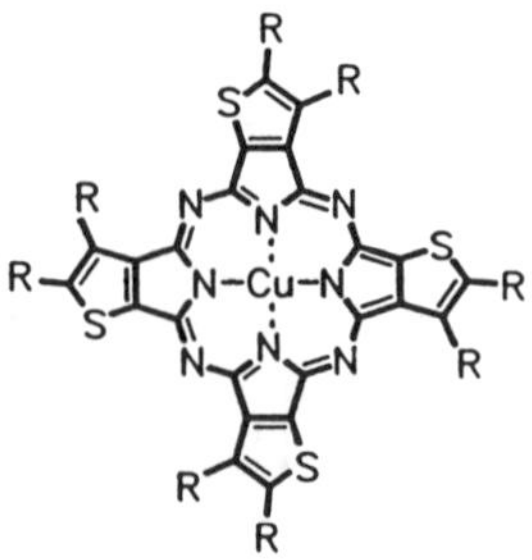

191 R = H
192 R,R = Benzo

Figure 16 Some metal-free and metallated polyaromatic and heteroaromatic phthalocyanine analogs (see also structures 65 and 103–107).

Table 6 Metal-Free and Some Metallated Polyaromatic and Heteroaromatic Phthalocyanine Analogs[a]

Comp[b]	Syn[c]	Pur[d]	Yield (%)	E.A.	MS	VIS	NMR	O	Metal[f]	Reference[g]
63	IIA	f	26	x	x	x		x	Cu,Pb,VO[h]	106,107
176	IIA	a,c	21	x		x			Cu,Zn,AlOH[i]	108
177	IIA	c	10	x		x			Cu,Zn,AlCl,VO	109
178	IIA	d	22	x		x	x			99
179	IIA	d	30	x		x	x			99
180a	IIB	a	69	x		x			Cu,Zn,Pb,Mg	70
180b	IIA	f	39	x		x	x			109a
181[j]	IXA	a	84	x[k]		x			VO,AlOH,Cu	110
182[j]	IXA[l]	f	52	x		x			VO,AlOH,Cu	110,111
183[j]	IXA	f	86	x[k]		x			VO,AlOH,Cu	110
184[j]	IXA[l]	a	20	x					VO	112
185[j]	IXA[l]	a	30	x		x			VO,AlOH	113
65[j]	IXA[l]	b	37	x					Cu,VO	59
186	IIA	a	67	x		x			Cu,Io,Zn,VO	71,114
187	IIA	a	72	x		x			Cu,Zn,Co,VO	71,114
187	IIIB	a	34	x		x			Mg,Cu,Zn,Co,AlOH	73
188	IXA	b,f	—					x[m]		115
189	IIA	a	80	x		x			Cu,Co,Zn,VO	71,114
189	IIIB	f	96	x				x		21
190	XIIA	f	97	x						51
103	XIIA	f	37	x		x			Cu,Zn,AlOH	73
191[j]	IIB	a	58	x					Cu	71
192[j]	IIB	a	59	x[k]					Cu	71

[a-g] These footnotes are the same as for Table 1.

[h] Other metal derivatives in ref. [106] are Mg, Zn, AlOH, GaOH, $SnCl_2$, Mn, Fe, Co, Ni, and Pd.

[i] Other metal derivatives in ref. [108] are $SnCl_2$, VO, Mn, Co, Ni, and Pd.

[j] These compounds were not made as metal-free phthalocyanines.

[k] These analyses are unsatisfactory.

[l] These syntheses use Method IXA, but use the o-dinitrile.

[m] Only method of characterization is the blue color of 188.

Table 7 Unusually Substituted Phthalocyanines[a]

Comp[b]	Syn[c]	Pur[d]	Yield (%)	Characterization[e]					Metal[f]	Reference[g]
				E.A.	MS	VIS	NMR	O		
33	VII	f	7	x	x					37
193	VII	f	7		x					37
35	VIII	d,i	3.5		x					39
36[b]	VII[b]	d,i	—		x					39
194[i]	IIA	f	60	x		x			Cu	74
195[i]	IIIB	d	34	x	x	x	x	x	Co,Ni,Pb	116
196	IIA	c	26	x	x	x	x	x	Cu	117
197[i]	IIA	c,f,g	35–49	x		x	x	x	Cu	118–120
198[i]	IIA	c	41	x		x		x	Cu	117
199[i]	IIA	g	15	x		x	x	x	Cu	91
200[i]	IIA	g	15	x		x	x	x	Cu	91
201[i]	IIA	g	15	x		x	x	x	Cu	91
202[i]	IIA	g	15	x		x	x	x	Cu	91

[a-g] These footnotes are the same as for Table 1.
[b] This compound was also made by Method VII by the author.
[i] These phthalocyanines were not made in their metal-free forms.

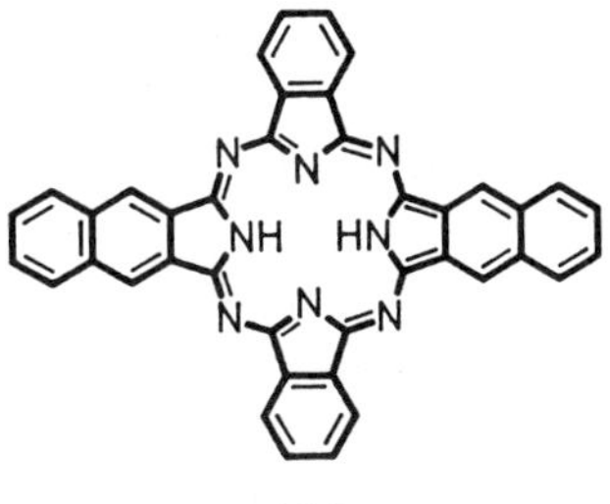

193

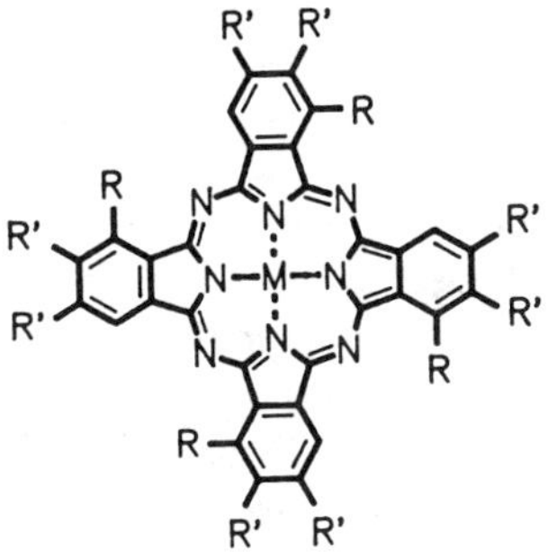

194 R = R′ = CF$_3$, M = Cu
195 R = H, R′ = CH$_2$OC$_8$H$_{17}$, M = Cu

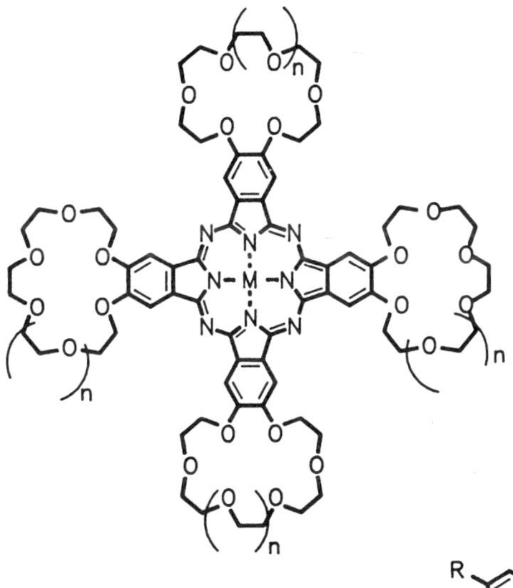

196 n = 1, M = H$_2$
197 n = 0, M = Cu
198 n = 2, M = Cu

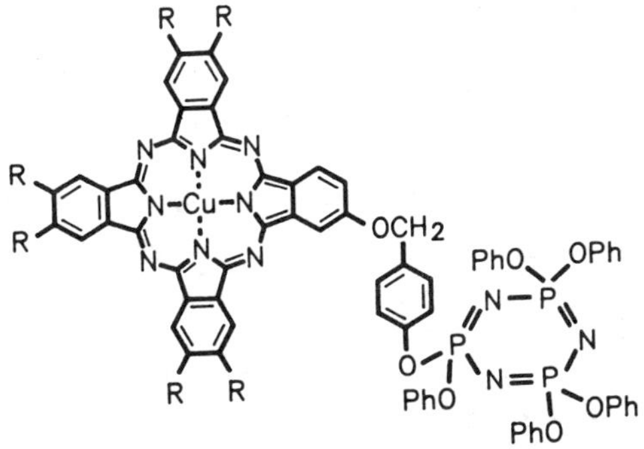

199 R = H
200 R = CH$_3$
201 R = CH$_2$OPh
202 R = CH$_2$OCH$_2$CH$_2$OCH$_3$

Figure 17 Monosubstituted, disubstituted, heptasubstituted, dodecasubstituted, and other unusual phthalocyanines (see also structures 33, 35, and 36).

cyanine (35) and the 2-phosphazenylphthalocyanine copper derivative 199 appear to be bonafide examples of this class of substituted phthalocyanines. Even 35 was isolated in a very small amount (2 mg) only after heroic chromatographic separations and mass spectral analysis of each fraction [39], while 199 was well-characterized but, unfortunately, its mass spectrum was unattainable and, therefore, some doubts about its purity may still remain.

Disubstituted phthalocyanines, such as 33 and 193, have been made in low yield by a unique method of synthesis (Method VII) and 36 was produced in low yield by Method VIII after extensive chromatographic purification. Although Method VII gave disubstituted phthalocyanines in only 7% yield, the products were pure when analyzed by mass spectroscopy. In our own laboratory, we have confirmed this patent literature result [37] by preparing 36 using Method VII and only traces of other phthalocyanines could be detected by mass spectroscopy using this procedure.

Some interesting tetra crown ethers of phthalocyanines (196–198) have been prepared while some unusual phosphazenylphthalocyanines (199–202) represent some very rare examples of heptasubstituted phthalocyanines. Although 200–202 are well-characterized, the unsymmetrical distribution of the substituents and their different substituents requires that 200–202 be analyzed by mass spectroscopy for confirmation of structure. Unfortunately, 200–202 did not give parent ions under the conditions described [91].

K. CONCLUSIONS

It has been our intention in this chapter to outline all of the available well-characterized, metal-free substituted phthalocyanines and some closely related analogs that have been synthesized since 1965. I have concentrated my efforts on the methods of synthesis, purification of these compounds, and level of characterization of these substituted phthalocyanines, but detailed discussions on the vast array of properties and applications of phthalocyanines will be the subject of the following chapters in this volume.

As metallated phthalocyanines have not been covered in depth in this chapter, metallated phthalocyanines substituted with axial ligands [77,121–124] have necessarily also been outside the scope of this chapter. This topic will be covered by other authors.

ACKNOWLEDGMENTS

I would like to thank Professor David Dolphin and the Department of Chemistry at the University of British Columbia for providing me with the encouragement and facilities to complete this chapter during a sabbatical leave from York University. I also thank Ms. Shafrira Greenberg of York University for valuable discussions and for proofreading the manuscript.

REFERENCES

1. A. Braun and J. Tcherniac, *Ber. Deut. Chem. Ges.*, 40 (1907) 2709.

2. G. T. Byrne, R. P. Linstead and A. R. Lowe, *J. Chem. Soc.*, (1934) 1017.

3. R. P. Linstead and A. R. Lowe, *J. Chem. Soc.*, (1934) 1022.

4. C. E. Dent, R. P. Linstead and A. R. Lowe, *J. Chem. Soc.*, (1934) 1033.

5. J. A. Elvidge and R. P. Linstead, *J. Chem. Soc.*, (1955) 3536.

6. C. E. Dent and R. P. Linstead, *J. Chem. Soc.*, (1934) 1027.

7. J. M. Robertson, *J. Chem. Soc.*, (1935) 615.

8. J. M. Robertson, *J. Chem. Soc.*, (1936) 1195.

9. J. M. Robertson and I. Woodward, *J. Chem. Soc.*, (1937) 219.

10. R. P. Linstead, *Ber. Deut. Chem. Ges.*, 72A (1939) 93.

11. A. B. P. Lever, *Adv. Inorg. Chem. Radiochem.*, 7 (1965) 27.

12. F. H. Moser and A. L. Thomas, *Phthalocyanine Compounds*, Reinhold, New York; Chapman and Hall, London, 1963.

13. K. Kasuga and M. Tsutsui, *Coord. Chem. Rev.*, 32 (1980) 67.

14. P. Sayer, M. Gouterman and C. R. Connell, *Acc. Chem. Res.*, 15 (1982) 73.

15. D. Wöhrle and G. Meyer, *Kontakte (Darmstadt)*, (1985) 38.

16. A. B. P. Lever, M. R. Hempstead, C. C. Leznoff, W. Lui, M. Melnik, W. A. Nevin and P. Seymour, *Pure Appl. Chem.*, 58 (1986) 1467.

17. A. H. Jackson, in *The Porphyrins*, Vol. 1, D. Dolphin, Ed., Academic, York, 1963, pp. 374–388.

18. B. D. Berezin, *Coordination Compounds of Porphyrins and Phthalocyanines*, Wiley, Chichester, 1981.

19. F. H. Moser and A. L. Thomas, *The Phthalocyanines*, Vols. 1, 2, CRC, Boca Raton, 1983.

20. P. A. Barrett, D. A. Frye and R. P. Linstead, *J. Chem. Soc.*, (1938) 1157.

21. P. J. Brach, S. J. Grammatica, O. A. Ossanna and L. Weinberger, *J. Heterocyclic Chem.*, 7 (1970) 1403.

22. V. M. Derkacheva and E. A. Luk'yanets, *Zh. Obshch. Khim.*, 50 (1980) 2313; *J. Gen. Chem. USSR*, 50 (1980) 1874.

22a. H. Tomoda, E. Hibiya, Nakamura, H. Ito and S. Saito, *Chem. Lett.*, (1976) 1003.

22b. H. Tomoda, S. Saito, S. Ogawa and S. Shiraishi, *Chem. Lett.*, (1980) 1277.

23. P. A. Barrett, C. E. Dent and R. P. Linstead, *J. Chem. Soc.*, (1936) 1719.

24. A. W. Snow, J. R. Griffith and N. P. Marullo, *Macromolecules*, 17 (1984) 1614.

25. U.S. Patent Appl. 768004 (1986); *Chem. Abstr.*, 105 (1986) 145325f.

26. C. C. Leznoff, S. M. Marcuccio, S. Greenberg, A. B. P. Lever and K. B. Tomer, *Can. J. Chem.*, 63 (1985) 623.

27. W. A. Nevin, W. Liu, S. Greenberg, M. R. Hempstead, S. M. Marcuccio, M. Melnik, C. C. Leznoff and A. B. P. Lever, *Inorg. Chem.*, 26 (1987) 891.

28. G. Pawlowski and M. Hanack, *Synthesis*, (1980) 287.

29. C. C. Leznoff and T. W. Hall, *Tetrahedron Lett.*, 23 (1982) 3023.

30. T. W. Hall, S. Greenberg, C. R. McArthur, B. Khouw and C. C. Leznoff, *Nouv. J. Chim.*, 6 (1982) 653.

31. C. C. Leznoff, S. Greenberg, S. M. Marcuccio, P. C. Minor, P. Seymour, A. B. P. Lever and K. B. Tomer, *Inorg. Chim. Acta*, 89 (1984) L35.

32. D. Wöhrle and G. Krawczyk, *Polym. Bull*, 15 (1986) 193.

33. S. Greenberg, S. M. Marcuccio, C. C. Leznoff and K. B. Tomer, *Synthesis*, (1986) 406.

34. J. A. Elvidge and N. R. Barot, in *The Chemistry of Double Bonded Functional Groups*, Part 2, S. Patai, Ed., Wiley, London, 1977, p. 1167.

35. N. H. Haddock and J. C. Woods, Brit. Pat. 762,783 (1956); *Chem. Abstr.*, (1957) 12891e.

36. Brit. Pat., 704595 (1954); *Chem. Abstr.*, 49 (1955) 7001.

37. E. M. Idelson, U.S. Pat., 4,061,654, (1977); *Chem. Abstr.*, 88 (1977) 171797m.

38. T. Wimmer, Austrian Pat., AT267711 (1969); *Chem. Abstr.*, 71 (1969) 712924f.

39. C. C. Leznoff, S. Greenberg, B. Khouw and A. B. P. Lever, *Can. J. Chem.*, 65 (1987) 1705.

40. W. C. Still, M. Khan and A. Mitra, *J. Org. Chem.*, 43 (1978) 2923.

41. N. M. Targett, J. P. Kilcoyne and B. Green, *J. Org. Chem.*, 44 (1979) 4262.

42. D. R. Boston and J. C. Bailar, Jr., *Inorg. Chem.*, 11 (1972) 1578.

43. H. Shirai, A. Maruyama, K. Kobayashi, N. Hojo and K. Urushidu, *Makromol. Chem.*, 181 (1980) 575.

44. J. Metz, O. Schneider and M. Hanack, *Inorg. Chem.*, 23 (1984) 1065.

45. J. H. Weber and D. H. Busch, *Inorg. Chem.*, 4 (1965) 469.

46. D. Bonderman, E. D. Cater and W. E. Bennett, *J. Chem. Eng. Data*, 15 (1970) 396.

47. T. Mishi, K. Arishima and H. Hiratsuka, *Jpn. Kokai Tokkyo Koho*, JP 61-32051 (1986); *Chem. Abstr.*, 105 (1986) 88745s.

48. S. Greenberg, A. B. P. Lever and C. C. Leznoff, *Can. J. Chem.*, 66 (1988) 1059.

49. G. F. Ficken and R. P. Linstead, *J. Chem. Soc.*, (1952) 4846.

50. G. E. Ficken, and R. P. Linstead, E. Stephen and M. Whalley, *J. Chem. Soc.*, (1958) 3879.

51. D. Wöhrle, G. Meyer and B. Wahl, *Makromol Chem*, 181 (1980) 2127.

52. N. Kobayashi, Y. Nishiyama, T. Ohya and M. Sato, *J. Chem. Soc. Chem. Commun.*, (1987) 390.

53. P. J. Duggan and P. F. Gordon, Eur. Pat. Appl., EP 155780 (1985); *Chem. Abstr.*, 105 (1987) 70242r.

54. P. A. Barrett, E. F. Bradbrook, C. E. Dent and R. P. Linstead, *J. Chem. Soc.*, (1939) 1820.

55. M. Hedayatullah, *Compt. Rend. Acad. Sci. Ser. II Mec. Phys.*, 296 (1983) 621.

56. B. Honigmann, H. U. Lenne and R. Schröeder, *Z. Kristallogr.*, 122 (1965) 185.

57. D. E. Games, A. H. Jackson and K. T. Taylor, *Org. Mass. Spectr.*, 9 (1974) 1245.

58. S. A. Mikhalenko, S. V. Barkanova, O. L. Lebedev and E. A. Luk'yanets, *Zh. Obshch. Khim.*, 41 (1971) 2735; *J. Gen. Chem. USSR*, 41 (1971) 2770.

59. S. Ya. Mikhalenko, L. A. Yagodina and E. A. Luk'yanets, *Zh. Obshch. Khim.*, 46 (1976) 1598; *J. Gen. Chem. USSR*, 46 (1981) 1557.

60. S. A. Mikhalenko, V. M. Derkacheva and E. A. Luk'yanets, *Zh. Obshch. Khim.*, 51 (1981) 1650; *J. Gen. Chem. USSR*, 51 (1981) 1405.

61. J. A. Anton, J. Kwong and P. A. Loach, *J. Heterocyclic Chem.*, 13 (1976) 717.

62. Y. C. Yang, J. R. Ward and R. P. Seiders, *Inorg. Chem.*, 24 (1985) 1765.

63. M. Abkowitz and A. R. Monahan, *J. Chem. Phys.*, 58 (1973) 2281.

64. I. Rosenthal, E. Ben-Hur, S. Greenberg, A. Concepcion-Lam, D. M. Drew and C. C. Leznoff, *Photochem. Photobiol*, 46 (1987) 959.

65. J. A. Elvidge and J. H. Golden, *J. Chem. Soc.*, (1957) 700.

66. V. F. Borodkin, *Zh. Prikl. Khim.*, 31 (1958) 813; *J. Appl. Chem. USSR*, 31 (1958) 803.

67. T. J. Hurley, M. A. Robinson and S. I. Trotz, *Inorg. Chem.*, 6 (1967) 389.

68. S. W. Oliver and T. D. Smith, *J. Chem. Soc. Perkin Trans II*, (1987) 1579.

69. S. Gaspard and Ph. Maillard, *Tetrahedron*, 43 (1987) 1083.

69a. I. G. Oksengendler, N. V. Kondratenko, E. A. Luk'yanets and L. M. Yagupol'skii, *Zh. Org. Khim.*, 13 (1977) 2234; *J. Org. Chem. USSR*, 13 (1977) 2085.

70. E. F. Bradbrook and R. P. Linstead, *J. Chem. Soc.*, (1936) 1744.

71. R. P. Linstead, E. G. Noble and J. M. Wright, *J. Chem. Soc.*, (1937) 911.

72. S. M. Marcuccio, P. I. Svirskaya, S. Greenberg, A. B. P. Lever, C. C. Leznoff and K. B. Tomer, *Can. J. Chem.*, 63 (1985) 3057.

73. D. Wöhrle, J. Gitzel, I. Okuro and S. Aono, *J. Chem. Soc. Perkin Tran. II*, (1985) 1171.

74. I. G. Oksengendler, N. V. Kondratenko, E. A. Luk'yanets and L. M. Yagupol'skii, *Zh. Org. Khim.*, 13 (1977) 1554; *J. Org. Chem. USSR*, 13 (1977) 1430.

75. I. G. Oksengendler, N. V. Kondratenko, E. A. Luk'yanets and L. M. Yagupol'skii, *Zh. Org. Khim*, 14 (1978) 1046; *J. Org. Chem. USSR*, 14 (1978) 976.

76. V. M. Derkacheva, O. L. Kaliya and E. A. Luk'yanets, *Zh. Obshch Khim.*, 53 (1983) 188; *J. Gen. Chem. USSR*, 53 (1983) 163.

77. M. Hanack, J. Metz and G. Pawlowski, *Chem. Ber.*, 115 (1982) 2836.

77a. C. C. Leznoff, S. Vigh, P. I. Svirskaya, S. Greenberg, D. M. Drew, E. Ben-Hur and I. Rosenthal, *Photochem. Photobiol.*, (1988) 49 (1989) 279.

78. G. Pawlowski and M. Hanack, *Syn. Commun.*, 11 (1981) 351.

79. H. Hopff and P. Gallegra, *Helv. Chim. Acta*, 51 (1968) 253.

80. S. A. Mikhalenko and E. A. Luk'yanets, *Zh. Obshch. Khim.*, 39 (1969) 2129; *J. Gen. Chem. USSR*, 39 (1969) 2081.

81. N. P. Marullo and A. W. Snow, *ACS Symp. Ser.*, 14 (1982) 325.

82. A. W. Snow and N. L. Jarvis, *J. Am. Chem. Soc.*, 106 (1984) 4706.

83. S. W. Oliver and T. D. Smith, *Hetrocycles*, 22 (1984) 2047.

84. R. P. Linstead and F. T. Weiss, *J. Chem. Soc.*, (1950) 2975.

85. L. I. Solov'eva, S. A. Mikhalenko, E. V. Chernykh and E. A. Luk'yanets, *Zh. Obshch. Khim.*, 52 (1982) 90; *J. Gen. Chem. USSR*, 52 (1982) 83.

86. M. J. Cook, M. F. Daniel, K. J. Harrison, N. B. McKeown and A. J. Thomson, *J. Chem. Soc. Chem. Commun.*, (1987) 1086.

87. Z. Witkiewicz, R. Dabrowski and W. Waclawek, *Mater. Sci. II*, (1976) 88.

88. M. J. Cook, M. F. Daniel, K. J. Harrison, N. B. McKeown and A. J. Thomson, *J. Chem. Soc. Chem. Commun.*, (1987) 1148.

89. W. A. Barlow, Eur. Pat. Appl., EP 157568 (1985); *Chem. Abstr.*, 104 (1985) 177813j.

90. E. A. Cuellar and T. J. Marks, *Inorg. Chem.*, 20 (1981) 3766.

91. H. R. Allcock and T. X. Neenan, *Macromolecules*, 19 (1986) 1496.

92. M. J. Camenzind and C. L. Hill, *J. Heterocyclic Chem.*, 22 (1985) 575.

93. C. Piechocki and J. Simon, *Nouv. J. Chim.*, 9 (1985) 159.

93a. I. Cho and Y. Lim, *Chem. Lett.*, (1987) 2107.

94. M. G. Gal'pern, V. K. Shalaev, T. A. Shatsskaya, L. S. Shishkanova, V. R. Skvarchenko and E. A. Luk'yanets, *Zh. Obshch. Khim.*, 53 (1983) 2601; *J. Chem. USSR*, 53 (1983) 2346.

95. D. Wöhrle and U. Hundorf, *Makromol. Chem.*, 186 (1985) 2177.

96. L. I. Solov'eva and E. A. Luk'yanets, *Zh. Obshch. Khim.*, 50 (1980) 1122; *J. Gen. Chem. USSR*, 50 (1980) 907.

97. C. Piechocki and J. Simon, *J. Chem. Soc. Chem. Commun.*, (1985) 259.

98. S. A. Mikhalenko and E. A. Luk'yanets, *Zh. Org. Khim.*, 11 (1975) 2216; *J. Org. Chem. USSR*, 11 (1975) 2246.

99. K. J. Harrison, M. J. Cook, S. D. Howe and A. J. Thomson, UK Pat. Appl., GB 2168372 Al (1986); *Chem. Abstr.*, 105 (1986) 228518x.

100. V. Manivannan, W. A. Nevin, C. C. Leznoff and A. B. P. Lever, *J. Coord. Chem.*, 19 (1988) 139.

101. H. Lam, S. M. Marcuccio, P. Svirskaya, S. Greenberg, A. B. P. Lever, C. C. Leznoff and R. L. Cerny, *Can. J. Chem.*, 67 (1989) 1087.

102. C. C. Leznoff, H. Lam, W. A. Nevin, N. Kobayashi, P. Janda and A. B. P. Lever, *Angew. Chem. Int. Ed. Eng.*, 26 (1987) 1021.

103. S. Vigh, H. Lam, A. B. P. Lever, C. C. Leznoff and R. L. Cerny, manuscript in preparation.

104. C. C. Leznoff, H. Lam, S. M. Marcuccio, W. A. Nevin, P. Janda, N. Kobayashi and A. B. P. Lever, *J. Chem. Soc. Chem. Commun.*, (1987) 699.

105. S. Gaspard, C. Giannotti, P. Maillard, C. Schaeffer and T-H. Tran-Thi, *J. Chem. Soc. Chem. Commun.*, (1986) 1239.

106. S. A. Mikhalenko and E. A. Luk'yanets, *Zh. Obshch. Khim.*, 39 (1969) 2254; *J. Gen. Chem. USSR*, 39 (1969) 2495.

107. M. L. Kaplan, A. J. Lovinger, W. D. Reents, Jr. and P. H. Schmidt, *Mol. Cryst. Liq. Cryst.*, 112 (1984) 345.

108. E. I. Kovshev and E. A. Luk'yanets, *Zh. Obshch. Khim.*, 42 (1972) 696; *J. Gen. Chem. USSR*, 42 (1972) 691.

109. M. G. Gal'pern, T. D. Talismanova, L. G. Tomilova and E. A. Luk'yanets, *Zh. Obshch Khim.*, 55 (1985) 1099; *J. Gen. Chem. USSR*, 55 (1985) 980.

109a. A. W. Snow and T. R. Price, *Synth. Metals*, 9 (1984) 329.

110. W. Freyer and L. Q. Minh, *Monatsh. Chem.*, 117 (1986) 475.

111. V. N. Kopranenkov and E. A. Luk'yanets, *Zh. Obshch. Khim.*, 41 (1971) 2341; *J. Gen. Chem. USSR*, 41 (1971) 2366.

112. W. Freyer, *Z. Chem.*, 26 (1986) 217.

113. W. Freyer, *Z. Chem.*, 26 (1986) 216.

114. M. G. Gal'pern and E. A. Luk'yanets, *Zh. Obshch. Khim.*, 39 (1969) 2536; *J. Gen. Chem. USSR*, 39 (1969) 2477.

115. *Jpn. Kokai Tokkyo Koho*, JP 60-95441 (1985); *Chem. Abstr.*, 104 (1985) 79155b.

116. M. Hanack, A. Beck and H. Lehmann, *Synthesis*, (1987) 703.

117. Ot. E. Sielcken, M. M. van Tilborg, M. F. M. Roks, R. Hendriks, W. Drenth and R. J. M. Nolte, *J. Am. Chem. Soc.*, 109 (1987) 4261.

118. A. R. Koray, V. Ansen and O. Bekaroglu, *J. Chem. Soc. Chem. Commun.*, (1986) 932.

119. N. Kobayashi and Y. Nishiyama, *J. Chem. Soc. Chem. Commun.*, (1986) 1462.

120. R. Hendriks, Ot. E. Sielcken, W. Drenth and R. J. M. Nolte, *J. Chem. Soc. Chem. Commun.*, (1986) 1464.

121. P. C. Krueger and M. E. Kenney, *J. Org. Chem.*, 28 (1963) 3379.

122. J. N. Esposito, J. E. Lloyd and M. E. Kenney, *Inorg. Chem.*, 5 (1966) 1979.

123. B. L. Wheeler, G. Nagasubramanian, A. J. Bard, L. A. Schechtman, D. R. Dininny and M. E. Kenney, *J. Am. Chem. Soc.*, 106 (1984) 7404.

124. E. Ciliberto, K. A. Doris, W. J. Pietro, G. M. Reisner, D. E. Ellis, I. Fragala, F. H. Herbstein, M. A. Ratner and T. J. Marks, *J. Am. Chem. Soc.*, 106 (1984) 7748.

2

Phthalocyanines in
Polymer Phases

Dieter Wöhrle

A. FUNDAMENTALS

Special porphyrins like phthalocyanines **1** are interesting worldwide as materials with unconventional properties. Variation of substituents at the ligand and of the metal ion in the core of the ligand is the prerequisite for constructing new materials that are active as sensitizers (e.g., in the photodynamic cancer therapy), catalysts (e.g., for thiol oxidation in gasoline fractions), electrocatalyst (e.g., for O_2 reduction in fuel cells), sensors, display devices, and information storage systems.

1

The combination of a phthalocyanine with a polymer or the incorporation of a phthalocyanine into a polymer matrix is another powerful tool for designing new materials with special properties. The combination of a metal ion, a ligand, and the chemical environment (such as polymer) determines the chemical and physical properties. Hemoglobin shows the result of this combination extraordinarily well. The reversible binding of dioxygen is the consequence of the combination Fe(II)/porphyrin–derivative/polymer. Preparative chemistry opens the way for synthesizing various combinations of phthalocyanines/polymers, which will be discussed in the following sections:

Section B, type A polymers: The ligand is part of a polymer network or a polymer chain. Polymers, insoluble in most solvents,
that exhibit high thermal stability and good catalytical and electrochemical activity are described.

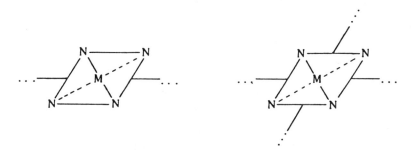

Section C, type B polymers: The metal atom of a phthalocyanine
as part of a polymer chain. The stacking of the phthalocyanines
results in high electrical conductivity of the polymers.

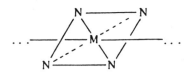

Section D, type C polymers: Phthalocyanines covalently bound
over the ligand to a polymer chain. Binding at linear polymers
leads to soluble polymers that were investigated, e.g., for electron transfer and photoelectron transfer reactions.

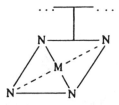

Section E, type D polymers: The interaction of a phthalocyanine
occurs coordinatively at the metal of a phthalocyanine with a
polymer donor ligand or electrostatically of a charged phthalocyanine with a charged polymer chain.

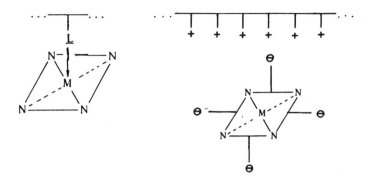

Section F, type E polymers: The simplest combination treats the physical incorporation into the matrix of an organic or inorganic polymer. These materials are of interest as dye stuffs, active electrode coatings, and catalysts.

In 1983, one paper described some results for the various combinations of phthalocyanine–polymers [1]. Since then several publications enlarged the field with new synthetic procedures for materials with fascinating properties.

B. POLYMERIC PHTHALOCYANINES THROUGH THE LIGAND

The phthalocyanine system contains four benzene rings at equal positions and of the same reactivity. Therefore the connection of phthalocyanine rings with each other by annulated benzene rings leads to polymers of type A existing in a two-dimensional structure [1–3]. Generally, these polymers are insoluble in organic solvents but some of them can be dissolved in conc. sulfuric acid. Two possibilities exist for preparing phthalocyanines covalently connected over the ligand. The well-known synthesis of low-molecular-weight phthalocyanines starts from 1,2-benzenedicarbonitriles (*o*-phthalodinitrile) or 1,2-benzenedicarboxylic acid derivatives (e.g., phthalic acid anhydride) [Eq. (1)]. These compounds are monofunctional in phthalocyanine synthesis because the cyclotetramerization results in the formation of low-molecular-weight compounds. Applying bifunctional reagents like aromatic tetracarbonitriles or tetracarboxylic acid derivatives in synthesis give polymeric phthalocyanine by polycyclotetramerization [Eq. (2)].

The second route utilizes tetrasubstituted low-molecular-weight phthalocyanine derivatives. By reacting with other bifunctional or higher functional compounds (or by self-condensation), polymeric phthalocyanines are prepared [Eq. (3)].

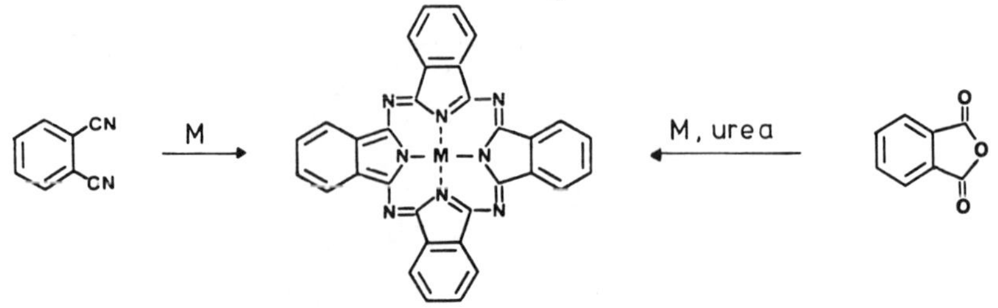

(1)

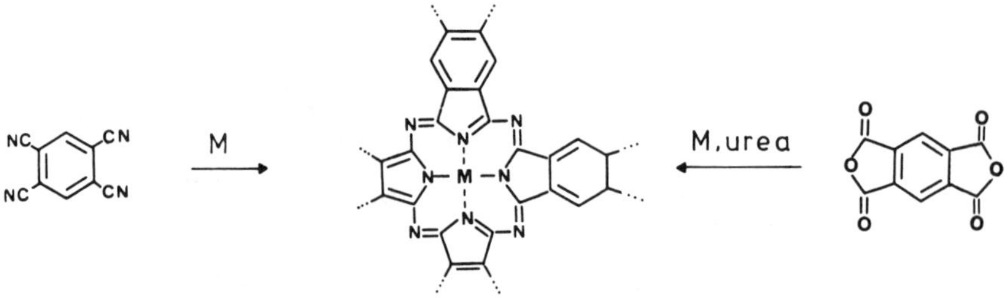

(2)

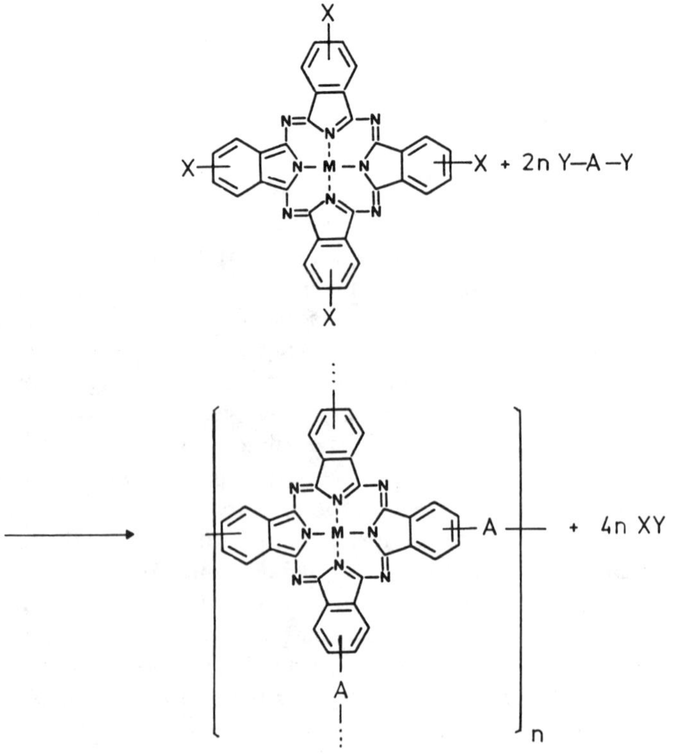

(3)

The synthesis of linear polymeric phthalocyanines connected by benzene rings was unsuccessful until now. The reaction of a mixture of, e.g., a 1,2-benzenedicarboxylic acid derivative with a 1,2,4,5-benzenetetracarboxylic acid derivative gives polymers that may consist of a mixture of several isomers. The direct synthesis of symmetrical disubstituted phthalocyanines for use as bifunctional monomers in polycondensation or polyaddition reactions also was not successful.

i. Polymeric Phthalocyanines from Aromatic Carbonitriles or Acid Derivatives

For the preparation of polymeric phthalocyanines, either tetracarbonitriles like 1,2,4,5-tetracyanobenzene, respectively various oxy-, arylenedioxy-, and alkylenedioxy-bridged diphthalonitriles (and other diphthalonitriles), or tetracarboxylic acid derivatives like 1,2,4,5-benzenetetracarboxylic acid dianhydride were employed mainly in the presence of metal salts or metals. The polycyclotetramerization of tetracarbonitriles exhibits the advantage that only a two-electron reduction occurs after cyclization to the phthalocyanine dianion, which gives a clear picture of the reaction pathway. Equations (4) and (5) show the stoichiometric equations for a two-dimensional network polymer starting from 1,2,4,5-tetracyanobenzene and Cu or $CuCl_2$. In contrast, the reactions of tetracarboxylic acid derivatives in the presence of urea are more complex.

$$2n \ (C_{10}H_2N_4) + n \cdot Cu^0 \rightarrow [(C_{20}H_4N_8)^{2-}Cu^{2+}]_n \qquad (4)$$

$$2n \ (C_{10}H_2N_4) + n \cdot CuCl_2 \rightarrow [(C_{20}H_4N_8)^{2-}Cu^{2+}]_n + n \cdot Cl_2 \qquad (5)$$

For complete characterization of the polymers, the following points must be considered: structural uniformity, nature of the end groups, metal content, and degree of polymerization (molecular weight). Most reports in the literature made insufficient statements regarding these important points. Equations (6)–(8) show the reactions for the most applied 1,2,4,5-tetracyanobenzene, oxy-, alkylenedioxy-, arylenedioxy-bridged diphthalonitriles, and 1,2,4,5-benzenetetracarboxylic acid derivatives, which will be considered later. These equations include some possible side products 3, 4, which, besides the phthalocyanine structure, can be covalently incorporated into the polymer (structurally nonuniform), the metal atom in the core of the ligand (degree of metallization), the nature of end groups, and the factor n (degree of polymerization).

(9)

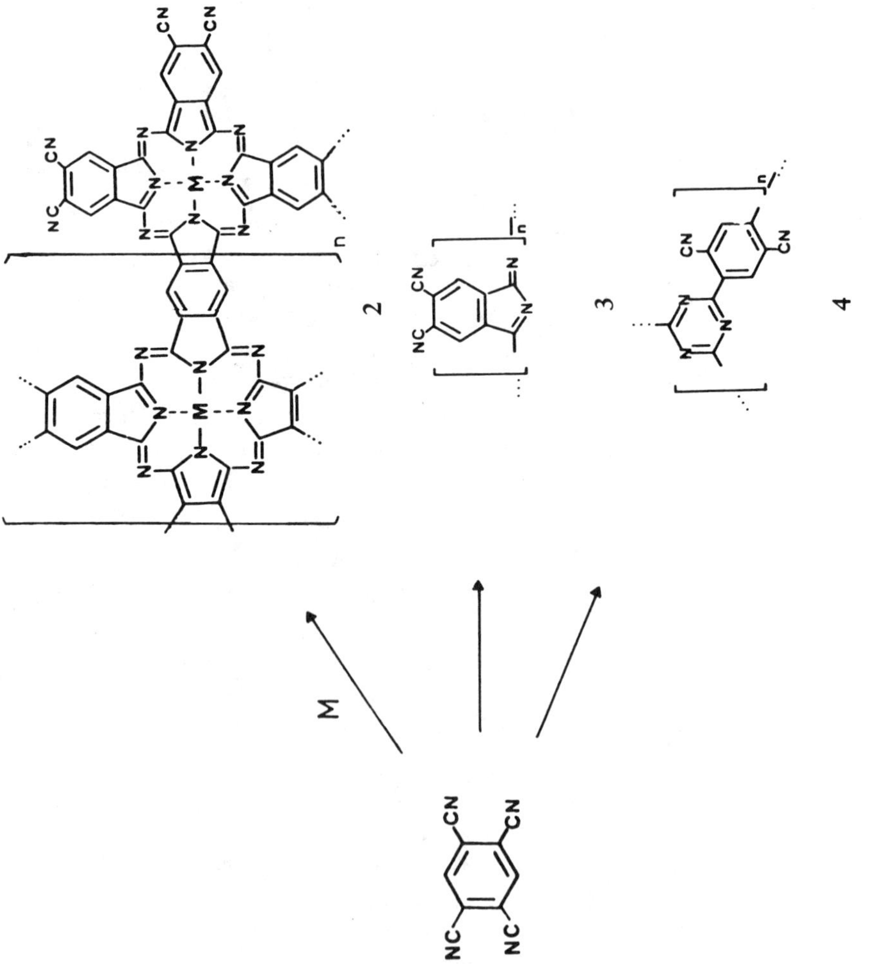

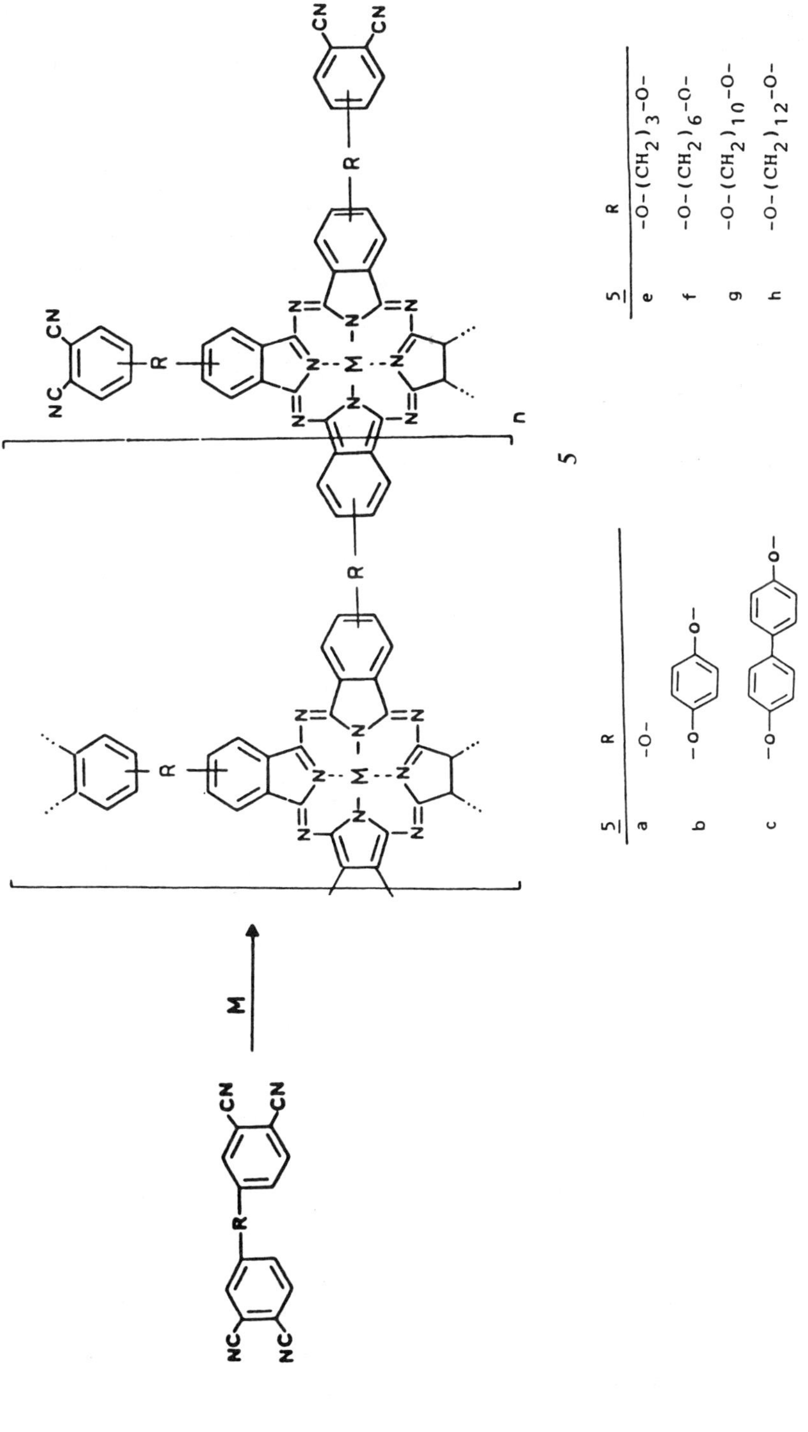

(7)

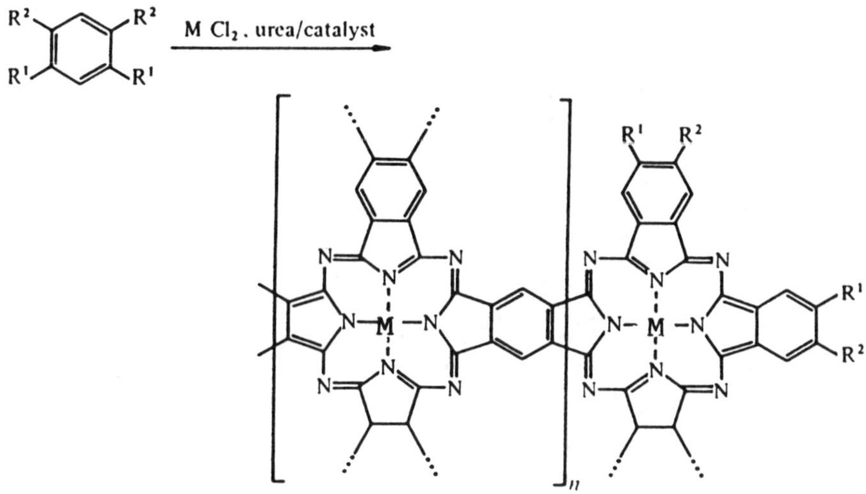

(8)

6

6a   R^1, R^2 = (anhydride structure)

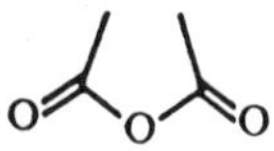

6b   R^1 = R^2 = COOH

6c   R^1, R^2 = (imide structure)

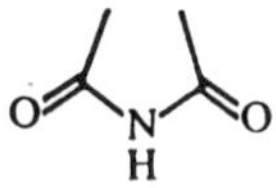

6d   R^1 = R^2 = CONH$_2$

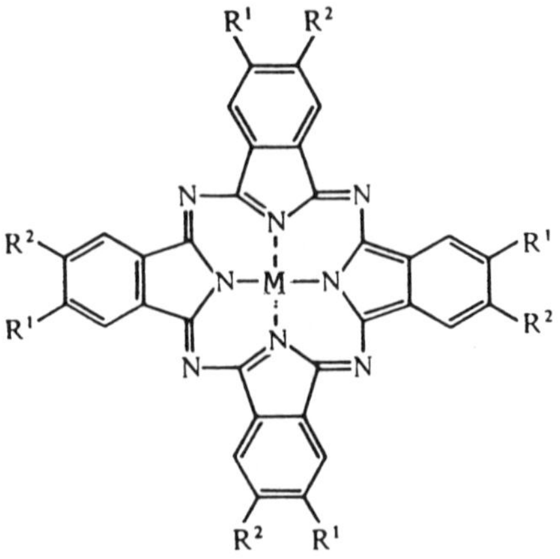

7	R^1	R^2
a	CN	CN
b	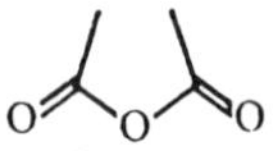 (anhydride)	
c	COOH	COOH
d	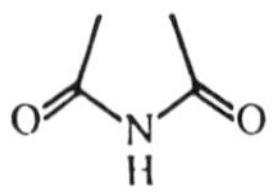 (imide)	
e	CONH$_2$	CONH$_2$

7

An exact structure determination is very difficult due to insolubility of the polymers in organic solvents. Therefore, a comparison to low-molecular-weight model compounds containing the same end groups as the polymers is helpful. 2,3,9,10,16,17,23,24-Octasubstituted phthalocyanines 7 can be converged to polymeric phthalocyanines and can be considered as model compounds for polymers 2, 6.

ii. Polymers from 1,2,4,5-Tetracyanobenzene and Heteroaromatic Tetranitriles

The reactions of 1,2,4,5-tetracyanobenzene (1,2,4,5-benzene tetracarbonitrile) with metals like Cu [4], Mg [4,6], Al [4], V [4], Cr [4], Mn [4], Fe [4], Co [4], Ni [4], Zn [4] or metal salts like Cu_2Cl_2 [4,7,9–13], $CuCl_2$ [4,5,8,9,14], CuF_2 [4], $CuBr_2$ [4], CuO [4], Cu_2O [4], $CuSO_4$ [4], Cu(acetylacetonate)$_2$ [4], $MgCl_2$ [4], $AlCl_3$ [4], VCl_3 [4], $CrCl_2$ [4], $MnCl_2$ [4], $FeCl_2$ [4,9], $FeCl_3$ [4,7,8], $CoCl_2$ [4], $NiCl_2$ [4], $ZnCl_2$ [4,7], $PbCl_2$ [4], $PtCl_2$ [4], and $SbCl_3$ [7] to polymers proceed in bulk in sealed tubes or in high-boiling solvents with $FeCl_3$ [8,15], $NiCl_2$ [16], Cu_2Cl_2 [17], and $CuCl_2$ [8,15,16]. Metal-free polymers were synthesized in bulk by reacting the tetranitrile with urea [18], Na acetate [4], Na_2S [19], and TaS [19], or demetallizing Li, Na, or Mg containing polymers [4,6,7]. Most papers do not consider structural uniformity, degree of metallization, the nature of the end groups, and degree of polymerization. It is interesting to note that the degree of ordering of metal-free polymers synthesized in the presence of urea varies over a wide range (0–100%) [18]. With respect to $C=O$ absorptions (saponification) and broad background absorption in the IR or UV-VIS (possible polynitrile 3 formation), the polymers do not seem to be uniform.

The bulk reactions were carefully investigated and special work-up procedures must be applied [4,5]:

Mixing of the tetranitrile with the anhydrous metal or metal salt (mole ratio 2/1) in dry inert gas → Heating in sealed glass tubes at 400°C for 4 h (metal salts) or 24 h (metal) →

Extended treatment in a Soxhlet apparatus with methylene chloride, acetone, and DMF → Handling with 1 M nitric acid. Washing with acetone. Drying at 200°C in vacuo

Simple methods such as UV-VIS spectra (in conc. sulfuric acid), IR spectra, and elemental analyses were used to determine the structure.

For the synthesis of structurally uniform polymeric phthalocyanines 2 from tetracyanobenzene, 400°C must be considered as a maximum reaction temperature [5]. Other structure elements are formed at >500°C. Nearly quantitative yields of the polymers are obtained depending on reaction temperature and reaction time. The best methods for obtaining structurally uniform metal-free polymeric phthalocyanines 2 (M = 2H) (determined by UV-VIS spectroscopy) are the reactions of tetracyanobenzene in bulk in the presence of 1,2,3,6-tetrahydropyridine as reducing agent or the reactions of the tetranitrile respectively octacyanophthalocyanine 7a in quinoline with the addition of sodium acetate [4,20,21]. The reactions of tetracyanobenzene in the presence of metal salts or metal in bulk lead to structurally uniform polymers 2 only when copper halides, copper, zinc chloride, or zinc are used. In the case of other metals or metal salts, the polymers also consists of other structural elements 3, 4 [Eq. (6)]. When polynitriles 3 and polytriazines 4 are covalently incorporated into the polymer network; they are identified by absorptions in their IR and UV-VIS spectra [4].

Structurally uniform polymeric phthalocyanines 2 are also obtained from octacyanophthalocyanine 7a (M = 2H) by heating with $CuCl_2$ and $NiCl_2$ in bulk [4,5]. Therefore 7a is considered as the low-molecular-weight model compound of the polymers 2. In high boiling solvents the reaction of 1,2,4,5-tetracyanobenzene with metal salts leads to other reaction products. The reaction with Cu_2Cl_2 in 1-methyl-2-pyrrolidone at 145–160°C for 2 h results in ~50% yield of dark-blue polymers consisting of structure elements 8 [17].

8

After work-up, reactions carried out carefully under inert conditions in the absence of dioxygen and water yield polymers 2 that contain cyano end groups as determined by IR spectroscopy at 2220 cm^{-1} [4,5,17]. Hydrolysis with KOH/triethyleneglycol/water converts the cyano end groups into carboxylic acid groups [4]. Elemental analyses and IR spectra (δ_{oop}N—H at

700 cm^{-1}) give a hint at the degree of metallization. Generally, reactions carried out in the presence of metal salts lead to higher degrees of metallization in comparison to the reactions with metals. A nearly quantitative metal content was achieved by the reaction of tetracyanobenzene with $CuCl_2$ in bulk [4]. Polymers with metals other than Cu(II) and Zn(II) contain more metal-free portions besides other structure elements [4]. One method of obtaining metal-containing **2** is the metallization of structurally uniform metal-free polymeric phthalocyanines with metal salts in high-boiling solvents. No detailed results are known.

Besides 1,2,4,5-tetracyanobenzene, some heteroaromatic tetranitriles (tetracyanothiophene, tetracyanofurane, tetracyanotetrathiafulvalene) [1,12,22,23], bis(1,2-dicyanoethylene-1,2-dithiolo) metal salts [1,2,24], and tetracyanoethylene [1–3] were used to prepare polymers mainly in the presence of metal acetylacetonates. By comparing the results of this older work with the more recent work of the reaction of tetracyanobenzene [4,5] it surely can be assumed that the polymers from other tetranitriles contain significant amounts of other structural units besides phthalocyanine.

The mechanism of the formation of polymeric phthalocyanines from tetracyanobenzene is of considerable interest. High temperatures and spontaneously proceeding reactions of the exothermic processes make isolation of intermediates difficult. First, tetracyanobenzene reacts to octacyanophthalocyanine, which reacts with itself or with other tetracyanobenzene molecules to form products of higher molecular weight [Eq. (9)] [4]. Metals or metal salts enhance the polymer yield and reaction rate as a result of the template effect.

$$\text{(9)}$$

The reactions of tetracyanobenzene and metals described above were carried out by mixing powders of both materials. Based on these experiences thin films of metal on various carriers or planar metal sheets were reacted with small amounts of tetracyanobenzene vapors (*in situ* synthesis) in sealed vessels at high temperatures (see route C in Fig. 1) [24–28]. Other routes for preparing thin films of octacyanophthalocyanine **7a** and polymeric phthalocyanines **2** are shown in Fig. 1 [27]. Thin films of uniform polymeric copper phthalocyanines **2** (thickness: 45–1200 nm) are obtained by the reaction of tetracyanobenzene with copper films (thickness: 1.5–30 nm) on quartz substrates at 400°C [28]. The ratio of the thickness of the polymer film to the employed copper film is ~30. It was shown that the reaction of copper films with gaseous tetracyanobenzene at $T < 350°C$ produces films of octacyano-

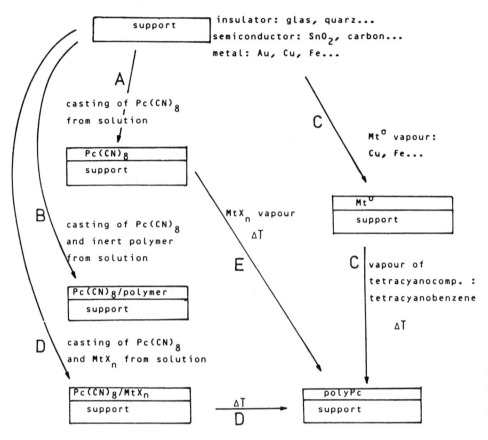

Figure 1 Schematic representation of several routes for preparing thin films of polymeric phthalocyanines **2** and octacyanophthalocyanine **7a**.

phthalocyanines **7a**, whereas at $T > 350°C$ films of polymeric phthalo-cyanines are obtained [29]. The molecular arrangements of **7a** in the stacks were investigated in detail [29,30]. Examination of the polymers' film growth mechanism demonstrates that after formation of the first monolayer or first few layers of the polymer, copper atoms diffuse from the copper film into the growing film surface in order to react with tetracyanobenzene. Initially, octacyanophthalocyanine **7a** is formed, which is then converted into the polymer via the peripheral cyano groups with cross-linking [see also Eq. (9)] [28]. The reactions of titanium foils with gaseous tetracyanobenzene lead to structurally uniform metal-free polymeric phthalocyanines **2** (M = 2H) [25]. In addition to the route C discussed, several other methods of obtaining thin films of **2, 7a** were investigated [27]. Other interesting methods for preparing thin films include double source evaporation of tetracyanobenzene and copper at $\sim 10^{-5}$ Torr argon atmosphere (film deposition rate 100 As^{-1}, thickness several microns) [31] and low-temperature plasma polymerization of evaporated low-molecular-weight phthalocyanines (thickness of polymer films 0.1–1 μm) [32]. The aim of all these papers is the study of the use of the coatings of polymeric phthalocyanines as devices in various areas of electron and photoelectron transfer processes.

iii. Polymers from Bridged Diphthalonitriles

In addition to 1,2,4,5-tetracyanobenzene, various diphthalonitriles were used as starting materials for the preparation of polymeric phthalocyanines. The group R in the bridged diphthalonitrile monomers in Eq. (7) consists of linkages such as oxygen [33–36], sulfur [35], alkylenedioxy respectively, arylendioxy [33,34,37–46], amide, imide, or ester [47–52], azomethine [53,54], perfluoroalkyl [55]. Conversion to metal-free brown-, green-, or blue-colored polymers is achieved by simple curing at temperatures ranging from 160 to 310°C [35,37–44,46,48,53,54]. Addition of special curing coreactants such as arylenediols, diamines, and 1,2,3,6-tetrahydropyridine results in higher conversion under milder conditions [35,38,41–43,45]. In some cases a metal such as copper bronze or a metal salt such as Cu_2Cl_2 and $SnCl_2$ were added [36,37,39,47,50–55]. Most papers investigate the thermal stability of these polymers, while some report on their mechanical properties [40,42,44,47,49] and electrical conductivity [33,45,46,53,54]. Only in a few cases, however, were detailed studies on the polymer structure included. It is assumed that curing of the diphthalonitriles in the presence of coreactants forms polymers containing mainly polyisoindolenines as structure elements [38,41,42,45]. UV-VIS spectroscopy of soluble polymers obtained by thermal cure of phenoxy-linked polymers showed that the content of phthalocyanine structure elements is less than 2%. Therefore, a very high degree of uncertainty on the structure of the polymers remains.

Only three papers [33–35] investigated the structure in detail. These polymers were prepared from oxy-, thio-, alkylenedioxy-, or arylendioxy-bridged diphthalonitriles [Eq. (7)]. Low-molecular-weight tetraphenoxy-, tetraphenylthio-, or tetraalkoxyphthalocyanines are used as low-molecular-weight model compounds for studying the structure by comparing UV-VIS and IR spectra with those of the polymers containing analogous structure elements.

Oxy- and thio-bridged metal-free polymeric phthalocyanines 5a (M = 2H) are prepared by heating 4,4-oxybis(phthalonitrile) or 4,4′-thiobis-(phthalonitrile) in the presence of 1,2,4,5-tetrahydropyridine (mole ratio 4:1) in evacuated sealed tubes at 275°C [35]. Quantitative yields of dark blue-purple powders are obtained. IR spectroscopy indicates that all detectable nitrile groups have reacted, significant amounts having been converted to phthalocyanine units, and very little triazine having been formed. The polymers are not dopable with iodine.

The reaction of oxy-, alkylenedioxy-, and arylenedioxy-bridged diphthalonitriles with $CuCl_2$ and Cu in bulk was investigated in detail [Eq. (7)] [33,34]. In addition, some reactions of oxy- and arylenedioxy-bridged phthalonitriles were carried out with other metal salts or metals [CuO, $CuSO_4$, Cu(acetate)$_2$, $MgCl_2$, $CoCl_2$, $PbCl_2$, Mg, Co, Pb] [33]. IR and UV-VIS spectroscopy were used to determine the structural uniformity (see Section B, i), metal content (absorptions in the IR spectra at 1050 cm^{-1} for the metal and at 1010 cm^{-1} for the metal-free polymers), end groups [$\nu(C\equiv N)$ at 2231–2225 cm^{-1}], and molecular weight (see Section B,v). It was found that analogous to the reactions of 1,2,4,5-tetracyanobenzene, the synthesis of structurally uniform phthalocyanines 5a–h was restricted by the reaction conditions and depended to a great extent on the kind of metal salt or metal. Mainly $CuCl_2$, applied in the reactions with tetranitriles in bulk, gave structurally uniform polymers (no triazines 4 or polynitriles 3 found) with a high degree of metallization of the ligand rings. The polymers were insoluble in organic solvents and the solubility in conc. H_2SO_4 decreased the higher the reaction temperature of the preparation. More flexible linkages between the two phthalonitrile units and improved electron-donation ability of the bridge permitted lower reaction temperatures with metal salts or metals while yielding higher degrees of polymerization (Table 1).

Easier protonation in acidic medium resulted in a more strongly polarized phthalocyanine. No longer wave length absorption was observed in the polymers in comparison to the low-molecular-weight model compounds tetraphenoxy- and tetrapropoxyphthalocyanine. As discussed in Section B, ii for tetracyanobenzene, some diphthalonitriles were reacted with thin copper films on quartz [28]. Thin films of 5a, b (M = Cu) were obtained.

Table 1 Some Data on Structural Uniformity of Polymeric Copper Phthalocyanines

Formula No.	Reaction temperature for synthesis (°C)	λ_{max} (nm) in conc. sulfuric acid	Maximum average molecular weight	References
6[a]	~270	744(>764)	>3,000	[60]
2[b]	~350	734	>4,000	[4]
5a[b]		814	>12,000	
5b[b]			>60,000	
5c[b]	250	832–838	∞	[33]
5d[b]				
5e[b]			≥140,000	
5f[b]	~200	844–848		[34]
5g[b]			∞	
5h[b]				

[a] Prepared from pyromellitic dianhydride, polymer with imido end groups [see Eq. (8)].
[b] Prepared from tetranitriles, polymers with cyano end groups [see Eqs. (6) and (7)].

An interesting new polymer was synthesized by the reaction of tetra-bromodibenzo-18-crown-6 with CuCN in pyridine [56]. Besides the rigid phthalocyanine the structurally uniform polymers additionally contain more flexible crown ether moieties. Alkali metal ion binding was investigated and showed higher selectivity for K^+ than for Na^+.

iv. Polymers from Tetracarboxylic Acid Derivatives

Tetracarboxylic derivatives are cheaper starting materials for polymeric phthalocyanines than tetranitriles. Therefore, the reactions are of considerable interest. Besides 3,3',4,4'-tetracarboxydiphenyl ether or sulfone, [57–59] mainly pyromellitic acid and its derivatives were used for the preparation of polymeric phthalocyanines [1–3]. For the reactions of the tetracarboxylic acid derivatives, metal salts, urea, and a catalyst (e.g., ammonium molybdate) have to be present [Eq. (8)]. The reactions were carried out either in bulk with Cu(I), Cu(II), Mg(II), Fe, Fe(II), Fe(III), Ni(II), Ga(III), and Ca(II) [57,60–67] or in high boiling organic solvents, usually nitrobenzene [68,69]. Metal-free polymers were produced by treating the calcium-containing polymer with an acidic solution [70]. It was shown that metal ions like Cu(II), Zn(II), Fe(II), and Ni(II) may be introduced into metal-free polymers very slowly [71]. It is impossible to compare the various preparations because reaction temperatures for reactions in bulk (T = 175–300°C), reaction times (t = 0.5–12 h), stoichiometric ratios of the reactants (pyromellitic acid derivative/metal salt/urea = 2/0.5–4/6–20), counterion of the metal salt, and especially purification methods are different. In most papers, yields are not reported. Complete elemental analyses were performed in only a few cases [60,62,64,66,69]. With one exception [60] detailed specification of the end group has not been done. The same holds for the degree of metallization.

A detailed discussion on synthesis and structure was carried out a few years ago. Pyromellitic acid, its dianhydride, diimide, and tetraamide were reacted with $CuCl_2$, $NiCl_2$, $FeCl_2$, and $MgCl_2$ in bulk in the presence of urea and ammonium molybdate [60]. Low-molecular-weight octasubstituted phthalocyanines **7b–e** with anhydride, carboxylic acid, imido, and amido groups were considered as model compounds [20,60]. The polymers were prepared and purified as follows:

Mixing of the tetracarboxylic acid derivative, metal salt urea, and catalyst in dry inert gas	Heating under inert gas → at 270°C for 60 min →
Extensive treating with DMF and acetone in a Soxhlet apparatus	Handling with 6 M HCl, water, → and acetone. Drying at 300°C in vacuo

After varying several parameters the following conditions leading to structurally uniform polymeric phthalocyanines **6c** in a yield of 45% were found to be the best: mole ratio pyromellitic dianhydride/CuCl$_2$/urea/ammonium molybdate = 1/0.5/18.86/0.007; reaction temperature 270°C; reaction time 60 min [60]. It was found that the ratio of the substrates influences the spin concentration of the paramagnetic centers [67]. 1,3,5-Triazine-2,4,6-diol dihydrate was found as side product [60]. Applying other pyromellitic acid derivatives or other metal salts, the yields are between 30 and 54%. In every case after work-up, imido end groups (polymer **6c**) were found by IR spectroscopy. The imido end groups were converted into polymers containing anhydride (polymer **6a**) and amido end groups (polymer **6d**). Triazines and polynitriles (see Section B,ii) were not detected by IR and UV-VIS spectroscopy. The polymers are insoluble in organic solvents but completely soluble in conc. sulfuric acid. Metal-containing polymers decompose slowly in acidic solutions [60,70,72]. The absorption ratios UV ($\lambda \sim 230$ nm):VIS ($\lambda \sim 740$ nm) in conc. sulfuric acid are between 0.79 and 1.38 in most cases and are only somewhat higher than those of low-molecular-weight octasubstituted phthalocyanines. Therefore, most polymers are structurally uniform. The positions of the UV-VIS absorption in conc. sulfuric acid are comparable for low-molecular-weight octasubstituted phthalocyanines **7b–e** and polymers **6a–d** with analogous end groups. No complete metallization of the phthalocyanine rings is achievable. The metal content ranges between 65 and 80% (in the case of the Mg containing polymer $\sim 40\%$).

(10)

Mössbauer spectra allow estimates on the purity of polymeric iron phthalocyanines [73,74]. The original polymer (from pyromellitic acid, $FeSO_4$, urea, and catalyst after reprecipitation from conc. sulfuric acid) shows four superimposed quadrupole doublets (low-molecular-weight Fe phthalo-cyanine, Fe in phthalocyanine units interior and periphal of the polymer, Fe in oxidized phthalocyanines) [73]. After chromatography over Sephadex LH 20 (to exclude oligomeric Pc and impurities with molecular weights lower than 5000) and heating to 300°C (desorption of O_2, assumed decarboxylation of COOH end groups) only one quadrupole doublet (Fe in phthalocyanine units) is present.

The mechanism for the formation of polymeric phthalocyanines using pyromellitic acid derivatives was studied [60] and is described in Eq. (10).

v. Molecular Weight Determination

Classical molecular weight determinations known in polymer chemistry are very difficult to conduct [1]. This difficulty is due to the insolubility or very poor solubility of the polymers in organic solvents. Additionally, different arrangements of connected phthalocyanine rings in a polymer must be taken into account. Several isomers exists for a polymer molecule with a definite molecular weight. The determination of the degree of polymerization of polymers from pyromellitic dianhydride was attempted by UV-VIS spectro-scopy [75]. The authors explained that with regard to π-electron delocaliza-tion in one conjugated system only one $S_0 \rightarrow S^*$ transition takes place. Therefore with increasing molecular weight extinction should decrease. On the other hand, π-electron delocalization should result in a long wavelength shift of the Q-band transition. No clear hints for reduced extinction and increased delocalization are found. Low molecular-weight model compounds exhibit similar absorption in comparison to the polymers: polymers with imido end groups **6c** $\lambda \sim 774$ nm with $\epsilon \sim 0.85 \cdot lg^{-1} cm^{-1}$ [20] (for polymers with other end groups and their analogous low-molecular-weight compounds see [4,20,33,34]. Impurities in structural nonuniform polymers (triazines, polynitriles) reduce Q-band intensities in comparison to the absorptions at shorter wavelength. Other studies tried to determine the degree of polymeriza-tion of polymers prepared from 1,2,4,5-tetracyanobenzene by IR [76] and from DMF-soluble polymers from pyromellitic acid by GPC [73]. Degrees of polymerization of 4–18 were found.

A possibility studied in detail is the determination of the degree of polymerization (number of connected phthalocyanine rings P_c, molecular weight M) by exact end group determination or elemental analysis [60]. This method was described [60] and applied to several polymers [4,33,34,60]. For a specific P_c different numbers of arrangements a of phthalocyanines exist (Table 2). The number of arrangements for a specific P_c is given by Eq. (11)

Table 2 Maximum Number of Arrangements for Different Numbers of Phthalocyanine Units

Number of phthalocyanine units P_c	Maximum number of arrangements a	Number of ring arrangements r formed by four phthalocyanine units				
		1	2	3	4	$\lim_{r\to\infty} r$
1	1					
2	1					
3	1					
4	2					
5	2					
6	3					
7	3					
8	4					
9	5					
$\lim_{P_c\to\infty} P_c$	$\lim_{r\to\infty} a$					

[60]. Each distinct arrangement *a* contains a certain number of rings *r* formed by annulation of four phthalocyanine rings.

$$a = \text{INT}(P_c + 2 - 2\sqrt{P_c}) \qquad (11)$$

Three limiting cases should be mentioned:

A. $P_c = 1$, $r = 0$. This case corresponds to the low-molecular-weight octasubstituted phthalocyanines **7**. The C, N, and Cu values of elemental analysis (for **6a–d**) attain their minimum and the H values (for **6a–d**) and the number of end groups (for **2, 6a–d**) their maximum.

B. $\lim_{P_C \to \infty} P_C$, $r = 0$ to $r = \infty$ all $P_C \geqslant 4$ for $r = 1$. This case corresponds to infinite chains. The structures with $r = 1$, $P_C > 4$ are identical to a circle, the ideal case of an infinite chain.

C. $\lim_{P_c \to \infty} P_C$, $\lim_{r \to \infty} r$: This case corresponds to a polymer with infinite network that is extended on the surface of an infinite sphere and does not possess any end group: Here C, N, and Cu values (for **6a–d**) reach their maximum, H (for **6a–d**) and the number of end groups (for **2, 6a–d**) their minimum.

For polymers with carboxylic end groups, analytical data and number of phthalocyanine rings are correlated in Table 3. The results in Tables 2 and 3 show that polymers with identical P_c (and different *r*) exhibit different analytical data while on the other hand polymers with different P_c and *r* can lead to identical analytical data. As a result, only the lower boundary values of P_c (and thus the molecular weight *M*) with a specific *r* can be determined from analytical data for structurally uniform polymeric phthalocyanines, Table 1 contains some data on the molecular weight. These values were obtained as follows: polymers **6c** with imide end groups by elemental analysis [60], polymers **6b** with carboxylic acid end groups by elemental analysis and titration of COOH groups [60], polymers **2, 5a–h** with cyano end groups by quantitative IR analysis of the cyano groups [4,33,34]. For end group analysis, correlation with analogous low-molecular-weight model compounds is necessary. The band structure calculation was discussed recently [246].

vi. Properties

Polymers obtained from different methods are black-, blue-, or brown-colored compounds. The black or brown color is probably due to impurities. The structurally uniform polyPc exhibit nearly the same color as the low-molecular-weight analogs. Impurities are easily determined by UV-VIS and IR spectra [4,5,33,34,60].

Solubility depends to a great extent on the kind of Pc: low-molecular-

Table 3 Correlation between Number of Phthalocyanine Units, P_c, and Analytical Data for 6b

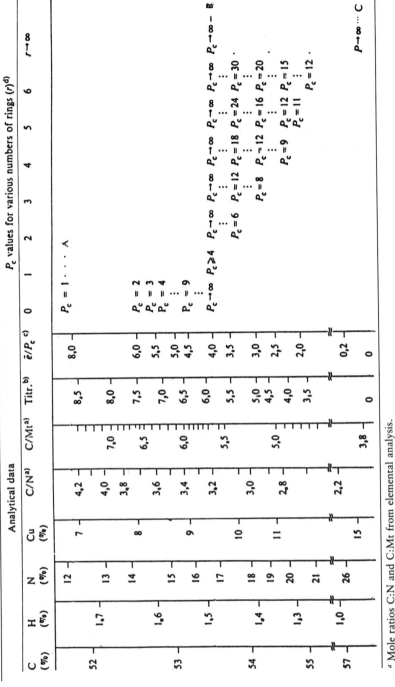

[a] Mole ratios C:N and C:Mt from elemental analysis.
[b] Titration data of end groups (mmol of COOH/g of 6b).
[c] Average number of end groups/phthalocyanine unit.
[d] For explanation of r, A, B, C, see text (Section B, v).

weight unsubstituted CuPc **1** is less soluble in organic solvents and well soluble in conc. H_2SO_4; octacyano CuPc **7a** has good solubility; polymers are less soluble or insoluble in organic solvents and conc. H_2SO_4.

Polymers containing, e.g., Co, Cu, Ni, Al are stable toward treatment with conc. H_2SO_4 for a short time while those with Mg, Cd, Pb, Sn, Fe decompose to give metal-free polymers. During this procedure, end groups may be saponified. Generally, hydrolytic stability of the Pc-ligand ring is lower under acidic conditions than that of low-molecular-weight Pc. The stability of metal-free and metal-containing polymers from PMDA and phthalic anhydride were investigated at various temperatures in 17 N H_2SO_4 [71,72]. Many papers describe the thermal stability of the polymers [1–3]. The results are difficult to compare. Heating rates in thermal analyses are different. In only a few cases was isothermal treatment, which really describes the long-term behavior use of such polymers, included in the investigations. Within one class of polymeric phthalocyanine, stability depends on the preparation procedure, metal atom, and end group.

One paper compares the thermal stabilities of **1**, polymers **2**, and the low-molecular-weight model compound **7** [5]. Thermal analysis and isothermal treatment both show that under inert conditions 500°C is the highest temperature **2** (M = Cu) can withstand while still maintaining the integrity of its structure, end groups, and metal content. Under oxidative conditions this polymer decomposes between 340 and 360°C. Under inert and oxidative conditions phthalocyanines with the same metal approximately show the following order of decreasing stability: **7** > **1** > **2**. Surprisingly the polymers are not the most stable. Under inert conditions, **7** (M = Ni) begins losing weight at 640°C and under oxidative conditions at 506°C. This low-molecular-weight phthalocyanine usually exhibits better thermal stability than various polymers **2–6**. For polymers from pyromellitic acid derivatives see [5] and [69].

Polymers **8** withstand temperatures of ~600°C [17] under inert conditions. Structurally uniform metal-free polymers **5a** begin losing weight at ~500°C under inert conditions [35]. Most of the polymers from diphthalo-nitriles with amide, imide, ester, azomethine, perfluoroalkyl, aryleneoxy, or alkyleneoxy linkages R [37–55] were prepared to study thermal stability and mechanical properties [40,42,44,47,49]. Since their structures are not fully proven, the results will not be discussed in detail. In general, thermal stabilities do not deviate significantly from the results discussed above. However, in some cases bridged diphthalonitriles cured in the presence of amine coreactants are extraordinarily stable [46].

Electronic conductivities of the polymers are also discussed in many papers [1–3]. It is difficult to compare the data since other impurities like adsorbed gases (dioxygen) influence conductivity besides structural uniformity, nature of the metal, and end group. All samples exhibit typical semiconducting behavior. The conductivity of pressed powders for some copper containing phthalocyanines are [1,77] $1 < 10^{-10}$, $2 < 10^{-2}$, $7a$ 2×10^{-2} S cm^{-1}. The fact that the low-molecular-weight model compound $7a$ exhibits high conductivity and, as discussed before, the polymers 2 show no long wavelength shift of the Q-band in comparison to $7a$ indicates that intermolecular charge transfer as the dominating factor. Thin films of 2 (M = Cu) (thickness 50–1261 nm) have a conductivity of 10^{-2}–10^{-7} S cm^{-1} [27,28; see also 31]. The conductivity of structurally uniform polymers $5a$, b is about $<10^{-11}$ (pressed powders) [33,35] respectively 10^{-4}–10^{-5} (thin films, thickness ~50 nm) [28]. The conductivity of pyrolyzed polymers can be varied and controlled as a function of pyrolysis temperature and annealing time [17,46]. As a result, conductivity ranges from that of an insulator to semiconductor to the near-metallic region. Additionally, two papers that deal with dielectric properties and the photoconductivity of copper-containing polymeric phthalocyanines should be mentioned [78,79].

The electrochemical and electrocatalytical properties of thin films of polymeric phthalocyanines prepared from vapors of 1,2,4,5-tetracyano-benzene with titanium sheets or copper films on pyrolytic graphite were investigated in aqueous electrolytes [24–28]. In the absence of a redox pair, the behavior of thin films of 2 (M = Cu) [28] is similar to that of thin films of the low-molecular-weight model compound $7a$ [80,81]. In contrast to thin films of 1 that are not electrochemically active under the applied conditions, thin films of 2 show an interesting behavior of reversible reduction and reoxidation. In 0.5 M sulfuric acid, the formal peak potential E^0 for reduction and reoxidation is only -0.064 V vs SCE. The reversible reduction requires the transport of electrons through the semiconductor interface and the intercalation of charge-compensating protons from the solution [Eq (12)]. With increasing pH, the potential shifts negative. The pH shift corresponds approximately to the hydrogen electrode.

$$\begin{bmatrix} \text{monomer unit} \\ \text{in polymer} \end{bmatrix} + e^{\ominus} + H^+ \rightleftarrows \begin{bmatrix} \text{monomer unit} \\ \text{in polymer} \end{bmatrix}^{\ominus} \cdot H^+ \qquad (12)$$

Coatings of the polymers 2 (M = 2H) in contact with an aqueous electrolyte containing the redox pair K_3, $K_4[Fe(CN)_6]$ have a surprisingly high electrocatalytic activity, which is similar to that of a platinum electrode [25]. Additionally, the polymer-coated electrodes are characterized by photoelectrocatalytic activities in the anodic oxidation of Br$^-$ and ascorbic acid [26]. The action spectra of the reactions correspond to the absorption spectra of 2.

In the solid state, polymeric phthalocyanines should have clathrate-like cavities or channels with different sizes between the phthalocyanine units depending on the length of the bridges (see annelated structure 2 and by R-linked structures 5) [34]. All polymers have low swelling factors and surface areas. But the absorption capacities investigated for organic solvents increase with increasing distance between the phthalocyanine units in **2, 5** [34]. Polymeric phthalocyanines connected by crown ether linkages show absorption of alkali metal ions [56].

The catalytic properties of polymeric phthalocyanines were described in a review elsewhere [1]. The catalytic activity of the polymers for the oxidation of thiols and the electrocatalytic activity for the reduction of dioxygen are described in Section F.

vii. Polymeric Phthalocyanines from Multifunctional Low-Molecular-Weight Phthalocyanines

This route for the preparation of polymeric phthalocyanines exhibits the advantage that well-characterized low-molecular-weight phthalocyanines are applied for polycondensation reactions. In general, tetrafunctional phthalocyanines like the tetraamines and tetracarboxylic acid derivatives of **9** are used as starting materials. Due to the high functionality (>2) network formation easily occurs, and difficult to characterize insoluble polymers are obtained (degree of network formation, molecular weight, end groups). The polymers were mainly investigated for their thermal stability.

9 (R = COOH, NH$_2$)

The simplest possibility uses polycondensation reactions of phthalocyanines by itself. The unsubstituted **1** (M = 2H, Cu, Mg, Zn, Ni) were evaporated and treated from the gas phase by low-temperature plasma [32]. The resulting greenish-colored smooth films are 60–800 nm thick. The insolubility shows polymeric character. Typical transitions for phthalocyanines at ~650–700 nm are observed in the VIS spectra of solid films. A cell of the construction ITO/polyPc/Al shows rectifying behavior and under irradiation photovoltaic responses (efficiency $< 10^{-3}\%$). In addition, the films are active in photoreductions. Another interesting procedure is the thermal reaction of 2,9,16,23-tetracarboxyphthalocyanine **9** (R = —COOH; M = Cu, Ni, Co) at 350–400°C in vacuum [82]. The polycondensation reactions were controlled by GC, MS, and IR. The structure of the polymers **10** are confirmed by spectroscopic investigations and determination of gaseous reaction products [Eq. (13)]. The polymers are stable in air up to ~ 400°C and under inert gas up to ~ 550°C.

$$n \cdot 9 \ (R = COOH) \xrightarrow{-CO_2, CO, H_2O, H_2}$$

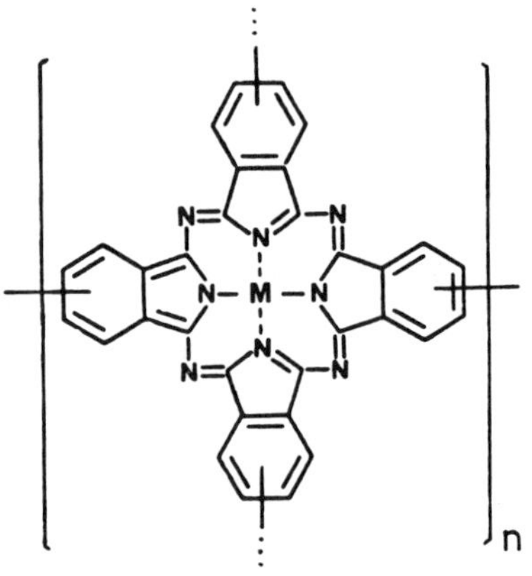

(13)

10

Tetracarboxyphthalocyanines **9** (R = —COOH; M = Cu, Co, Ni, Zn) were converted in the reaction with 3,3′-diaminobenzidine to polymers **11** [83]. The reactions were conducted in melt at 200–400°C or polyphosphoric acid at 185°C (mole ratio 9:benzidine = 1:2). The structures of the insoluble greenish to bluish-black polymers were confirmed by elemental analysis and IR spectra. Decomposition occurs in air at ~ 450°C and under an inert gas at ~ 650°C. Low-molecular-weight phthalocyanines with two 3,4-dicarboxy-benzoyl anhydride groups were prepared by a statistical synthesis of 3,3′,4,4′-benzophenone tetracarboxylic acid dianhydride, phthalic anhydride, metal salt and urea [84–86]. This "bifunctional monomer" (several possible structural isomers) was reacted with various aromatic diamines in N-methylpyrrolidone to soluble polyamic acids that react to insoluble polyimides **12** by heating. The electronic conductivity is $< 10^{-6}$ S cm^{-1}·**9** (R = —CONH$_2$; M = Co, Ni) is converted with formaldehyde to water-soluble **9** (R = —CONHCH$_2$OH), which reacts with urea at 100–150°C to form polymers [87].

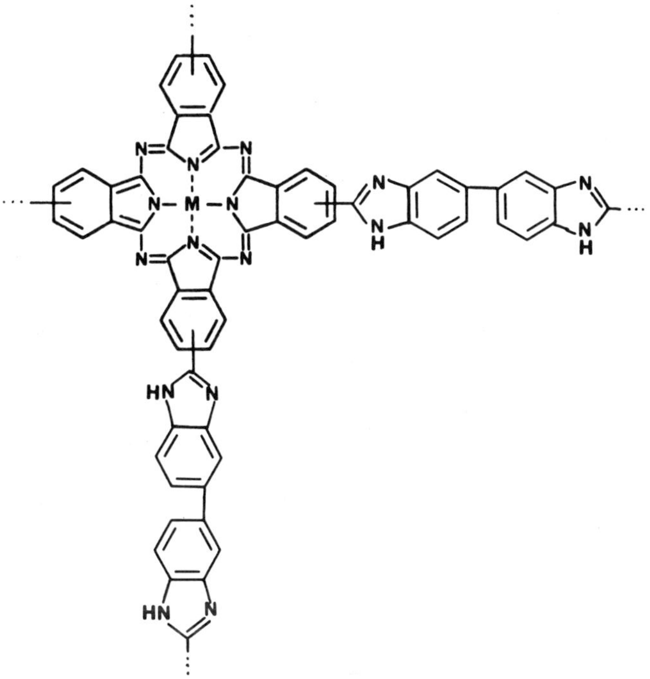

11

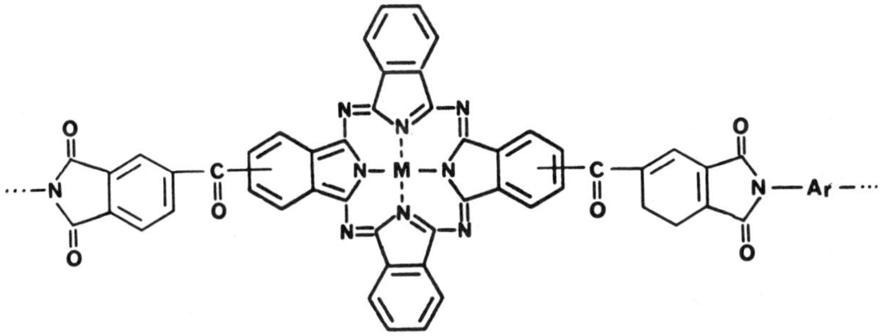

12

2,9,16,23-Tetraaminophthalocyanines **9** [88] were employed in the synthesis of polyimides and as curing agents for epoxy resins. **9** (M = Cu, Co, Ni, Zn) and pyromellitic dianhydride resp. benzophenone tetracarboxylic dianhydride were stirred in DMSO at 25–70°C to get the polyamic acids [89]. These precursors are soluble and films can be cast. Fairly flexible, tough films of the corresponding polyimides were then obtained by heating to 300°C. The reactions were monitored by IR spectra. Another procedure uses the cocon-

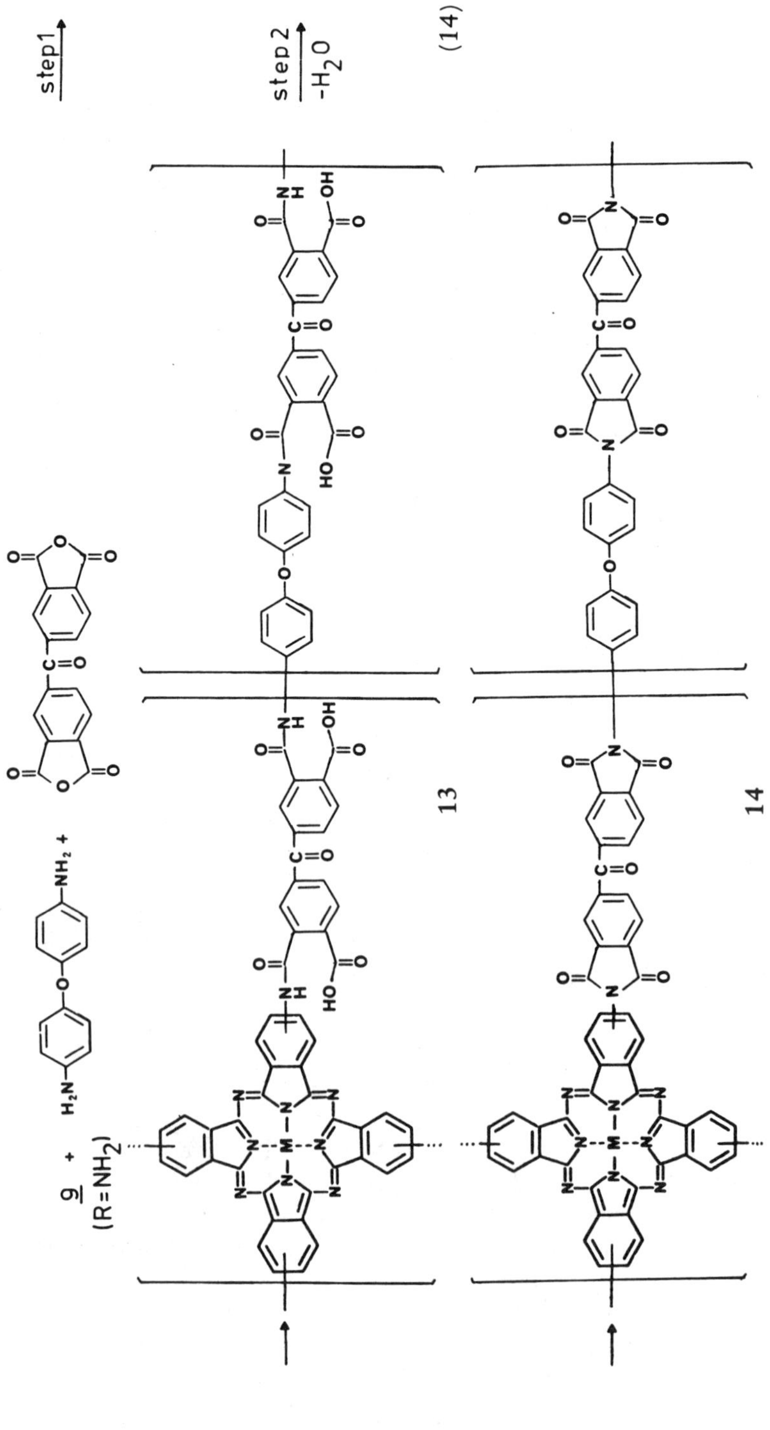
9 + (R=NH2)
step1
step 2
-H2O
13
14
(14)

densation of the tetraamine **9** in the presence of various aromatic diamines and the above-mentioned tetracarboxylic acid dianhydrides [90]. Variables such as molar concentrations of the reagents, solvents, and temperature were investigated to optimize the conditions of the polycondensation. IR spectra and inherent viscosity were used to characterize the polymers. Solutions of the polymeric acid copolymers can be used to fabricate films or fibers. The polyimide copolymers obtained in the second step of the reactions are insoluble. Excellent thermal stabilities up to 500°C in air and 600°C under an inert gas are reported. A typical description for the procedure is as follows. Equation (14) shows the reaction leading to the polymeric acid **13** and the polyimide **14**.

1.25 mmol **9** (R = NH₂)　　　10 mmol
and 7.5 mmol　　　　　　　　benzophenonetetra-
4,4′-diaminodiphenyl ether → carboxylic dianhydride →
dissolved in 123 ml DMSO　added dropwise
under inert gas

　　　　　　　　　Casting of
2 h stirring ↗ films,　　　↘ Heating at
at room T　drawing of　　　325–350°C
　　　　　　fibers

　　　　↘　　　　　　　　in vacuo
　　　Precipitation ↗ or under
　　　by toluene　　inert gas

In addition, **9** (R = —NH₂) was used to cure epoxy resins in homogeneous reactions [91,92]. At first, **9** (R—NH₂) and the epoxy resins are dissolved in DMSO. After removal of the solvent, the material was cured at 180–250°C. Chemically resistant, thermally stable, and tough materials are obtained. Considerable improvement of the heat resistance over those cured with other commonly used curing agents was observed.

C. COFACIALLY STACKED POLYMERIC PHTHALOCYANINES

The formation of an octahedral configuration with additional ligands at the metal atom of a metal phthalocyanine is the fundamental requirement for the formation of polymers with chelates linked cofacially in a face to face orientation:

These stacked and bridged phthalocyaninato metal complexes can be obtained by the reaction of metal phthalocyanines with other bifunctional inorganic or organic reagents. Research in this field grew during the past years due to the good electronic conductivity of cofacially stacked macrocycles in the undoped and even better conductivity in the doped state. The rigid one-dimensional connection of the macrocycles combined with an excellent interaction that switches between a lower and a higher conducting state is the foundation for seeking more practically interesting applications in the near future.

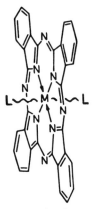

low molecular
octahedral
configuration

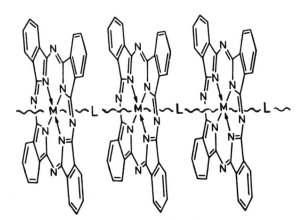

polymeric octahedral configuration

In addition to the phthalocyanines **1**, the following dianionic ligands **15–20** surrounding the metal in a square planar arrangement were used to prepare cofacially stacked polymeric macrocycles [93]:

2,3-naphthalocyanine

15

tetrabenzoporphyrin

16

5,10,15,20-tetraphenylporphyrin
17

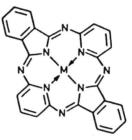

hemiporphyrazine
18

dibenzo-1,4,8,11-
tetraaza-14-annulene
19

2,3,9,10-tetramethyl-1,4,8,11-
tetraaza-1,3,8,10-cyclotetradecatetraene
20

Depending on the oxidation state of the metal ions in the core of the macrocycles, polymers with different bonds in the main chain can be designed:

B1: Covalent–covalent bonds between the macrocycles containing tetravalent metals and a bifunctional ligand L capable of covalent bonds.

B2: Covalent–coordinative bonds between the macrocycles containing trivalent metals and an inorganic or organic ligand capable of one covalent and one donor function.

B3: Coordinative–coordinative bonds between the macrocycles containing bivalent metals and a bifunctional ligand L capable of coordinative bonds.

$$\underline{B1} \quad \cdots\!-\!M\!-\!L\!-\!M\!-\!L\!-\!M\!-\!\cdots$$

$M = Si, Ge, Sn \quad L = -O-, \ -S-, \ -O-R-O-, \ -C\equiv C-, \ -N=C=N-$

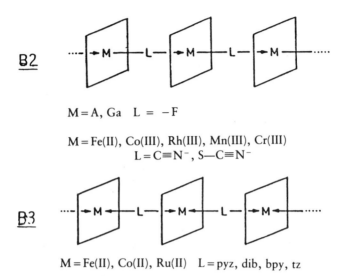

$$M = A, Ga \quad L = -F$$

$$M = Fe(II), Co(III), Rh(III), Mn(III), Cr(III)$$
$$L = C \equiv N^-, S - C \equiv N^-$$

$$M = Fe(II), Co(II), Ru(II) \quad L = pyz, dib, bpy, tz$$

Since unsubstituted phthalocyanines generally are poorly soluble in organic solvents, it can be expected that stacked and bridged chelates are even less soluble. Therefore, information on the solid-state structure must be derived in most cases from several lines of evidence: IR, UV-VIS, ^{13}C-NMR, X-ray, Mössbauer (for Fe-containing macrocycles), ESCA, etc. In some cases polymers containing bulky substituents on the annulated benzene rings of the ligand were prepared that show excellent solubility in organic solvents.

The thermal and chemical stability of the polymers are in the following order: type B1 > type B2 > type B3. Processing of the polymer depends on its nature: type B1 is dissolved in some conc. acids, advantageous mainly in the presence of another thermoplastic polymer, followed by extruding into an aqueous coagulation bath; type B2: with M = Al, Ga and L = F sublimation under formation of thin films. For all polymers, spin casting in the presence of an inert polymer binder can be a technology to obtain thin films. Thin films of soluble polymers can be prepared by the Langmuir–Blodgett technique.

i. Covalent–Covalent Bonds in the Main Chain

Simple oxo-bridged group IVB polymers **21** [Eq. (15)] were first prepared by Kenney et al. [94–96], followed by additional papers. Besides phthalocyanine **1** the Si(IV), Ge(IV), and Sn(IV)-containing ligand rings of the polymers are [94–106] hemiporphyrazine **18** [97,98], tetrabenzoporphyrin **16** [107], and porphyrins like tetraphenylporphyrin **17** or etioporphyrin [98,99,108].

The group IVB precursor phthalocyanines **1** $Si(Pc)Cl_2$, $Ge(Pc)Cl_2$, and $Sn(Pc)Cl_2$ were synthesized as described elsewhere [94,101,109,110]. Conversion of the dichlorides to the corresponding dihydroxides Si(Pc)(OH)$_2$, Ge(Pc)(OH)$_2$, and SnPc(OH)$_2$ was achieved by hydrolysis in a pyridine/NaOH solution [94,101,109,110].

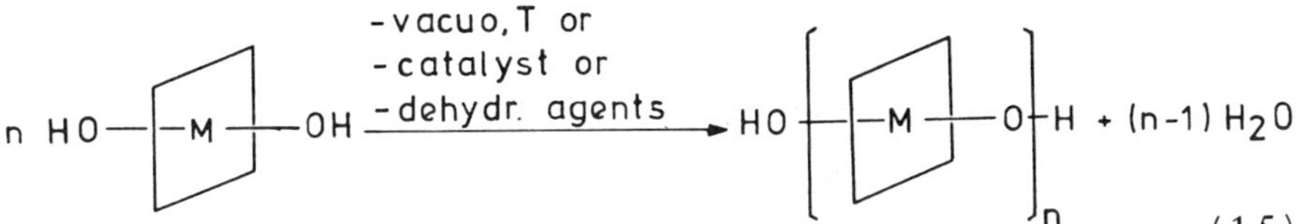

$$\text{n HO} - \boxed{-\text{M}-} - \text{OH} \xrightarrow[\substack{-\text{catalyst or} \\ -\text{dehydr. agents}}]{-\text{vacuo,T or}} \text{HO} \left[\boxed{-\text{M}-} - \text{O} \right]_n \text{H} + (n-1)\,\text{H}_2\text{O}$$

(15)

21

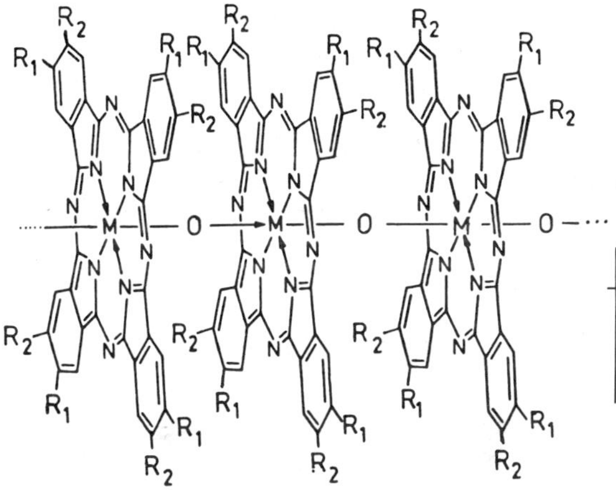

R$_1$	R$_2$
H	H
tert-butyl	H
(CH$_2$)$_n$CH$_3$	O(CH$_2$)$_n$CH$_3$

21

The oldest method for the preparation of catena-poly[phthalocyaninato-silicon(IV)-μ-oxo-] **21** and the analogous germanium and tin polymers is thermal dehydration in bulk at 325–440°C *in vacuo* for several hours [94–96,100,101,103,104]. The unsubstituted polymers are insoluble in organic solvents and soluble in conc. sulfuric acid. A general concept was used to improve the solubility. The oxo-bridged substituted polymers exhibit solubility in organic solvents [104,106].

The metallized oxo-bridged polymers **21** are exceptionally stable materials. They are unaffected by aqueous HF at 100°C, aqueous 2 *M* NaOH at reflux, and conc. H$_2$SO$_4$ at room temperature. Two methods were used to determine the degree of polymerization $\bar{P}_n$. In the IR spectra the ratio of the areas under the absorption peaks of the Si—OH end group at 836 cm^{-1} was compared quantitatively to the phthalocyanine frame mode at 911 cm^{-1} [101,103]. A second method uses tritium labeling after applying M(Pc)-(O^{3}H)$_2$ in the dehydration process [101]. Polymers **21** with $\bar{P}_n$ ~ 100 (M_n ~

Table 4 Degree of Polymerization of Some Polymers with Covalent/Covalent Bonds in the Chain

Polymer	Polymerization conditions	$\bar{P}_n$ (method)	Reference
$+$Si(Pc)—O$+_n$	Bulk, 440°C, 12 h	120 (IR end groups) 102 (tritium labeling)	[101]
	Bulk, 420°C, 10 h	100 (IR end groups)	[103]
	Suspension nitro benzene, ZnCl$_2$	11 (VIS data)	[100]
	Quinoline	20 (IR end groups)	[103]
	Tributylamine, 2% FeCl$_3$	100 (IR end groups)	[104]
$+$Ge(Pc)—O$+_n$	Bulk, 440°C, 10 h	70 (IR end groups) 62 (tritium labeling)	[101]
$+$Sn(Pc)—O$+_n$	Bulk, 325°C, 0.5 h	100 (IR end group) 15 (tritium labeling)	[101]

55,000) can be obtained. Table 4 shows an increase of $\bar{P}_n$ in the sequence of Sn ~ Ge < Si, which is also valid for organometallic Group IV chemistry. Poly(organosilicones)(siloxanes) exhibit higher molecular weight in comparison to the analogous poly(organogermanium) and poly(organotin) compounds. For Pc polymers, $\bar{P}_n$ increases with longer reaction time and higher reaction temperature. It is important to note is that $\bar{P}_n$ increases also with higher polymer yield [103]. Therefore the conditions of a normal equilibrium polycondensation (Carother equation) are fulfilled. The polycondensation occurs by random reaction of all end groups with each other and not by addition of the monomer to the growing surface of a polymer crystal. The reaction process in the solid state is unknown.

Besides the simple bulk reaction, heating as a suspension in a high-boiling solvent, partly in the presence of dehydrating agents (ZnCl$_2$, benzoyl-chloride) was conducted ($\bar{P}_n$ ~ 11, Table 4) [100,103,111]. Progress was achieved by conducting the reaction in a high-boiling solvent like tributyl-amine in the presence of metal chlorides like FeCl$_3$ as catalysts [104]. $\bar{P}_n$ of **21** is approximately $\leqslant 100$.

Interesting information on polymers **21** is drawn from X-ray powder diffraction pattern and both scanning and transmission electron microscopy [101,102,112–115]. In contrast to the fibrillar morphology of polyacetylene, the results indicate, e.g., for $+$Ge(Pc)—O$+_n$ a crystalline, lath-like morphology [113]. The chains of the polymer are parallel to the large surface of the lath and lie in most crystals parallel to the lengthwise direction of the lath. But deviations from these orientations are observed. The pattern can be indexed according to the tetragonal crystal system. This is very similar to the columns or crystal structure of the doped [NiPc]I$_{1.10}$ [116]. The stacking intervals schematically shown in Fig. 2 increase as follows: Ni—Ni (in [NiPc]I$_{1.0}$)

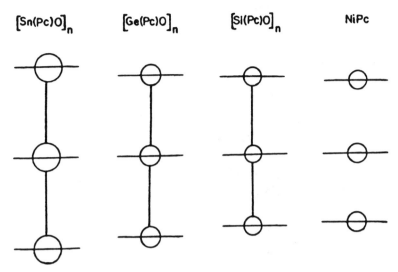

Figure 2 Scale drawing of the interplanar relationships in cofacially stacked polymeric phthalocyanines **21** and in $[Ni(Pc)]I_{1.0}$.

0.3244 nm, Si—O—Si 0.333 nm, Ge—O—Ge 0.351 nm, Sn—O—Sn 0.395 nm. In contrast to the doped NiPc, the polymers have the great advantage of permanent cofacial stacking. The polymers exhibit the typical transitions of the phthalocyanine system in the UV-VIS spectra [101,106]. However, the Q-band transition is shifted from ~680 nm for $M(Pc)(OH)_2$ to ~620 nm for $+M(Pc)—O+_n$, which is interpreted in terms of exciton coupling of neutral-excitation transitions between adjacent macrocycles [106,117]. The kinetic stability of the oxo-bridged polymers in conc. sulfuric acid was studied [100]. The simple thia-bridged catena-poly[phthalocyaninatogermanium(IV)-μ-thia] were prepared from $Ge(Pc)(OH)_2$ [118,119]. Heating of this compound with H_2S in an autoclave at 130°C leads to polymers. Molecular weights are unknown.

Hydroxy groups in the $M(Pc)(OH)_2$ exhibit great reactivity. This was demonstrated by reactions of these compounds [M = Si(IV), Ge(IV), Sn(IV)] with an excess of alcohols, phenols, carboxylic acids, and thiophenols that yield the corresponding bisalkoxy, bisphenoxy, bisesters, and bisphenylthio derivatives [98,99,111]. If the reactions are conducted now, with equimolar amounts of bifunctional alcohols, phenols, and carboxylic acids [98,99,111] or an excess of 1,4-benzodithiol [118] in high boiling organic solvents, the corresponding polymers are obtained [Eq. (16)]. Besides Pc, other systems like hemiporphyrazines **18**, tetraphenylporphyrins **17**, and etioporphyrins were applied in the polycondensation process [98,99,108,111]. The reactions with thiols show that the substitution is of S_N1 character.

$$(16)$$

The introduction of the comonomer $+M(Pc)-O+_n$ into other polymer chains has been reported. $Si(Pc)(OH)_2$ was incorporated into the chain of polyurethanes prepared from diisocyanates and dihydroxymethyl carboranes [120]. The reaction of modified silicon phthalocyanine disilanols with bis(ureido)dimethylsilane or bis(ureido)siloxanes yields blue SiPc-siloxane polymers with the general formula $+Si(Pc)-O-Si(CH_3)(C_6H_5)+O-Si(CH_3)_2+_mO-Si(CH_3)-(C_6H_5)-O+_n$ (n = 2–4). $\bar{M}_n$ is approximately 1.2–1.4×10^4 [110]. The aim of this work was to enhance the thermal stability and to influence the transition temperature of the polymers. Phthalocyanine complexes of silicon and germanium were incorporated covalently via their core metal atom in poly(ethylene terephthalate), poly(ethylene adipate), poly(amid-6), and polyurethanes [121]. Quantities of 10^{-4} M complexes color the plastics, intensively relative to a monomer unit in the polymer. Catena-poly[phthalocyaninatometal)-μ-(ethynyl)] compounds **25** [M = Si(IV), Ge(IV), Sn(IV)] were successfully synthesized by the reaction of $MPcCl_2$ with bisbromomagnesium acetylene in THF [Eq. (17)]. The use of bulky substituents resulted in easily soluble polymers [122,123]. Analogous to the oxo-bridged polymers **21**, the ethynyl-bridged polymers **25** also exhibit a bathocromic shift of the Q-band. Several of the low-molecular-weight trans-bis(1-alkynyl)-metal-IVB-phthalocyanines were prepared [124].

$$n\ M(Pc)Cl_2 + n\ BrMg-C\equiv C-MgBr \longrightarrow$$

$$+\ 2n\ MgBrCl \quad (17)$$

25   M = Si, Ge, Sn
R = H, *tert*-butyl, Si(CH₃)₃

The electronic conductivity of the polymers containing phthalocyanines and other cofacially linked macrocycles were studied in detail (Table 5). "Undoped" polymers exhibit semiconducting behavior in the order of 10^{-6}–10^{-10} S cm^{-1}. A great increase in conductivity was observed by treating the macrocycles with electron acceptors like iodine, bromine, nitrosonium salts, and *o*-chloranil. Halogen-doped polymers [Eq. (18)] were prepared by stirring the powdered samples with solutions of the halogen in organic solvents or exposing the powders to halogen vapor. $+Si(Pc)O+_n$ **21** can also be doped by dissolving in sulfuric acid and precipitating with an aqueous I_3^- solution [102]. A cell for electrochemical doping of pressed polymer powders was described [103]. Bu₄N$^\oplus$BF₄$^\ominus$ in CH₂Cl₂ or LiClO₄ in H₂O were used as electrolytes. Electrochemical oxidation was carried out galvanostatically (0.5 F per mole of constitutive unit).

$$+M(Pc)-O+_n + ny/2 \cdot X_2 \rightarrow (+M(Pc)-O+(X_y)+)_n \qquad (18)$$

$$(M = Si, Ge, Sn;\ X = Br, I)$$

Doping with *o*-chloranil was simply achieved by mixing it with the polymers in a mortar [125,126]. Principally, the influence of doping on the conductivity is an old technique [125–128] that was investigated more intensively by several analytical methods (resonance Raman, IR, UV-VIS, X-ray, susceptibility, NMR, ESR, etc.) a few years ago [102,112,129,130 and references cited therein]. It was shown that partial oxidation of the cofacial array occurs. The nature of $+M(Pc^{\delta+})$—O+ is consistent with π radical cations predominantly ligand in character. Doping does not significantly alter the interplanar separations. The resistivity mechanisms of the polymeric conductors were discussed [131]. It can be seen from Table 5 that doping—mainly with iodine—significantly increases electrical conductivity. The temperature dependence of the powder conductivities is thermally activated as in semiconductors. Powder conductivities are affected by interparticle contact resistance and averaging over all crystallographic directions. Powder data of the iodine doped polymers show that higher, "metallic" conductivities, in the chain direction of $+\!\!+M(Pc)$—O$+\!(I_x)\!+\!-_n$ (M = Si, Ge) are likely. Conductivity values obtained from optical data suggest a conductivity of 10^2–10^3 S cm^{-1} at 300°C along the stacking axis [130]. It is interesting to note that the conductivities are stable at room temperature *in vacuo*, too. Furthermore, it has been reported that polymers—of the $+Ge(Pc)$—S$+_n$-type—exhibit photoconductivity [119].

An interesting technology was developed by extruding solutions of $+Si(Pc)$—O$+_n$ **21** in trifluoromethansulfonic acid into an aqueous coagula-

Table 5 Pressed Powder Electrical Conductivity Data for Cofacially Linked Phthalocyanines with Covalent–Covalent Bonds (CA: *o*-Chloranil)

Compound	σ_{RT} (S cm^{-1})	ΔE (eV)	Reference
Si(Pc)(OH)$_2$	6×10^{-9}	0.79	[125]
[Si(Pc)(OH)$_2$]CA	9×10^{-4}	0.23	[125]
[Si(Pc)O]$_n$	5×10^{-6}	0.29	[102]
	3×10^{-7}	0.17	[125]
([Si(Pc)O)]I$_{1.55}$)$_n$	1.4	0.035	[102]
([Si(Pc)O]CA)$_n$	2×10^{-3}	0.17	[125]
[Si(Pc)OC$_2$H$_4$O]$_n$	4×10^{-11}	1.4	[125]
([Si(Pc)OC$_2$H$_4$O]CA)$_n$	9×10^{-9}	0.9	[125]
([Si(Pc)(t$-$bu)$_4$O]I$_{2.0}$)$_n$	2×10^{-3}		[106]
([Si(Pc)O](BF$_4$)$_{0.36}$)$_n$	2×10^{-1}		[103,130]
[Ge(Pc)O]$_n$	2×10^{-10}		[102]
	5×10^{-10}	1.4	[125]
([Ge(Pc)O]I$_{1.08}$)$_n$	1×10^{-1}	0.042	[102]
([Ge(Pc)O]CA)$_n$	1×10^{-5}	0.20	[125]
[Sn(Pc)O]$_n$	1×10^{-9}	0.56	[102]
	2×10^{-6}	0.6	[125]
([Sn(Pc)O]I$_{1.76}$)$_n$	7×10^{-7}	1.28	[102]
([Sn(Pc)O]CA)$_n$	3×10^{-9}	1.0	[125]
[Si(Pc)C≡C]$_n$	2×10^{-11}		[123]
[Ge(Pc)S]$_n$	2×10^{-8}	1.78	[119]

tion bath [132,133]. The silicon polymer (7–30 wt%) is dissolved at 80°C in the acid. Extrusion into a bath (25°C) results in air-stable dark-colored fibers. Iodine doping was achieved by adding the halogen to the spinner solution prior to spinning, to the coagulation bath as $I_3^{\ominus}$ salt, or by immersing the dried fibers in a benzene solution of the halogen. Anodic electrochemical doping is also possible. The mechanical properties of the $\{\!\{\text{Si(Pc)}\!-\!\text{O}\}\!(I_y)\}_n$ fibers, which are brittle, can be improved by adding aramide polymer [poly(1,4-diiminophenylene terephthalyl)] (1–7 wt%) to the spinning solution. By X-ray the presence of preferentially oriented polyamide and doped $\{\text{Si(Pc)}\!-\!\text{O}\}_n$ crystalline regions of the fibers were shown. Some conductivities are $\{\!\{\text{Si(Pc)O}]I_{1.76}\}_n$ 1.8 S cm^{-1}, $\{\!\{\text{Si(Pc)O}\}$ (polyamide)$_{0.59}I_{1.45}\}_n$ 9.1 × 10^{-2} S cm^{-1}. A similar result was obtained applying doped NiPc or H$_2$Pc in the spinning process [133].

At this point it is interesting if the doped polymers have advantageous properties compared to doped NiPc or H$_2$Pc. All samples show a well-studied high electronic conductivity as a result of similar orientation in the solid state. The polymers consisting of covalently linked phthalocyanines possess a permanent molecular structure with a stable orientation. Therefore, the properties of the polymers are stable over a longer period of time.

ii. Covalent–Coordinative Bonds in the Main Chain

Polymers of the type B2 contain one σ-bond and one coordinative bond to the central metal ion in the main chain. Two different kinds of polymers were described. In one case the phthalocyanine ligand contains the main group elements Al(III) or Ga(III) covalently bound to (the small and strong electronegative) F$^{\ominus}$. In the other case, trivalent transition metals occupy the core of the macrocycle and are covalently attached to an anion like C≡N$^-$ that has numerous, easily delocalizable π-electrons.

Fluoroaluminium and fluorogallium phthalocyanines **26** are polymeric compounds with the formula $\{M(Pc)\!-\!F\}_n$ (M = Al, Ga). The linear $\{M\!-\!F\}$ backbone is surrounded by a stack of attached planar phthalocyanine rings [134–136]. The fluoro compounds are prepared from M(Pc)(OH)·xH$_2$O (M = Al, Ga) by repeated evaporation to dryness with concentrated aqueous hydrofluoric acid [134,135].

The products were purified by repeated sublimation at an oven temperature of 510–530°C for [Al(Pc)F]$_n$ and 470–490°C for [Ga(Pc)F]$_n$. The fact that both compounds sublime at elevated temperatures is understandable in terms of a depolymerization process involving M(Pc)F monomers or [M(PcF)]$_x$ oligomers. [Al(Pc)F]$_n$ forms needle-like crystallites with broadened, blue-shifted visible spectra whereas [Ga(Pc)F]$_n$ has block-like crystallites with a broadened, red-shifted visible absorbance spectrum [137]. Crystallographic studies had shown [134,135] the cofacial fashion of **26** in a geometry similar

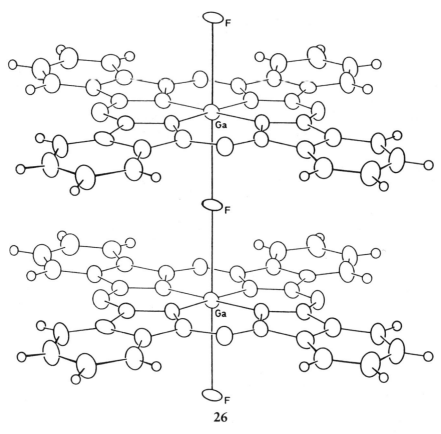

26

Figure 3 Structure of +Ga(Pc)—F+$_n$ **26**.

to the silicon and germanium polymers **21** (Fig. 3). The M—F bonds possess a dipole strong enough for favoring electrostatic interaction of the following type:

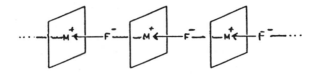

The molecules are linear and are stacked parallel to the long axes of the crystals with ring–ring separation of 0.366 nm in the Al compound and 0.386 nm in the Ga compound. In contrast, the analogous AlCl, AlBr, AlI, GaCl, and InCl compounds of phthalocyanines appear to be ordinary square-pyramidal phthalocyanine complexes [134,137].

Table 6 Pressed Powder Electrical Conductivity Data for Cofacially Linked Phthalocyanines with Covalent–Coordinative and Coordinative–Coordinative Bonds

Compounds	σ_{RT} (S cm^{-1})	ΔE (eV)	Reference
[Al(Pc)F]$_n$	$< 10^{-7}$ ($\sim 10^{-4}$)		[135[140]]
([Al(Pc)F]I$_{3.3}$)$_n$	3.4	0.017	[135]
([Al(Pc)F]I$_{1.0}$)$_n$	0.13	0.05	[135]
([Ga(Pc)F]$_n$	4.6×10^{-10}	0.85	[135]
([Ga(Pc)F]I$_{2.1}$)$_n$	0.15	0.04	[135]
([Al(Pc)F](PF$_6$)$_{0.7}$)$_n$	~ 4		[139]
([Ga(Pc)F](BF$_4$)$_{0.9}$)$_n$	~ 10	~ 0.003	[139]
[Co(Pc)CN]$_n$	2×10^{-2}	0.10	[153]
[Co(TBP)CN]$_n$^{a}	4×10^{-2}	0.11	[153]
[Co(Pc)SCN]$_n$	6×10^{-3}	0.22	[153]
Fe(Pc)(pyz)$_2$	2×10^{-12}		[160]
[Fe(Pc)pyz]$_n$	7.8×10^{-8}	0.404	[159]
([Fe(Pc)pyz]I$_{2.1}$)$_n$	1.28×10^{-1}	0.033	[159]
([Fe(Pc)pyz](PF$_6$)$_{0.5}$)$_n$	3.5×10^{-2}		[165]
Fe(Pc)(tz)$_2$	$< 10^{-9}$		[167]
[Fe(Pc)tz]$_n$	2×10^{-2}	0.10	[167]
[Fe(2,3-Nc)tz]$_n$^{b}	3×10^{-1}	0.07	[167]

^{a} TBP, tetrabenzoporphyrin.
^{b} 2,3-Nc, 2,3-naphthalocyanine.

As described for the silicon and germanium phthalocyanines **21**, [Al(Pc)F]$_n$ and [Ga(Pc)F]$_n$ **26** can be doped with electron acceptors such as iodine and nitrosonium salts NO$^+$Y$^-$ (Y = BF$_4$, PF$_6$) [135,136,138–140]. Doping results in greatly improved electronic conductivity (Table 6). The reaction with iodine was carried out in inert organic solvents or with iodine vapor and leads to purple black powders. The rate and extent of the reactions of [MPcF]$_n$ (M = Al, Ga) with iodine depend on the purity of the phthalocyanine compound: (a) when suspended in pentane, sublimed [Al(Pc)F] is converted within 5 min into ([Al(Pc)F]I$_{3.3}$)$_n$ and (b) unsublimed [Al(Pc)F]$_n$ after 24 h into ([Al(Pc)F]I$_{1.5}$)$_n$ [135]. The samples slowly lose iodine over a period of days to weeks, *in vacuo* finally giving stable compositions: sample (a) I$_{3.3}$ to I$_{2.4}$, sample (b) I$_{1.5}$ to I$_{1.0}$. At temperatures of $T \sim 250°C$ under vacuum, complete removal is effected leaving undoped [M(Pc)F]$_n$ [135]. The doping–dedoping process is performed at 50°C by repeatedly flushing with I$_2$/N$_2$ and N$_2$ gas [140]. The reversible change in conductivity is ~ 10 with a rising time of about 1 min. The iodine-treated samples were investigated by IR, TGA, and MS [135]. In comparison with [Si(Pc)O]$_n$ **21**, the iodine-doped [M(Pc)F]$_n$ **26** have lower thermal stability with regard to loss of iodine and the conductivity is equal or greater. The [M(Pc)F]$_n$ **26** exhibit the advantage of sublimation, e.g., to thin films that can be doped afterwards.

The reactions of [M(Pc)F]$_n$ **26** with nitrosonium salts were conducted in nitromethane or dichloromethane to yield ([M(Pc)F]Y$_x$)$_n$ (M = Al, Ga; Y =

PF_6, BF_4; $x = 0.38-0.90$) [139]. Weight loss for the doped compounds starts at $\sim 150\,°C$. Compressed tablet conductivities are about $0.2-10$ S cm^{-1}. The conductivity increases with increasing dopant concentration. The number of unpaired spins increases approximately with dopant concentration. The unpaired electrons are delocalized on the ligand. A remarkable feature of these materials is the invariant conductivity on exposure to ambient air. The stability of these doped compounds to thermal stress and exposure to air contrasts the behavior of iodine-doped $[M(Pc)F]_n$ and many other conducting polymers and nonmetallic materials.

Phthalocyanines containing trivalent transition metal ions [Mn(III), Cr(III), Fe(III), Co(III), Rh(III), Ru(III)] are able to form polymers linked by cyanide [141–148], thiocyanate [149,150], and azide [150]. Also polymeric tetrabenzoporphyrin–Co(III) with cyanide as bridging ligand was prepared [150].

Some methods exist for obtaining catena-poly[phthalocyaninato-metal(III)-μ-cyano] $+M(Pc)—CN+_n$ (Eq. 19) [141–148]. One route for the synthesis of these polymers is the displacement of an axial axion $X^{\ominus}$ (halide) by $CN^{\ominus}$ in coordinatively unsaturated compounds M(Pc)X. This method was sucessfully applied to the synthesis of polymers containing Fe(III) and Mn(III). Fe(Pc)Cl and Mn(Pc)Cl were synthesized by oxidative chlorination of FePc and MnPc [151,152]. Treatment of both chlorides with aqueous alkali metal cyanide solutions yields the polymers $+M(Pc)—CN+_n$ **27**. In the case of Co- and Cr-containing phthalocyanines the corresponding $M(Pc)Cl_2$ compounds were used. The $M^{3+}Pc^{1-}$ compounds are simultaneously reduced and polymerized by adding them to an aqueous alkali metal cyanide solution. A general route leading to the cyano bridged polymers is the cleavage of alkali metal (M') cyanide from alkali metal–dicyano(phthalocyaninato)–transition metal(III) complexes $M'[M(Pc)(CN)_2]$. These complexes can be synthesized by *in situ* oxidation of MPc (M = Fe, Co, Mn, Cr, Rh) with dioxygen in the presence of an excess of cyanide or by the reaction of $M(Pc)Cl_2$ (M = Co, Cr) respectively, M(Pc)Cl (M = Mn, Fe, Rh) with alkali metal cyanide in ethanol. The polymers are formed during extraction of the dicyano complexes with water or by addition of a solution of the dicyano complex in an organic solvent into hot water. Furthermore, polymers with substituents at the phthalocyanine macrocycle were prepared: 2,3,9,10,16,17,23,24-octa(octyloxymethyl) and octamethyl [141], and 2,9,16,23-tetranitro [142].

The polymers are poorly soluble in organic solvents. However, polymers with eight octyloxymethyl substituents are excellently soluble in various organic solvents [141]. In comparison to the low-molecular-weight $M'[M(Pc)(CN)_2]$, no shift of the Q-band transition is observed in polymeric **27** ($\lambda \sim 670$ nm) [141]. The C≡N vibrations in their IR spectra are shifted by around $15-30$ cm^{-1} to higher energy values in the polymers compared to those observed in monomeric $M'[M(Pc)(CN)_2]$. The increase of the CN frequency denotes the formation of a cyano bridge. The polymeric character is also

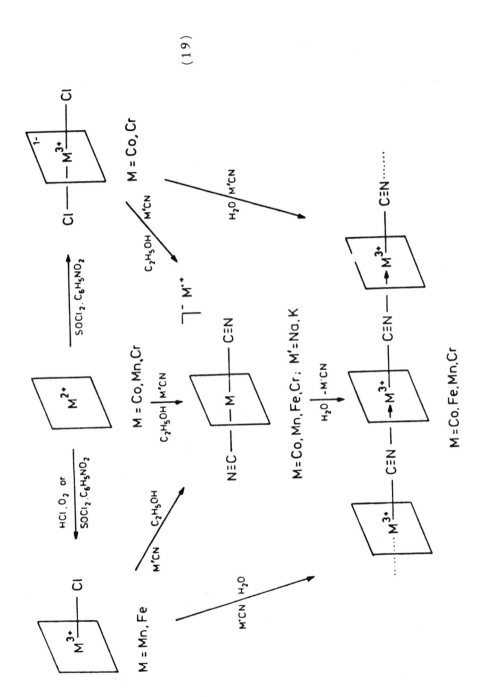

(19)

established by its conversion to the low-molecular-weight adducts M(Pc)(CN)-
(L) with bases (L = pyridine, piperidine etc.). The polymer $+$Co(Pc)—CN$+_n$ can
be dissolved in CF_3COOH and is recovered almost unchanged if the solution is
kept below 0°C for a short time [153]. The preparation of the thiocyanato
bridged polymer $+$Co(Pc)—SCN$+_n$ is very similar to that of cyano-bridged
polymers [149,150]. Polymers unsubstituted at the benzene rings are insoluble
in organic solvents and decompose in the presence of organic bases to give
Co(Pc)(SCN)(L) or Co(Pc)(L)$_2$ under reduction of M(III) to M(II).

27

Most papers include conductivity data. Table 6 contains the highest
values. Without doping, the polymeric cyano- and thiocyanato-bridged
polymers can show high values of $<4 \times 10^{-2}$ S cm^{-1}. In addition, the
photoconductivity of [Co(Pc)—CN]$_n$ 27 [154] and liquid crystalline behavior
of substituted soluble cyano-bridged polymers are described [141].

iii. Coordinative–Coordinative Bonds in the Main Chain

Polymers of the type B3 contain macrocycles with transition metal ions in
the oxidation state $+2$ capable of hexacoordination and neutral organic
donors with two groups or heteroatoms capable of coordination. Examples of
organic donors are pyrazine, 1,2,4,5-tetrazine, 4,4′-bipyridyl, and 1,4-
diisocyanobenzene. The macrocycles applied are phthalocyanines [with M =
Ru(II), Fe(II), Co(II)] [155–161], naphthalocyanines 15 [with M = Fe(II)]
[156,162], and 2,3,9,10-tetramethyl-1,4,8,11-tetraaza-1,3,8,10-cyclotetra-
decatetraene 20 [with M = Fe(II)] [163].

Some examples for the preparation of the polymers are given. The
catena-poly[phthalocyaninatoiron(II)-pyrazino] $+$Fe(Pc)—pyz$+_n$ 28 can be

prepared as dark violet insoluble solids from FePc and pyrazine via the method of Eq. (20) [159–161]. The materials have been characterized by elemental analysis, TGA/DTA, IR, UV-VIS, ESR spectroscopy, magnetic susceptibility, and others. Quantitative IR spectroscopic investigations show that the degree of polymerization is ∼ 20 [164].

Powder diffraction data indicate high crystallinity [159]. The polymers are not isostructural with $+M(Pc)-O+_n$ 21 or $+M(Pc)-F+_n$ 26. The general trend for thermal stability of the adducts of nitrogen base to transition metal phthalocyanines is an increase from the bisadduct to the polymer and the monoadduct [155]:

$$Co(Pc)(pyz)_2 < [Co(Pc)pyz]_n < CoPcpyz$$

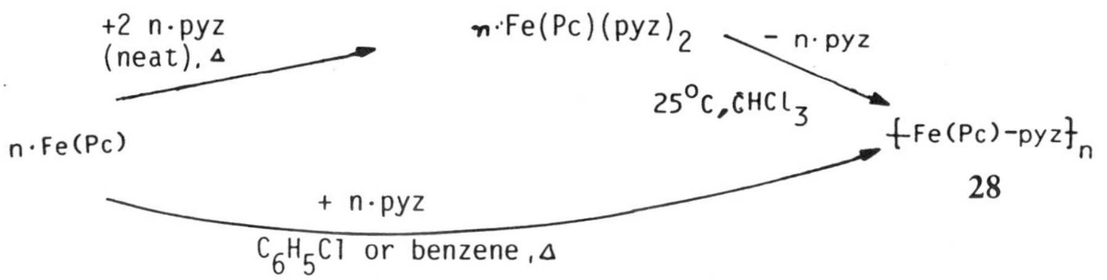

$$n \cdot Fe(Pc) \xrightarrow[\substack{+2\ n \cdot pyz \\ (neat),\,\Delta}]{} n \cdot Fe(Pc)(pyz)_2 \xrightarrow[\substack{25^\circ C,\, CHCl_3}]{-\ n \cdot pyz} +Fe(Pc)-pyz+_n$$

$$n \cdot Fe(Pc) \xrightarrow[\substack{+\ n \cdot pyz \\ C_6H_5Cl\ or\ benzene,\,\Delta}]{} +Fe(Pc)-pyz+_n$$

28

(20)

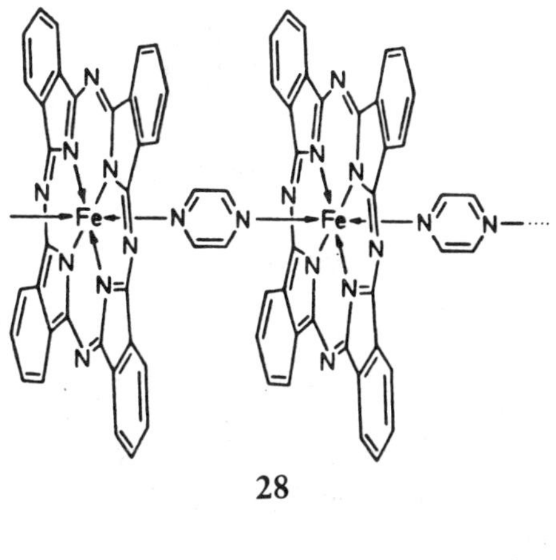

28

Catena-poly[phthalocyaninatoiron(II)-μ-(1,4-diisocyano-benzene] $+Fe(Pc)-bzNC+_n$ 29 is obtained by the reaction of FePc and 1,4-bis(isocyano-benzene) (small excess) in refluxing acetone [160,161]. Another example is the preparation of catena-poly[phthalocyaninatoiron(II)-μ-(1,2,4,5-tetrazine)] $+Fe(Pc)-tz+_n$ 30 from FePc and 1,2,4,5-tetrazine (excess) in chlorobenzene at 110–120°C [161]. The polymeric character of the reaction products was determined by various methods, however, the degree of polymerization was not determined. The solid state UV-VIS reflection spectra of the bisadducts and the polymers exhibit similar transitions. In some cases, additionally to the Q-band, a transition at λ > 700 nm was observed [160].

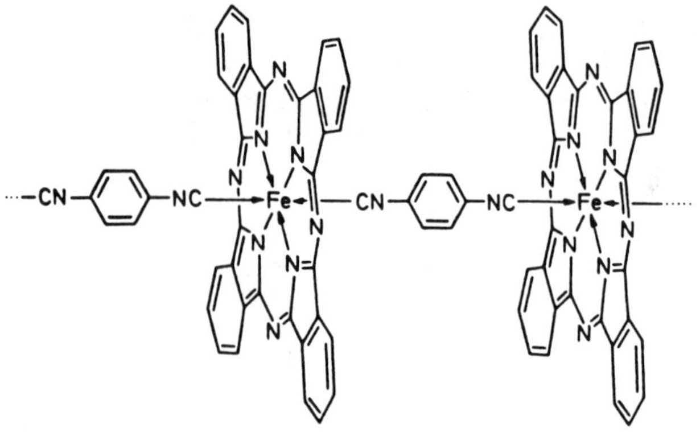

29

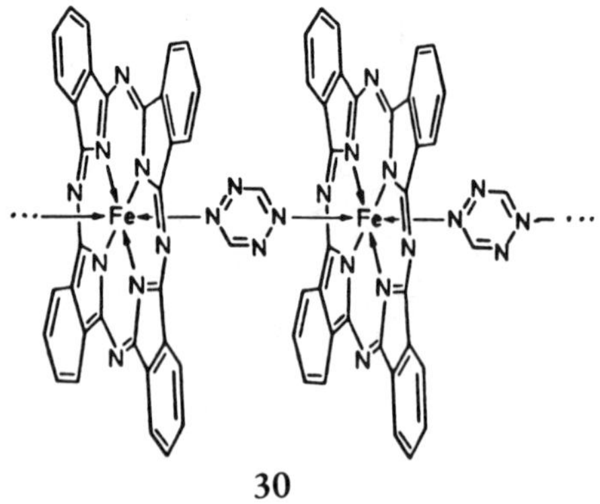

30

Most of the papers previously mentioned include results on measurements of electronic conductivity [see also 165–168] (Table 6). Undoped polymers **28–30** exhibit conductivities up to 10^{-1} S cm^{-1}. The conductivities increase from the bisadducts of the N-donors to the corresponding polymers. Doping with iodine (e.g., from benzene, [159]) or with nitrosonium salts (electrochemically [165]) results in increased conductivity. In comparison to iodine doped $[M(Pc)O]_n$ **21** and $[M(Pc)F]_n$ **26**, the interaction of iodine with polymers **28–30** is weak, and iodine can be washed out with organic solvents [159]. In contrast to **21** and **26**, the conductivity of the polymers **28–30** with a coordinative–coordinative bond in the main chain is assumed to be not ligand but metal/N-donor centered.

D. COVALENTLY BOUND POLYMERIC PHTHALOCYANINES

The covalent immobilization of phthalocyanines at a polymer chain combines the properties of the polymer carrier such as its solubility and thermal and mechanical behavior with that of the chelate. Principally, the properties of the bound phthalocyanines are preserved but in addition influenced by the polymer environment, e.g., the polarity of the polymer. The covalent bond is distinguished from a coordinative bond by its long-term stability.

Various possibilities exist for realizing a covalent bond between a polymer and phthalocyanine:

polymer-analogous reactions between a polymer and a phthalo-cyanine (both containing reactive groups),

synthesis of phthalocyanines directly on a modified polymer,

polymerization reactions of vinyl-substituted phthalocyanines.

The phthalocyanine-containing polymers described in this chapter and in Section E were investigated mainly for their behavior as oxidation catalysts and in electron transfer or photoelectron transfer reactions. The study, as an oxidation catalyst, is interesting for the oxidation of thiols which is widely applied industrially in the chemical treatment of petroleum distillates for removing thiols (Merox extraction) and for converting thiols to disulfides by oxidation with air (Merox sweetening). The overall reaction for the oxidation of thiolate anions with dioxygen is shown in Eq. (21).

$$4 \, RS^{\ominus} + O_2 + 2 \, H_2O \rightarrow 2 \, R—S—S—R + 4 \, OH^{\ominus} \qquad (21)$$

i. Covalent Binding of Phthalocyanines

A general route that allows binding of different porphyrins at linear polymers was described recently [169–171]. The substituted porphyrins 31–33 contain nucleophilic amino groups of similar reactivities. Therefore, an identical synthetic procedure can be applied to conduct the covalent binding to a polymer. Besides the binding of one porphyrin, the addition of different porphyrins to the reaction mixture allows the fixation of two or three porphyrins at one polymer system in a one-step procedure. Mainly a method was selected where a diluted solution of the polymer was added dropwise to a diluted solution to the porphyrins. The reaction of poly(styrene-co-chloro-methylstyrene) in the presence of a small amount of triethylamine in DMF results in the formation of polymers **34**, which are soluble only in organic

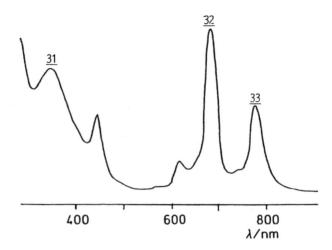

Figure 4 UV-VIS spectrum of the positively charged polymer **35** in water containing covalently bond moieties of **31, 32, 33**.

solvents such as DMF and toluene [169,172,173]. If the reaction of poly(chloromethylstyrene) is proceeding in the presence of an excess of triethylamine, the covalent binding of the porphyrin and a quaternization reaction occur simultaneously. Positively charged polymers **35** soluble in water were obtained [171]. Negatively charged polymers **36** containing porphyrin moieties are easily synthesized by the reaction of poly(methacrylic acid) (activation of the carboxylic acid group by carbodiimides or triphenylphosphine/CCl_4) with the porphyrins [170,173]. Uncharged water-soluble polymers **37** containing the porphyrin moieties are obtained by the reaction of poly(N-vinylpyrrolidone-co-methacrylic acid) with the low-molecular-weight substituted porphyrins in the presence of the same activating agents for the carboxylic acid groups. Residual carboxylic acid groups were converted to methyl esters [170]. The porphyrin content in the polymers was determined by quantitative VIS spectra and analysis of metal. The polymers prepared contain porphyrin moieties up to ~10 wt %. The porphyrins **31–33** contain four reactive functional groups. Therefore inter- and intramolecular cross-linking may occur in the reaction with the polymers employed. Intermolecular cross-linking could be avoided up to an amount of 2 mol% of applied porphyrins corresponding to one unit of the polymers. Higher amounts of porphyrins result in the formation of gels due to intermolecular cross-linking. Viscosity measurements indicate intramolecular cross-linking (micro gel formation) in some cases.

Figure 4 gives an example of a VIS spectrum of a polymer **35** with combined moieties of all porphyrins **31–33**.

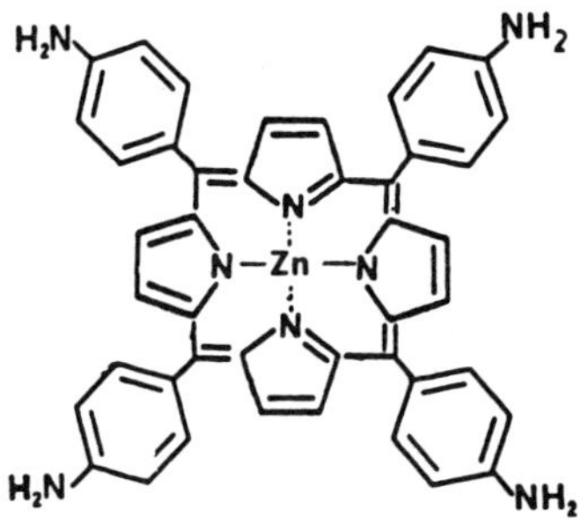

tetraaminotetra-
phenylporphyrin
31

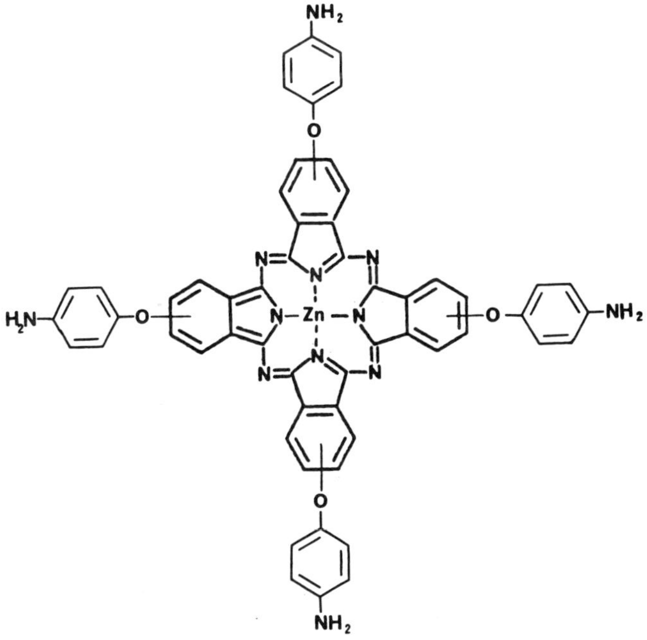

tetra(aminophenoxy)-
phthalocyanine
32

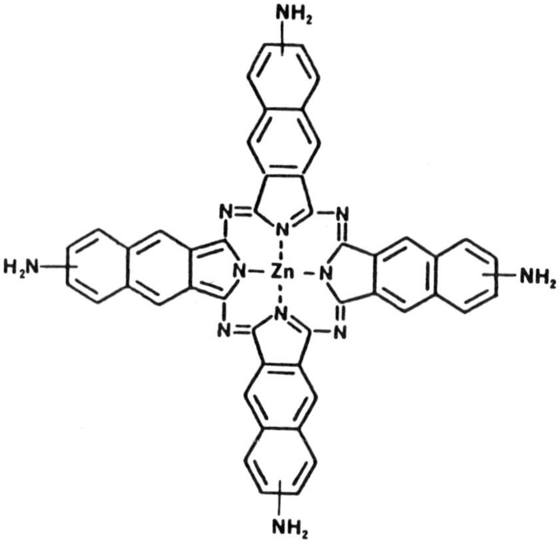

tetraaminonaphthalocyanine
33

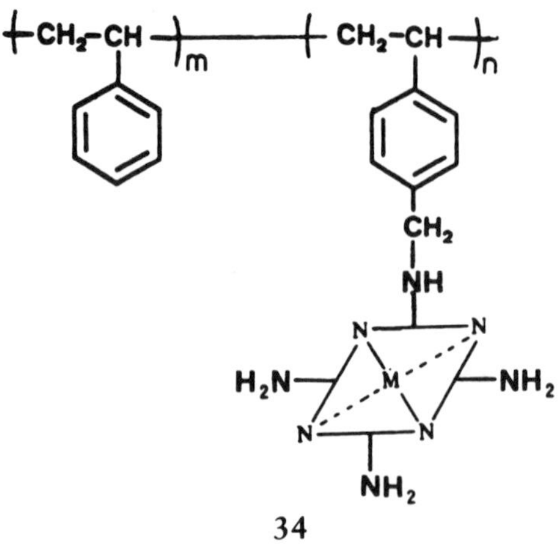

34

It is known that water-soluble phthalocyanines aggregate strongly in aqueous solutions. This is indicated by a shift of the Q-band to shorter wavelengths. The cationically charged polymers **35** show no aggregation whereas the other polymers **36, 37** aggregate slightly in water. Therefore the polymer backbone may inhibit a strong intermolecular porphyrin–porphyrin interaction. Also the positions of the Soret and fluorescence bands of the polymers **34–37** give hints at the influence of the polymer environment on the porphyrins [174]. The strongest long wavelength shift exists in the positively charged polymers **35** and the shortest wavelength shift in the negatively charged ones **36**. For the other polymers, no significant influence was observed. Mainly, the positive charges of the polymer backbone interact strongly with the electron-rich metal chelates, which prevents the porphyrin–porphyrin interaction. The porphyrin moieties in the polymers can act as antenna for reactions in a variable time scale:

   nanosecond to microsecond range: singlet life time,

   microsecond to millisecond range: triplet life time,

   second to millisecond range: electron transfer,

   minute range: photoelectron transfer,

   hour range: oxygen transfer.

By studying these reactions information concerning the polymer environment can be obtained.

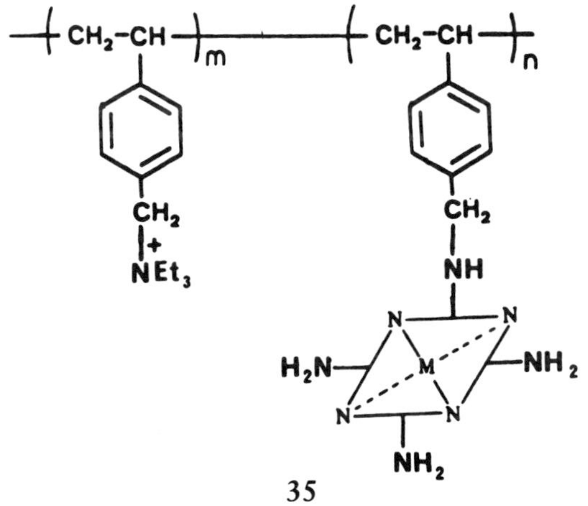

35

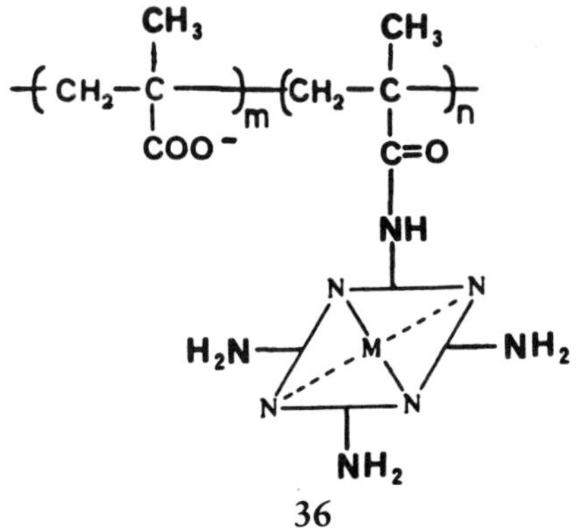

36

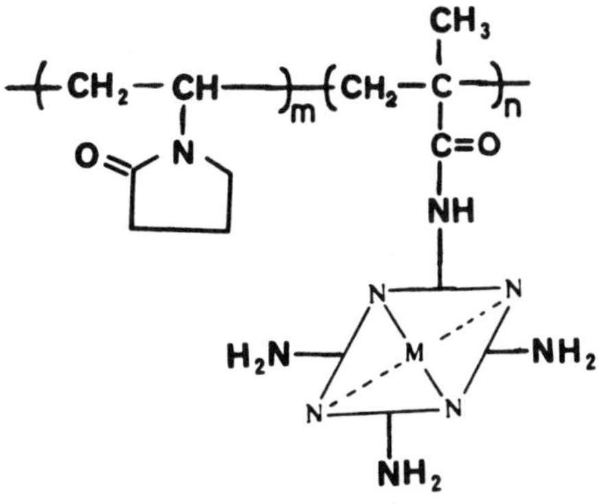

37

Determination of the triplet life time in the microsecond to millisecond range results in the fundamental observation that all polymer-bound porphyrins exhibit a higher lifetime in comparison to low-molecular-weight porphyrins [174]. Generally, the lifetimes increase in the following order: low-molecular-weight porphyrins < porphyrins at uncharged polymers < porphyrins at negatively charged polymers ≪ porphyrins at positively charged polymers. Therefore the polymer environment results in a strong to very strong shielding effect.

Electron-transfer reactions as model reactions for metallo-enzymes studied in the second to millisecond range depend strongly on the kind of charge of the polymer [172–174]. The experiments were performed by mixing a reducing agent such as dithionite and the polymer containing as Mn(III)–OH porphyrin and then directly measuring the electron transfer by the stopped-flow technique. The positively charged polymer 35 especially enhances the rate with anionic dithionite. In this case, much lower activation energy values and negative activation entropy values indicate a highly ordered transition state for the electron transfer. The rapid electron transfer may run efficiently by changing the conformation of the polymer. Compared to low-molecular-weight salts, polyelectrolytes can accelerate or inhibit an electron transfer much more efficiently.

The photoelectron-transfer reactions by irradiation with visible light in the presence of methylviologen as acceptor and, e.g., mercaptoethanol as donor were carried out in a different manner [169,174]. Irradiation, which induces electron transfer, was started about 10 min after mixing the reagents. Therefore formation of random coils with partly included donor and acceptor molecules occurs in solution. These molecules may be bound to the polyelectrolyte by both electrostatic and nonelectrostatic forces. Surprisingly, the photoelectron transfer taking place in a minute time range is relatively independent of the polymeric charge. No general trend is obvious. In addition, the covalently bound porphyrins are more efficient sensitizers in comparison to the low-molecular-weight ones, which is partly due to the higher triplet life time of the polymer-incorporated moieties. Therefore the polymer coil, which changes its conformation in the minute time scale, does not strongly attenuate the diffusion of the low-molecular-weight reagents.

A recently investigated reaction involves the time scale for oxygen transfer. The epoxidation of 2,5-dihydrofuran in water with hypochlorite in the presence of porphyrins was studied [174,175]. In this case, the polymer environment does not play a significant role. This is due to the fact that the kinetics of the reaction are not influenced by the polymer coil or its charge but by another reaction that includes oxygen transfer. The polymer is only important for achieving solubility in a special solvent. The employment of Mn(III)–porphyrins bound to insoluble polymers would offer the possibility of easy separation of the reaction products from the reaction mixture.

Tetraaminophthalocyanine [9, R = NH_2, M = Co(II)] was coupled to the amino groups of cross-linked aminated polystyrene or polyacrylamide (with aniline substituted groups) with cyanuric chloride yielding **38** and **39**. [176]. The polymers were tested as catalysts in thioloxidation. Polymer **39** shows a higher activity than the low-molecular-weight cobalt phthalocyanine due to a high concentration of the mononuclear superoxo complex of the phthalocyanine with O_2, which is more active than a dinuclear peroxo complex. Polymer binding reduces dinuclear phthalocyanine formation.

38   **39**

2,9,16,23-Phthalocyaninetetracarboxylic acid and its derivatives were also employed as starting materials for the preparation of covalently bound polymeric phthalocyanines. The tetrachloride of the phthalocyanine [9; R = COCl; M = Fe(III), Co(II), Cu(II), Ni(II)] was attached to linear polystyrene or poly(2-vinylpyridine-co-styrene) in nitrobenzene with $AlCl_3$ to give polymers such as **40** [177,178]. The copolymer with styrene contains ⩽4 mol% bound phthalocyanines. Gel formation occurs at >4 mol% of phthalocyanine content. The reason for this is the formation of some phthalocyanine-cross-linked polymer chains. In contrast, copolymers containing the less reactive vinylpyridine unit can bind up to ~15 mol% phthalocyanine without cross-linking. Investigation of the catalase-like activity shows that polymer binding leads to a lower activation energy, due to a higher concentration of unaggregated active centers, than with low-molecular-weight phthalocyanines [177]. The polymer **40** was used as a cathodic catalyst on platinum for the reaction of oxygen with hydrogen in a secondary fuel cell [179]. Dioxygen produced by electrolysis of water is stored during the charging process in the polymer matrix and then electrochemically reduced in the discharging process. The open circuit voltage (vs. Zn) of the cell is 1.2 V and the discharge capacity is 46 A·h/kg. No significant decay of these properties was observed after more than 30 cycles. Doping of films of **40** with various dopants such as HCl, H_2SO_4, and I_2 yields highly conducting materials ($\sigma_{20°C} = 10^{-4}$–10^{-1} S cm^{-1}) [180].

40

The reaction of the tetracarboxylic acid of phthalocyanine [9; R = COOH; M = Ni(II), Cu(II), Co(II)] with partially chloromethylated polystyrene in DMSO/pyridine affords water-soluble polymers **41**. These polymers were investigated in the sensitized reduction of methylviologen by visible light [181].

Poly(vinylamine) and macroporous highly cross-linked styrene–divinylbenzene (grafted with poly(vinylamine)) were reacted with 9 [R = COOH; M = Co(II)] [182,183]. The condensation of the tetracarboxylic acid with poly(vinylamine) was carried out in THF in the presence of dicyclohexylcarbodiimide. The resulting polymer **42** contains ~0.013% of the phthalocyanine. The activity of the polymers in mercaptoethanol oxidation is comparable to the activity of coordinatively bound 9 [R = COOH; M = Co(II)] at polyvinylamine (Section E) but higher than the activity of the low-molecular-weight 9 [R = SO$_3$H; M = Co(II)].

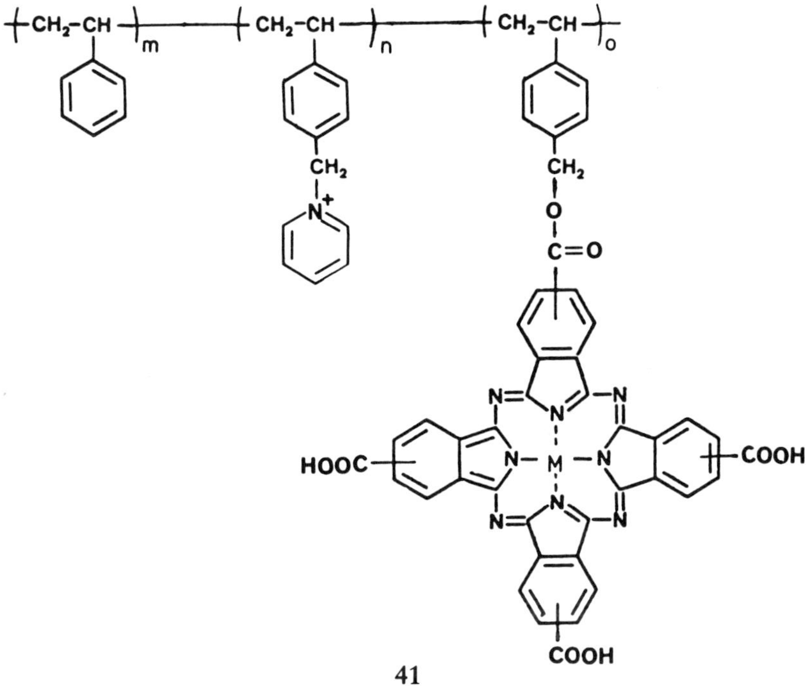

41

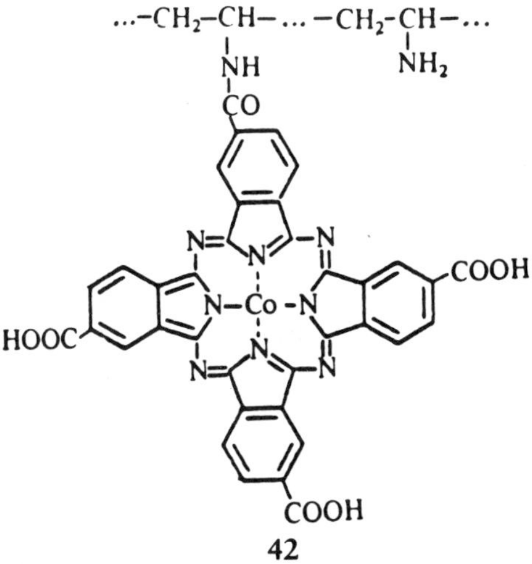

42

Attachment of chlorosulfonated phthalocyanines (9; R = SO$_2$Cl; M = Co(II)) to macroreticular polystyrene or poly(4-aminostyrene) led to polymers **43**, **44** containing 0.0027–0.035% bound phthalocyanines [184]. The phthalocyanines are mainly aggregated at the edges of the beads. The activity of the polymers in the oxidation of cyclohexene at 78°C compared with low-molecular-weight phthalocyanine is only higher after grinding the polymer.

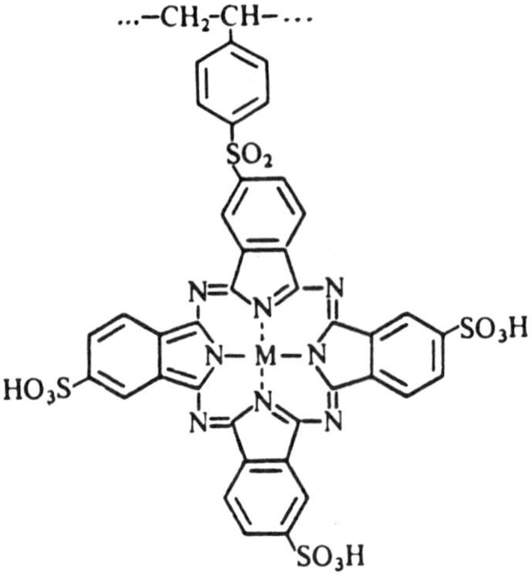

...−CH₂−CH−···

...−CH₂−CH−···

43

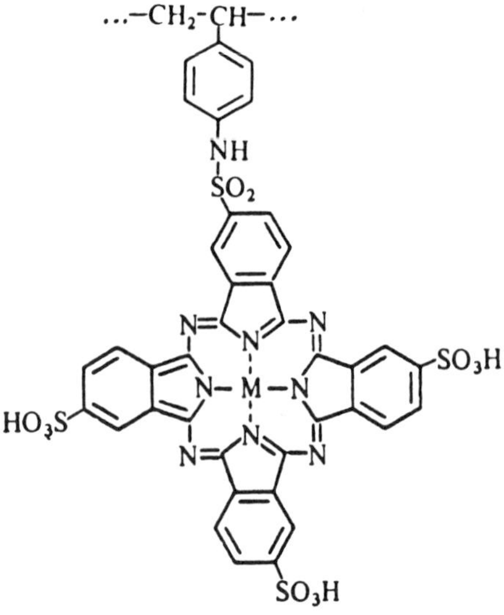

44

Phthalocyanines 9 [R = COCl; M = Fe(III), Co(II), Ni(II), Cu(II)] were polymer bound by a Friedel–Crafts reaction with poly(γ-benzyl-L-glutamate) [185]. The polymers contain about 0.2–2.8 mol% of phthalocyanine groups. They are soluble in various organic solvents and have an α-helical conformation.

ii. Synthesis of Phthalocyanines at Polymers

The first step for the synthesis of phthalocyanines on a polymer carrier is the immobilization of an aromatic o-dinitrile. In a second step the conversion of the covalently polymer-bound dinitrile with dissolved low-molecular-weight o-dinitrile molecules to the polymer-bound phthalocyanines is carried out. The bound chelates may be cleaved to obtain a low-molecular-weight phthalocyanine containing only one reactive functional group (unsymmetrical phthalocyanines). For the synthesis of unsymmetrical phthaloyanines this method is favored above a statistical reaction of different low-molecular-weight substituted phthalonitriles [186].

The pioneer work in this field was described in 1982 [186]. Cross-linked substituted divinylbenzene–styrene copolymers were used. First, a polymer bound trityl chloride was reacted with diols such as 1,6-hexanediol followed by reaction with 4-nitrophthalonitrile to a polymer bound dinitrile (0.23–0.54 mmol/g) [Eq. (22)]. For easier phthalocyanine synthesis the polymer-bound dinitrile was converted with NH₃ to the polymer-bound diiminoisoindoline.

The following condensation with 4-isopropoxydiiminoisoindoline in DMF/ dimethylaminoethanol results in the formation of the dark green polymer bound phthalocyanines **45a**. Hydrolysis with HCl in dioxane leads to the monohydroxy substituted phthalocyanines **45b**.

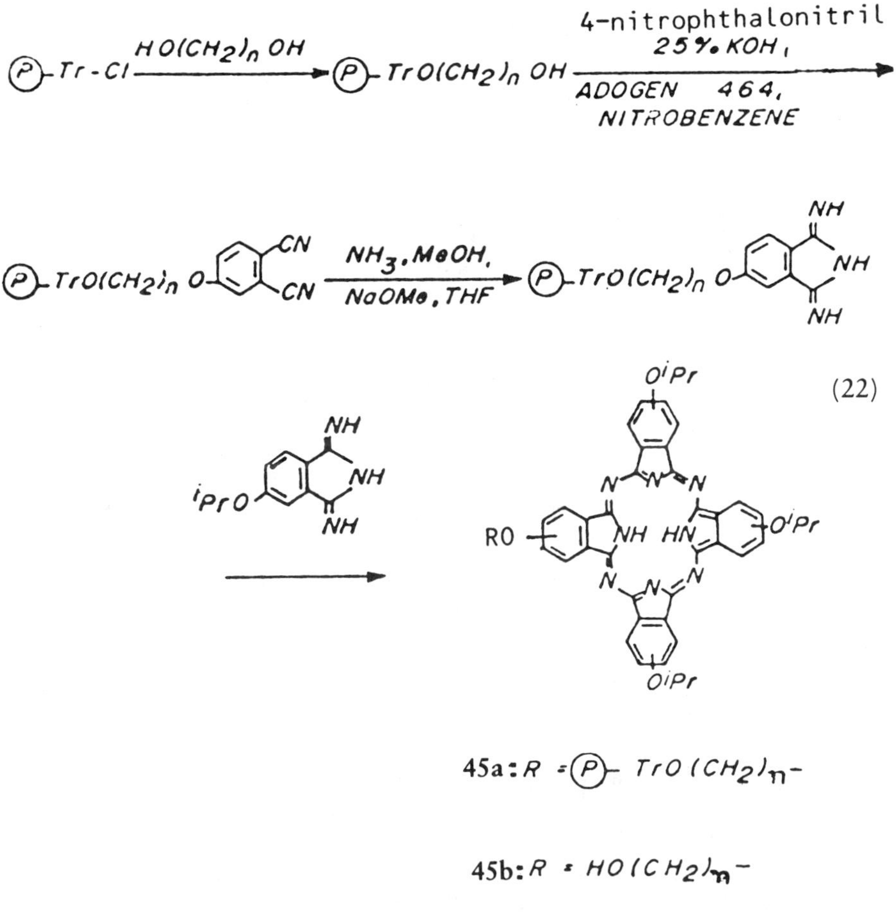

(22)

45a: $R = \text{(P)} - Tr O (CH_2)_n -$

45b: $R = HO (CH_2)_n -$

n = 4,6

(P) = divinylbenzene-
    styrene copolymer

A second paper describes the synthesis of polymer-bound 2-[4-(4'-hydroxyphenoxy]-9,16,23-triphenoxyphthalocyanine (M = Zn) [187]. First 4-(4'-hydroxyphenoxy)phthalodinitrile was bound on a macroreticular divinylbenzene–styrene copolymer to obtain the polymer-bound dinitrile. Then, reaction with an excess of 4-phenoxyphthalodinitrile in dimethylaminoethanol in the presence of Zn(II) leads to **46** (0.227 mole phthalocyanine groups/g polymer, quantitative conversion). The low-molecular-weight monohydroxy-phenoxy-substituted phthalocyanine is obtained in ~20% yield by treating **46** with HBr in glacial acid.

46

Soluble poly(organophosphazenes) bearing covalently bound phthalo-cyanine (M = Cu) side groups have been synthesized [Eq. (23)] [188]. First poly(dichlorophosphazene) was converted into a polymer containing pendant dinitriles in a three-step procedure. The reactions with an excess of various dinitriles lead to soluble polymers **47** with 2.5–10% phthalocyanine side groups. In order to avoid cross-linking (and insolubilization) the loading with dinitriles was limited to 10% of total side groups, and a very large excess (~50 equiv.) of low-molecular-weight dinitriles was employed for the conversion into the chelate. In addition dilute reaction conditions and lower coupling temperatures than normal minimize cross-linking. The paper of interest here includes the synthesis of various phthalocyanines on phosphazene trimers. The electronic absorption spectra in benzene show that the polymer chain effectively inhibits phthalocyanine aggregation. Iodine-doped polymers exhibit electrical conductivity in the range of 10^{-5}–10^{-6} S cm^{-1}.

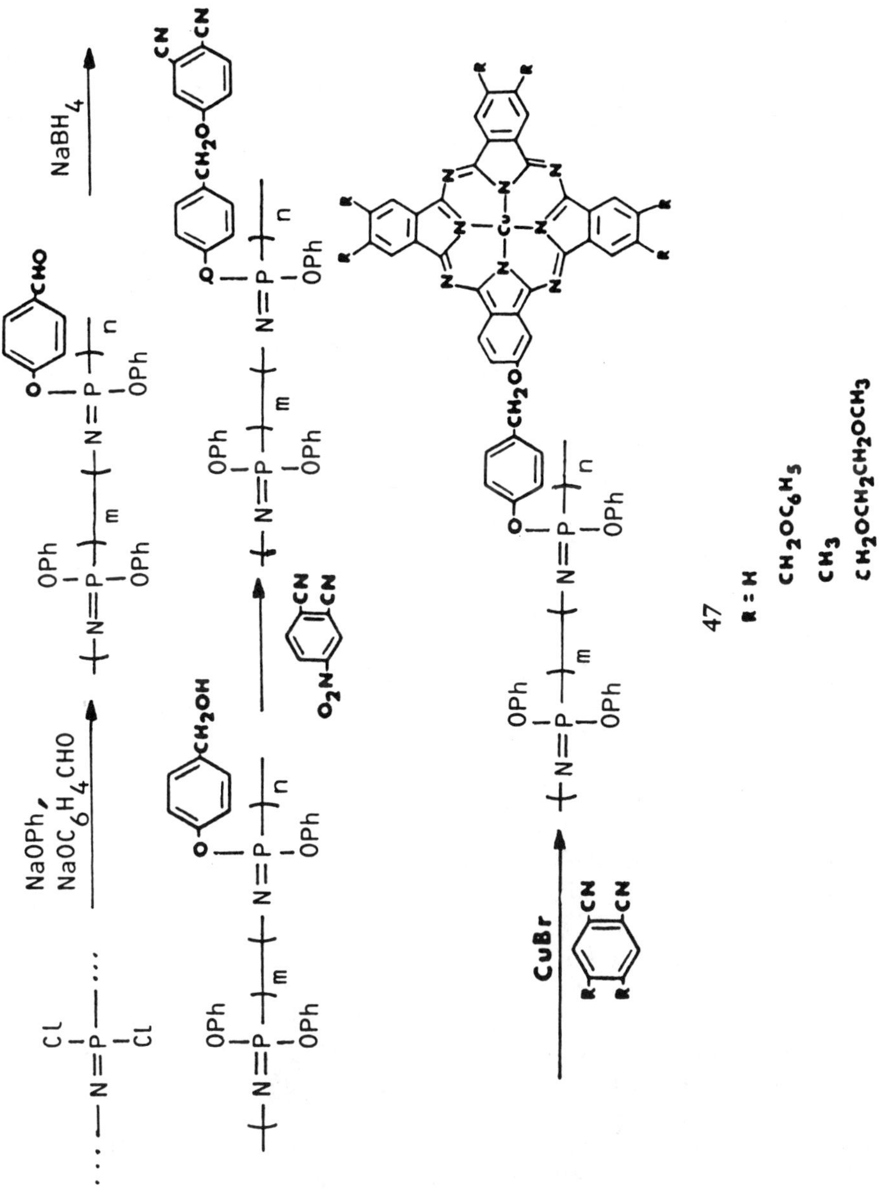
NaOPh,
NaOC6H4CHO
NaBH4
CuBr
47
R = H
CH2OC6H5
CH3
CH2OCH2CH2OCH3

(23)

iii. Polymerization of Vinyl Groups Containing Phthalocyanines

The use of vinyl-substituted phthalocyanines in polymerization reactions is of substantial interest because many copolymers may be prepared. Until now only a few papers are known.

2-Hydroxyethylmethacrylate reacts with **9** [R = COCl; M = Co(II)] to give the tetravinyl compounds **48** [189]. Radical copolymerization with an excess of poly(N-vinylcarbazole) in benzene results in a copolymer with the composition phthalocyanine:vinylcarbazole = 0.0083. Results on the polymerization of 2,9,16,23-tetravinylphenoxyphthalocyanine metal chelates oriented in thin films obtained by the Langmuir–Blodgett technique will be published soon [190].

48

E. COORDINATIVE AND ELECTROSTATIC INTERACTIONS OF PHTHALOCYANINES WITH POLYMERS

Coordinative polymer binding of phthalocyanines was examined using polymer ligands such as poly(ethylenimine) [191–193], poly(vinylamine) [191–196], amino group-modified poly(acrylamide), or silica gel [191] phthalocyanine derivatives **9** (R = —SO₃H, —COOH; M = Co) [191,192,194–196] and **7b** [M = Fe(III)] [193] were employed. The above-mentioned thiol oxidation [190,191,193–195] and the oxidation of guaiacol [192] were investigated.

Solutions or suspensions of the polymers and the water-soluble phthalocyanines are mixed for later investigations. For **9** (R = —COOH, M = Co), conclusive evidence of axial coordination was obtained from ESR showing a 5-coordinative complex structure [194]. It was shown that increasing concentrations of poly(vinylamine) shifted the equilibrium between the monomer and the dimer form to the monomeric phthalocyanine. Therefore a high concentration of polymer ligands separates the chelate molecules in the polymer coil (shielding effect).

The oxidation of thiols, especially 2-mercaptoethanol, with coordinatively bound phthalocyanines **9** (R = SO_3H, COOH; M = Co) as catalysts in water starts with the coordination of the nucleophilic thiolate anion to the phthalocyanine [191,195]. The coordination of dioxygen at Co(II) may occur either on the opposite side of the thiolate-containing phthalocyanine molecule [191] or after splitting off a thiol radical at the reduced phthalocyanine [195]. The thiol radicals dimerize to disulfides and reduced dioxygen forms H_2O_2 that later may oxidize thiolate anions [Eq. (21)].

Some important results are as follows:

Increase of activity of the polymer systems compared to low-molecular-weight phthalocyanines by, e.g., a factor of 50.

Higher turnover numbers with the polymer system of 10^6–10^7 compared the low-molecular-weight phthalocyanine with 10^4–10^5.

Increase of activity of the polymer system with increasing polyamine content, decreasing concentration of **9** (R = SO_3H) and decreasing molecular weight of poly(vinylamine) (for some results see Fig. 5).

The oxidation of guaiacol with **7b** [M = Fe(III)] in the presence of several polymers and H_2O_2 as oxidizing agent was studied in water at different pH [194]. For example with poly(vinylamine) at pH > 6.0 the free amino group coordinates with the axial position of Fe(III) at the chelate ring and inhibits catalysis. At pH < 6.0, the environment in poly(vinylamine) is positively charged and induces the formation of the reactive $OOH^\ominus$. This results in an increased rate of oxidation.

Positively charged polymers such as ionenes $+N^\oplus(CH_3)_2—(CH_2)_x—N^\oplus(CH_3)_2—(CH_2)_y+_n$ have a different interaction with **9** [R = SO_3H; M = Co(II)] [192,197,198]. This interaction must be purely electrostatic. It was observed that the ionenes strongly enhance aggregation of the cobalt complexes. A stoichiometric complex between the tetrasulfonic acid and the ionone appeared to be formed with a $N^\oplus$/Co ratio of 4:1. The complexes are resistant toward oxygen adduct formation (e.g., no μ-peroxo bridged adduct formed). For the oxidation of thiols these polymers are preferred over poly(vinylamine) since their cationic charge, required for increasing local

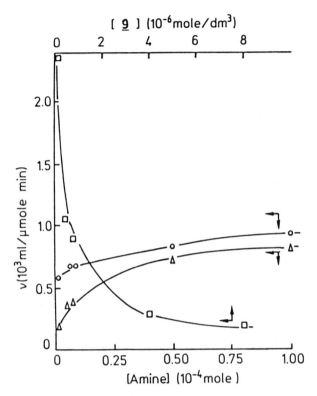

Figure 5 Catalytic activity of poly(vinylamine) **9** [R = SO₃H; M = Co(II)] complexes in the oxidation of 2-mercaptoethanol (14.25 × 10⁻³ mol in 11 ml H₂O) at 25°C as a function of the poly(vinylamine) (○: $\bar{P}_n$ = 50; △: $\bar{P}_n$ = 1680) content with **9** 10⁻⁸ mol and as a function of content of **9** (□) with poly(vinylamine) ($\bar{P}_n$ = 570) 4 × 10⁻³ mol. *ν* corresponds to O₂ consumption in ml/min per μmol of Co.

thiolate anion concentration, is independent of pH. As a consequence, catalytic activity is less influenced by pH. The maximum activity is even somewhat higher for ionenes and the catalytic activity decreases less in successive thiol oxidation runs. The kinetics of the thiol oxidation follow the two-substrate Michaelis–Menten rate law. Cationic latexes were prepared by emulsion copolymerization of styrene, divinylbenzene, and a vinyl monomer containing a quaternary amino side group [199]. Oxidation of 1-decanethiol at pH 9 occurs rapidly, even though the thiol is immiscible with water.

    An interesting method is using phthalocyanines as models for hemoproteins. Here an attempt was made to obtain a coordinative bond between **9** [R = SO₃H; M = Fe(II), Co(II)] at natural globin [200]. The interaction between globin or hemoglobin and the phthalocyanine leads to green crystals. Spectroscopic methods indicate that heme is partially displaced [Eq. (24)].

Some results are as follows:

The phthalocyanine is bound as monomer from the solution containing dimers.

Like heme, the phthalocyanine is located deeply inside the protein.

The Fe-chelate shows reversible dioxygen uptake. Cyanide ions are bound irreversibly.

$$\sim N \rightarrow FeP \quad + \quad FePc \quad \longrightarrow \quad \sim N \rightarrow FePc \quad + \quad hem \quad (24)$$
$$\underline{9} \ (R= SO_3^-)$$

The binding of water-soluble phthalocyanines on proteins and proteides is the fundamental requirement to use these chelates in photodynamic cancer therapy and diagnosis [201]. Until now a mixture of porphyrins (HPD) was employed successfully. However, these porphyrins exhibit the disadvantage of absorbing laser light at too short a wavelength and having low extinction coefficients. In order to avoid adsorption by thermal tissue photosensitizers absorbing at $\lambda > 640$ nm are necessary. Water-soluble phthalocyanines and naphthalocyanines are attractive candidates for increasing the application of photodynamic cancer therapy and diagnosis (see Rosenthal and Ben-Hur, this volume).

F. PHYSICAL INCORPORATION OF PHTHALOCYANINES IN ORGANIC AND INORGANIC POLYMERS (CARRIERS)

The commercial use of phthalocyanines as pigment dye stuffs for organic polymers is not discussed here [for reviews see 202–204]. Some new special developments are mentioned.

Phthalocyanines are either homogeneously dissolved or dispersed in a polymer. This depends on the concentration of the phthalocyanine and substituents on the ligand. Bulky or long-chain substituents R in **9** enhance solubility.

Photochemical hole burning is known as an optical information storage system with high storage density $(10^8–10^{11}$ bit/cm^2) [205,206]. The storage system is prepared by dissolving $\geqslant 10^{17}$ photoreactive molecules in a solid transparent amorphous material. The metal-free phthalocyanine dissolved in poly(methylmethacrylate) is characterized by a broad absorption of the Q-band region due to the different environments present. Equation (25) schematically depicts tautomerization of the central protons. Because both tautomers absorb at different wavelengths, it is possible to write and read optical informations by using light of different wavelengths. If one phthalocyanine tautomerizes, holes are burned in the broad absorption peak. The main disadvantage is that this storage system works at very low temperatures. If, for example, holes are burned at 4.2 K and the system is heated to 100 K, the holes are totally destroyed and do not reappear even if the system is recooled to 4.2 K. Thermal tautomerization of the central protons occurs at elevated temperatures. High-density storage of information seems obtainable at 77 K as long as the hole-burned information is written and read at lower temperatures. The use of substituents on the ligand with different electronic or steric effects may be a possibility for using this storage system at higher temperatures. However, no information is available yet.

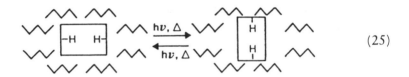

$$\text{(25)}$$

Dye-in-polymer films consisting of 20 wt% **9** [R = *tert*-butyl; M = V(0)] and a vinyl chloride–vinyl acetate copolymer have been prepared by the spin-coating technique [207]. The vanadyl phthalocyanine precipitates in different glassy solids in the polymer. The background of this work is that a so-called phase II of the vanadyl phthalocyanine is the most photoactive form with good IR absorption, compatible with the low-cost GaAs diode laser [207–209]. This is interesting for various laser-addressed marking technologies. The long-chain tetracarboxylic esters **9** [R = —COOC$_{10}$H$_{21}$; M = Fe(III), Co(II), Ni(II), Cu(II)] were dispersed in polystyrene by spreading a 20% benzene solution of the polymer and the phthalocyanine on plates [210]. Conductivity was investigated. Smooth green films containing the octacyano-derivative **7a** (M = 2H) in poly(N-vinylcarbazole) respectively polyimide were prepared on gold substrates by casting from solution [211]. The polymer films (thickness 140–2600 nm) contain 50–10% wt% of **7a**. These films function as stable electrochromic redox systems in the absence of redox couples in an electrolyte. Due to a combination of high intrinsic electronic conductivity and electron-accepting properties of the cyano groups, films of **7a**, without a polymer, are easily reduced and reoxidized at 230 mV vs. NHE in a weakly acidic electrolyte [212]. Unsubstituted **1** is electrochemically inert under these

Table 7 Performance of Some Photovoltaic Cells (Nesa Glass/H_2Pc
(7)-Polymer/Metal$\leftarrow h\nu$) at Room Temperature[a]

Cell	Thickness of organic film (μm)	Metal films	I_T (mW cm^{-1})[b]	$\bar{V}_{oc}$ (V)[c]	J_{sc} (μA cm^{-2})[d]	$\eta'(\%)$[e]
1	0.5	In	1	0.40	179	2.4
2	0.9	In	1	0.40	110	1.42
3	1.8	In	1	0.38	48	0.63
4	1.8[f]	In	1.1	0.40	130	1.58
5	1.8[f]	In	0.135	0.35	27	2.30
6	1.8	In	0.135	0.39	7	0.64
7	1.8	Al	1	0.97	6	0.19
8	1.8	Sn	1	0.34	26	0.29
9	1.8	Pb	1	0.53	7.2	0.13

[a] Content of x$-$$H_2$Pc in the organic film 60%. Thickness of the metal film $\sim$900 Å.
Irradiation with visible light of 80 mW/cm^2.
[b] I_T, light energy after traversing the metal film.
[c] $\bar{V}_{oc}$, open circuit voltage.
[d] J_{sc}, short circuit current.
[e] Conversion efficiency $\eta' = (J_{sc} \cdot \bar{V}_{oc}/I_T) \cdot$ FF (FF = fill factor).
[f] Organic film doped with 14% trinitrofluorenone.

conditions. In the polymer, dispersed **7a** forms conducting pathways through the polymer film which leads to behavior similar to that of films of pure **7a**. The electrochemical activity for reduction and reoxidation [Eq. (26)] of mechanically stable films of **7a** in a polymer depends partly on the kind of polymer, the amount of the phthalocyanine, the layer thickness, and the conditions of film preparation.

$$Pc(CN)_8 + n \cdot H^{\oplus} + n \cdot e^{\ominus} \rightleftharpoons Pc(CN)_8H_n \qquad (n \leqslant 3) \qquad (26)$$

Photovoltaic cells using Schottky junctions composed of thin films of an organic dye have been reported [213–216]. Besides merocyanines, phthalocyanines are the most promising compounds for higher conversion efficiencies. The adsorption of dioxygen is necessary to cause a substantial increase of conversion efficiency [215]. One explanation for the influence of dioxygen is an increased efficiency of minority carrier injections. The metal-free **1** was transformed into the x-form by milling (diameter of the particles 80 nm) [217]. A mixture of 2.3% **1** and 1.5% polymer (polycarbonate, polyvinylacetate, polyvinylcarbazole) in methylene chloride was cast on Nesa glass. After evaporating the solvent, a metal film was deposited on the organic film. The composition of some cells is described in Table 7. The cells exhibit good rectifying behavior and diode parameters. Current flows with a negative potential at the metal (e.g., indium) electrode. The conversion efficiencies under irradiation with visible light in the cell Nesa glass/**1**/metal$\leftarrow h\nu$ can reach 3%. Disadvantages are the very low transmission of the metal film and the decreasing quantum efficiency with increasing light intensity. Comparing cells 1–3 (Table 7) it is seen that thinner layers show more promising properties. Thicker layers favor recombination. The addition of acceptors such as trinitrofluorenone also leads to higher conversion due to a higher J_{sc} (cells 4,5). Higher temperatures increase the efficiency by 4–5% K^{-1}. No clear picture of the function of the cells can be given but some models are mentioned in the literature [215,216,221]. One model states that by contacting the p-conductor 1 with indium (metal with low work function of 4 eV), the negative charge carriers generated by irradiation flow into the metal and the positive charge carriers into the Nesa glass (Fig. 6). In the cells composed of particles of 1 dispersed on a polymer binder, the effect of the polymer was studied. Polar polymers such as poly(vinylidene fluoride), poly(acrylonitrile), and others give a high photocurrent and conversion efficiency [218]. This is attributed to the enhancement of exciton dissociation, which in turn is due to a large electric field formed by polar groups of the polymer. The effects of recrystallization of the phthalocyanine were studied in the cell Nesa glass/1 in poly(N-vinylcarbazole)/Au [219]. For the lower resistant 1, carriers are generated in the polymer. The MIS (metal/insulator/semiconductor)-type cell: Nesa glass/1 (M = Cu)/polyethylene/Au shows higher efficiencies than the corresponding MS (metal/insulator)-type cell [220]. Films of phthalocyanines form liquid junctions (a kind of Schottky junction) when dipped into an electrolyte solution. This is a simple method of constructing photovoltaic cells. Promising cells consist of 1 (M = 2H) in polymers on Nesa glass [222]. These cells are dipped into an aqueous $K_4/K_3[Fe(CN)_6]$ resp. I^-/I_2 redox-pair solution. Under irradiation with visible light, cathodic photocurrents J_{sc} of < 0.63 mA/cm^2 with open circuit voltages V_{oc} of < 0.33 V were observed. The conversion efficiency (at 75 mW/cm^2 of white light) is < 0.1%. Films of 1 (M = Zn) in poly(N-vinylcarbazole) on Nesa glass in contact with an aqueous electrolyte can be used as a sensitive dioxygen sensor [223]. By irradiation with visible light in the presence of dioxygen (redox couple O_2/reduced O_2) cathodic photocurrents are observed. The device can be switched either by irradiation/darkness or O_2/inert gas cycles (Fig. 7). The polymer improves the stability and performance of the cells. Results on the investigations of phthalocyanines in combination with "inorganic polymers" such as active carbon, Al_2O_3, SiO_2, and others are discussed briefly because these carriers are indeed polymeric but normally classified as inorganic compounds.

Unsubstituted and substituted low-molecular-weight phthalocyanines 1, 9, tetraphenylporphyrins 17, and tetraazaannulenes 19 and also polymeric phthalocyanines 5, 6 [M = Fe(II), Co(II)] on active carbon were studied as catalysts for dioxygen reduction in the fuel cell [224–232] and for thiol oxidation in the chemical treatment of petroleum distillates (Merox sweetening) [233–236]. Phthalocyanines on γ-Al_2O_3 and SiO_2 were studied in thiol oxidation [236,237], dioxygen chemisorption [238,239], and NH_3 adsorption [240]. Other inorganic carriers are CdS and ZnO (photochemistry) [241,242].

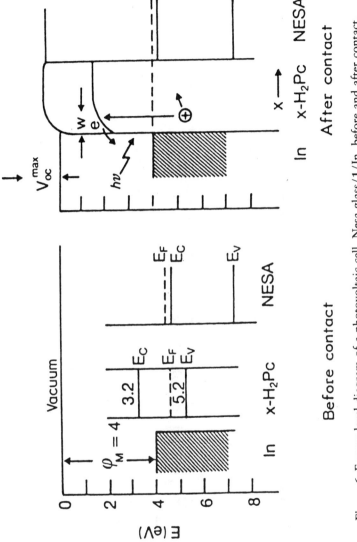

Figure 6 Energy level diagram of a photovoltaic cell, Nesa glass/1/In, before and after contact.

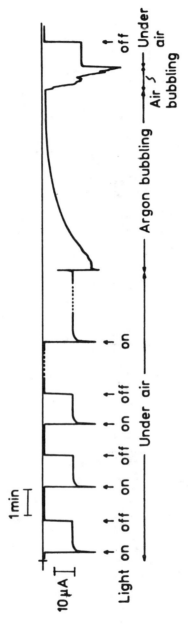

Figure 7 Current changes [film of **1** (M = Zn)/poly(N-vinylcarbazole) in contact with an aqueous electrolyte] induced by switching of the irradiation on and off, and by argon and air bubbling.

Another interesting route is the synthesis of faujasite zeolite-encaged metal phthalocyanines by the reaction of metal ion-exchanged faujasites with phthalonitrile [243–245]. The phthalocyanines are monomolecularly encaged in the supercages.

G. SHORT CONCLUSION

It was shown that many kinds of polymer/phthalocyanine combination can be prepared. Due to the high functionality of the benzene rings of the ligand, cross-linked polymers are easily obtained by reaction of low-molecular-weight chelates or their precursors to polymers and on the other side by the reaction of highly substituted phthalocyanines with reactive linear polymer chains. As is generally known, cross-linked polymers are difficult to characterize. Therefore, to synthesize a special polymer/phthalocyanine combination, much experience in both fields (polymer chemistry and phthalocyanine chemistry) is needed.

The polymer–phthalocyanine combinations exhibit many interesting properties such as energy transport, energy conversion, and energy-saving memory storage. The phthalocyanine is stabilized by the surrounding polymer matrix, and, in addition, the interaction of a polymer molecule with a phthalocyanine molecule can improve some properties. Therefore, efforts should be undertaken to realize more of these systems for practical applications.

REFERENCES

1. D. Wöhrle, *Adv. Polym. Sci.*, 50 (1983) 46.
2. D. Wöhrle, *Adv. Polym. Sci.*, 10 (1972) 35.
3. A. A. Berlin and A. I. Sherle, *Inorg. Macromol. Rev.*, 1 (1971) 235.
4. D. Wöhrle, U. Marose and R. Knoop, *Makromol. Chem.*, 186 (1985) 2209.
5. D. Wöhrle and B. Schulte, *Makromol. Chem.*, 186 (1985) 2229.
6. L. G. Cherkashina and A. A. Berlin, *Vysokomol. Soedin.*, 8 (1966) 627.
7. C. J. Norrell, H. A. Pohl, M. Thomas and K. D. Berlin, *J. Polym. Sci., Polym. Phys. Ed.*, 12 (1974) 913.
8. H. Inoue, Y. Kida and E. Imoto, *Bull. Chem. Soc. Jpn.*, 40 (1967) 184.
9. T. Hara, Y. Ohkatsu and T. Osa, *Bull. Chem. Soc. Jpn.*, 48 (1975) 85.
10. L. G. Cherkashina, Ye. L. Frankevich, I. V. Yeremina, Ye. I. Balabanov and A. A. Berlin, *Vysokomol. Soedin.*, 7 (1965) 1264.
11. A. A. Berlin, L. G. Cherkashina, Ye. L. Frankevich, Ye. M. Balabanov and Yu. G. Aseyov, *Vysokomol. Soedin.*, 6 (1964) 832.
12. W. Hanke, *Z. Anorg. Allg. Chem.*, 347 (1966) 67.
13. A. Epstein and B. S. Wildi, *J. Chem. Phys.*, 32 (1960) 324.

14. R. Bannehr, G. Meyer and D. Wöhrle, *Polym. Bull.*, 2 (1980) 841.

15. D. R. Boston and J. C. Bailar, *Inorg. Chem.*, 11 (1972) 1578.

16. W. Hanke, *Z. Chem.*, 6 (1966) 69.

17. J. W.-P. Lin and L. P. Dudek, *J. Polym. Sci., Polym. Chem. Ed.*, 23 (1985) 1579, 1589.

18a. A. I. Sherle, V. V. Promyslova, N. I. Shapiro, V. P. Epshtein and A. A. Berlin, *Vysokomol. Soedin. Ser. A*, 22 (1980) 1258.

18b. M. Ya. Kushnerev and A. I. Sherle, *Vysokomol. Soedin. Ser. A*, 23 (1981) 1187.

19. J. Bellido, J. Cardoso and T. Akachi, *Makromol. Chem.*, 182 (1981) 713.

20. D. Wöhrle and U. Hündorf, *Makromol. Chem.*, 186 (1985) 2177.

21. D. Wöhrle, G. Meyer and B. Wahl, *Makromol. Chem.*, 181 (1980) 2127.

22. G. Manecke and D. Wöhrle, *Makromol. Chem.*, 102 (1967) 1.

23. G. Koßmehl and M. Rohde, *Makromol. Chem.*, 178 (1977) 715.

24a. G. Manecke and D. Wöhrle, *Makromol. Chem.*, 116 (1967) 36; 120 (1968) 176.

24b. R. Bannehr, G. Meyer, D. Wöhrle and N. Jaeger, *Makromol. Chem.*, 182 (1981) 2633.

25. D. Wöhrle, R. Bannehr, N. Jaeger and B. Schumann, *Angew. Makromol. Chem.*, 117 (1983) 103.

26. D. Wöhrle, R. Bannehr, B. Schumann and N. Jaeger, *J. Mol. Catal.*, 21 (1983) 255.

27. D. Wöhrle, B. Schumann, V. Schmidt and N. Jaeger, *Makromol. Chem. Macromol. Symp.*, 8 (1987) 195.

28. D. Wöhrle, V. Schmidt, B. Schumann, A. Yamada and K. Shigehara, *Ber. Bunsenges. Phys. Chem.*, 91 (1987) 975.

29a. H. Yanagi, S. Maeda, S. Hayashi, and M. Ashida, *Cryst. Growth*, 92 (1988) 498.

29b. H. Yanagi, S. Maeda, Y. Ueda and M. Ashida, *Electron Microsc.*, 37 (1988) 177.

30. J. R. Fryer and T. A. Kinnaird, *Inst. Phys. Conf. Ser.*, 68 (1983) 27.

31. M. Yudasaka, K. Nakanishi, T. Hara, M. Tanaka and S. Kurita, *Synth. Met.*, 19 (1987) 775.

32. Y. Osada, A. Mizumoto and H. Tsuruta, *J. Macromol. Sci.-Chem. A*, 24 (1987) 403.

33. D. Wöhrle and B. Schulte, *Makromol. Chem.*, 189 (1988) 1167.

34. D. Wöhrle and B. Schulte, *Makromol. Chem.*, 189 (1988) 1229.

35. A. W. Snow, J. R. Griffith and N. P. Marullo, *Macromolecules*, 17 (1984) 1614.

36. C. S. Marvel and M. M. Martin, *J. Am. Chem. Soc.*, 80 (1958) 6600.

37. T. M. Keller and J. R. Griffith, *Org. Coat. Plast. Chem.*, 40 (1979) 781.

38. T. M. Keller, T. R. Price and J. R. Griffith, *Org. Coat. Plast. Chem.*, 43 (1980) 804.

39. T. M. Keller and J. R. Griffith, *Resins for Aerospace. ACS Symp. Ser.*, 132 (1980) 25.

40. R. Y. Ting, T. M. Keller, N. P. Marullo, P. Peyser, T. R. Price and C. F. Poranski, Jr., *Polym. Prepr. (Am. Chem. Soc., Div. Polym. Chem.)*, 22 (I) (1981) 50.

41. T. M. Keller and T. R. Price, *J. Macromol. Sci., Chem. A*, 18 (1982) 931.

42. R. Y. Ting, T. M. Keller, T. R. Price and C. F. Poranski, Jr., *ACS Symp. Ser.*, 195 (1982) 337.

43. N. P. Marullo and A. W. Snow, *Copolymerization of Polymers with Chain-Ring Structure. ACS Symp. Ser.*, 195 (1982) 325.

44. J. A. Hinkley, *J. Appl. Polym. Sci.*, 29 (1984) 3339; J. A. Hinkley, *ACS Symp. Ser.*, 282 (1985) 43.

45. T. M. Keller, *J. Polym. Sci., Polym. Lett. Ed.*, 24 (1986) 211.

46. T. M. Keller, *J. Polym. Sci., Polym. Chem. Ed.*, 25 (1987) 2569.

47. T. R. Walton, J. R. Griffith and J. G. O'Rear, *Polym. Sci. Technol.*, 96 (1975) 665.

48. J. R. Griffith, J. G. O'Rear and T. R. Walton, *Adv. Chem. Ser.*, 142 (1975) 458.

49. W. D. Bascom, R. L. Cottingham and R. Y. Ting, *J. Mater. Sci.*, 15 (1980) 2097.

50. B. N. Achar, G. M. Fohlen and J. A. Parker, *J. Polym. Sci. Polym. Chem. Ed.*, 23 (1985) 1677.

51. B. N. Achar, G. M. Fohlen and J. A. Parker, *J. Appl. Polym. Sci.*, 29 (1984) 353.

52. B. N. Achar, G. M. Fohlen and J. A. Parker, *J. Polym. Mat.*, 2 (1985) 16.

53. T. R. Walton, J. R. Griffith, J. G. O'Rear and J. P. Reardon, *Coat. Plast, Prep. Pap. Meet. (Am. Chem. Soc., Div. Org. Coat. Plast. Chem.)*, 37 (1977) 180.

54. T. R. Walton, J. R. Griffith and J. P. Reardon, *J. Appl. Polym. Sci.*, 30 (1985) 2921.

55. T. M. Keller and J. R. Griffith, *Org. Coat. Plast. Chem.*, 39 (1978) 546.

56. A. Gül and O. Bekaroglu, *Makromol. Chem., Rapid Commun.*, 8 (1987) 243.

57. C. S. Marvel and J. H. Rassweiler, *J. Am. Chem. Soc.*, 80 (1958) 1197.

58. A. S. Akopov, T. N. Lomova and B. D. Berezin, *Izv. Vyssh. Uchebn. Zaved., Khim. Khim. Tekhnol.*, 19 (1976) 1177; 21 (1978) 663.

59. B. A. Zhubanov and A. K. Zharmagambetor, *Izv. Akad. Nauk SSSR, Ser. Khim.*, 23 (1973) 71.

60. D. Wöhrle and E. Preußner, *Makromol. Chem.*, 186 (1985) 2189. For details on molecular weight determination see also G. Knothe and D. Wöhrle, *Makromol. Chem.*, 190 (1989) in press.

61. W. C. Drinkard and J. C. Bailar, *J. Am. Chem. Soc.*, 81 (1959) 4795.

62. Ye. I. Balabanov, Ye. L. Frankevich and L. G. Cherkashina, *Vysokomol. Soedin*, 5 (1962) 1684.

63. B. D. Berezin and L. P. Shormanova, *Vysokomol. Soedin, Ser. A*, 10 (1968) 384.

64. D. R. Boston and J. C. Bailar, *Inorg. Chem.*, 11 (1972) 1578.

65. J. Blomquist, L. C. Moberg, L. Y. Johansson and R. Larsson, *J. Inorg. Nucl. Chem.*, 43 (1981) 2287.

66. J. Blomquist, V. Helgeson, L. C. Moberg, L. Y. Johansson and R. Larsson, *Electrochim. Acta*, 27 (1982) 1445.

67. L. Kreja and A. Plewka, *Electrochim. Acta*, 25 (1980) 1283; *Angew. Makromol. Chem.*, 102 (1982) 45.

68. H. S. Nalwa, J. M. Sinha and P. Vasudevan, *Makromol. Chem.*, 182 (1981) 811.

69. B. N. Achar, G. M. Fohlen and J. A. Parker, *J. Polym. Sci., Polym. Lett. Ed.*, 20 (1982) 1785.

70. L. P. Shormanova and B. D. Berezin, *Vysokomol. Soedin. Ser. A*, 10 (1968) 1154.

71. B. D. Berezin and L. P. Shormanova, *Izv. Vyssh. Uchebn. Zaved., Khim. Khim. Tekhnol.*, 16 (1973) 442.

72. B. D. Berezin and L. P. Shormanova, *Vysokomol. Soedin. Ser. A*, 11 (1969) 1033.

73. J. Blomquist, L. C. Moberg, L. Y. Johansson, and R. Larsson, *Inorg. Chim. Acta*, 53 (1981) L39.

74. H. Meier, H. U. Tschirwitz, E. Zimmerhackl, W. Albrecht, and G. Zeitler, *J. Phys. Chem.*, 81 (1973) 712.

75. B. D. Berezin and A. N. Shlyapova, *Vysokomol. Soedin Ser. A*, 15 (1973) 1671.

76. A. I. Sherle, V. V. Promyslova, M. I. Shapiro, V. P. Epshtein, and A. A. Berlin, *Vysokomol. Soedin. Ser. A*, 22 (1980) 1258.

77. V. A. Zhorin, S. I. Beshenko, L. I. Makarova, A. I. Sherle, Yu. A Berlin and N. S. Yenikolopyan, *Vysokomol. Soedin. Ser. A*, 25 (1983) 551.

78. H. S. Nalwa, L. R. Dalton and P. Vasudevan, *Eur. Polym. J.*, 21 (1985) 943.

79. H. Meier, W. Albrecht and E. Zimmerhackl, *Polym. Bull.*, 13 (1985) 43.

80. B. Schumann, D. Wöhrle and N. Jaeger, *Electrochem. Soc.*, 132 (1985) 2144.

81. D. Wöhrle, H. Kaune and B. Schumann, *Makromol. Chem.*, 187 (1986) 2947.

82. B. N. Achar, G. M. Fohlen and J. A. Parker, *J. Polym. Sci., Polym. Chem. Ed.*, 21 (1983) 589.

83. B. N. Achar, G. M. Fohlen and J. A. Parker, *J. Polym. Sci., Polym. Chem. Ed.*, 20 (1982) 269, 2073.

84. H. Shirai, S. Yagi, A. Suzuki and N. Hojo, *Makromol. Chem.*, 178 (1979) 1889.

85. H. Shirai, K. Kobayashi, Y. Takemae, A. Suzuki, O. Hirabaru and N. Hojo, *Makromol. Chem.*, 180 (1979) 2073.

86. H. Shirai, Y. Takemae, K. Kobayashi, Y. Kondo, O. Hirabaru and N. Hojo, *Makromol. Chem.*, 189 (1984) 1395.

87. H. Itoh, K. Hanabusa, E. Masuda, H. Shirai and N. Hojo, *J. Polym. Sci., Polym. Lett.*, 25 (1987) 413.

88. B. N. Achar, G. M. Fohlen, J. A. Parker and J. Keshavaya, *Polyhedron*, 6 (1987) 1463.

89. B. N. Achar, G. M. Fohlen and J. A. Parker, *J. Polym. Sci., Polym. Chem. Ed.*, 20 (1982) 2773, 2781.

90. B. N. Achar, G. M. Fohlen and J. A. Parker, *J. Polym. Sci., Polym. Chem. Ed*, 21 (1983) 1025, 3063; 22 (1984) 319; 23 (1985) 801.

91. B. N. Achar, G. M. Fohlen and J. A. Parker, *J. Polym. Sci., Polym. Chem. Ed.*, 21 (1983) 1505.

92. B. N. Achar, G. M. Fohlen, M. S. Hsu and J. M. Parker, *J. Polym. Sci., Polym. Chem. Ed.*, 22 (1984) 1471.

93. T. Zipplies and M. Hanack, *Methoden der Organischen Chemie*, Vol. E20, G. Thieme-Verlag, Stuttgart, 1987, p. 2237.

94. R. J. Joyner and M. E. Kenney, *Inorg. Chem.*, 82 (1960) 5790.

95. R. D. Joyner and M. E. Kenney, *Inorg. Chem.*, 1 (1962) 717.

96. J. E. Owen and M. E. Kenney, *Inorg. Chem.*, 1 (1962) 334.

97. D. Wöhrle and G. Meyer, *Makromol. Chem.*, 175/3 (1974) 715.

98. G. Meyer, and M. Hartmann and D. Wöhrle, *Makromol. Chem.*, 176 (1975) 1919.

99. M. Hartmann, G. Meyer and D. Wöhrle, *Makromol. Chem.*, 176 (1975) 831.

100. B. D. Berezin and A. S. Akopov, *Vysokomol. Soedin. Ser. A*, 16 (1974) 450, 2334.

101. C. W. Dirk, T. Inabe, K. F. Schoch and T. J. Marks, *J. Am. Chem. Soc.*, 105 (1983) 1539.

102. B. N. Diehl, T. Inabe, J. W. Lyding, K. F. Schoch, C. R. Kannewurf and T. J. Marks, *J. Am. Chem. Soc.*, 105 (1983) 1551.

103. E. A. Orthmann, V. Enkelmann and G. Wegner, *Makromol. Chem., Rapid Commun.*, 4 (1983) 687.

104. E. A. Orthmann and G. Wegner, *Makromol. Chem., Rapid Commun.*, 7 (1986) 243.

105. M. Hanack, J. Metz and G. Pawlowski, *Chem. Ber.*, 115 (1982) 2836.

106. J. Metz, G. Pawlowski and M. Hanack, *Z. Naturforsch.*, 38b (1983) 378.

107. M. Hanack and T. Zipplies, *J. Am. Chem. Soc.*, 107 (1985) 6127.

108. G. Meyer and D. Wöhrle, *Z. Naturforsch.*, 32b (1977) 723.

109. T. J. Marks, K. F. Schoch and B. R. Kundalkar, *Synth. Met.*, 1 (1980) 337.

110. J. B. Davison and K. J. Wynne, *Macromolecules*, 11 (1978) 186.

111. G. Meyer and D. Wöhrle, *Makromol. Chem.*, 175/3 (1974) 728.

112. C. W. Dirk, E. A. Mintz, K. F. Schoch and T. J. Marks, in *Advances of Organometallic and Inorganic Polymer Science*, C. E. Carraher, J. E. Sheats, and C. U. Pittman, Eds., Marcel Dekker, New York, 1982, p. 275.

113. X. Zhou, T. J. Marks and S. H. Carr, *J. Polym. Sci., Polym. Chem. Ed.*, 23 (1985) 305. *Mol. Cryst. Liq. Cryst.*, 118 (1985) 357.

114. C. W. Dirk, T. Inabe, K. F. Schoch and T. J. Marks, *J. Am. Chem. Soc.*, 105 (1983) 1539.

115. S. H. Carr, X. Zhou, T. Inabe and T. J. Marks, *Polym. Mater. Sci. Eng.*, 49 (1983) 94.

116. C. J. Schramm, R. P. Scaringe, D. R. Stojakovic, B. M. Hoffman, J. A. Ibers and T. J. Marks, *J. Am. Chem. Soc.*, 102 (1980) 6707.

117. N. S. Hush and I. S. Woolsey, *Mol. Phys.*, 21 (1971) 465.

118. K. Fischer and M. Hanack, *Chem. Ber.*, 116 (1983) 1860.

119. H. Meier, W. Albrecht, E. Zimmerhackl and K. Fischer, *J. Mol. Electron.*, 1 (1985) 47.

120. A. D. Delman, J. J. Kelly and B. B. Simms, *J. Polym. Sci. A-1*, 8 (1970) 111.

121. G. Meyer, P. Plieninger and D. Wöhrle, *Angew. Makromol. Chem.*, 72 (1978) 173.

122. K. Mitulla and M. Hanack, *Z. Naturforsch.*, 35b (1980) 1111.

123. M. Hanack, J. Metz and G. Pawlowski, *Chem. Ber.*, 115 (1982) 2836. M. Hanack, K. Mitulla and O. Schneider, *Chem. Scripta*, 17 (1981) 139.

124. M. Hanack, K. Mitulla, G. Pawlowski and L. R. Subramanian, *J. Organomet. Chem.*, 204 (1981) 315; *Angew. Chem.*, 91 (1979) 343.

125. G. Meyer and D. Wöhrle, *Materials Sci.*, 7 (2-3) (1981) 265.

126. H. Uth and D. Wöhrle, *Bremer Briefe z. Chem.*, 1 (2/3) (1977) 23.

127. J. Curry and E. Cassidy, *J. Chem. Phys.*, 37 (1962) 2154.

128. H. Meier, W. Albrecht and U. Tschirwitz, *Angew. Chem.*, 22 (1972) 1077.

129. P. J. Toscano and T. J. Marks, *Mol. Cryst. Liq. Cryst.*, 118 (1985) 337.

130. T. Inabe, M. K. Moguel, T. J. Marks, R. Burton, J. W. Lyding and C. R. Kannewurf, *Mol. Cryst. Liq. Cryst.*, 118 (1985) 349.

131. W. J. Pietro, T. J. Marks and M. A. Ratner, *J. Am. Chem. Soc.*, 107 (1985) 5387.

132. T. Inabe, J. W. Lyding and T. J. Marks, *J. Chem. Soc., Chem. Commun.*, (1983) 1084.

133. T. Inabe, J. F. Lomax, J. W. Lyding and C. R. Kannewurf, T. J. Marks, *Synth. Met.*, 9 (1984) 303.

134. J. P. Linsky, T. R. Paul, R. S. Nohr and M. E. Kenney, *Inorg. Chem.*, 19 (1980) 3131.

135. R. S. Nohr, P. M. Kuznesof, K. J. Wynne, M. E. Kenney and P. G. Siebenmann, *J. Am. Chem. Soc.*, 103 (1981) 4371.

136. P. M. Kuznesof, R. S. Nohr, K. J. Wynne and M. E. Kenney, *J. Macromol. Sci. A*, 16 (1981) 299.

137. T. J. Klofta, P. C. Rieke, C. A. Linkous, W. J. Buttner, A. Nanthakumar, T. D. Mewborn and N. R. Armstrong, *J. Electrochem. Soc.*, 132 (1985) 2134.

138. P. M. Kuznesof, K. J. Wynne, R. S. Nohr and M. E. Kenney, *J. Chem. Soc., Chem. Commun.*, (1980) 121.

139. P. Brant, D. C. Weber, S. G. Haupt, R. S. Nohr and K. J. Wynne, *J. Chem. Soc. Dalton Trans.*, (1985) 269.

140. G. Berthet, D. Djurado, C. Fabre, F. Faury, C. Maleysson and H. Robert, *Mol. Cryst. Liq. Cryst.*, 118 (1985) 345.

141. M. Hanack, A. Beck and H. Lehmann, *Synthesis*, (1987) 703.

142. M. Hanack and R. Fay, *Recl. Trav. Chim. Pays-Bas*, 105 (1986) 427.

143. M. Schwartz, W. E. Hatfield, M. D. Joesten, M. Hanack and A. Datz, *Inorg. Chem.*, 24 (1985) 4198.

144. M. Hanack and X. Münz, *Synth. Met.*, 10 (1985) 357.

145. O. Schneider and M. Hanack, *Z. Naturforsch.*, 39b (1984) 265.

146. A. Datz, J. Metz, O. Schneider and M. Hanack, *Synth. Met.*, 9 (1984) 31.

147. J. Metz and M. Hanack, *J. Am. Chem. Soc.*, 105 (1983) 828.

148. C. Hedtmann-Rein, M. Hanack, K. Peters, E.-M. Peters and H. G. von Schnering, *Inorg. Chem.*, 26 (1987) 2647.

149. M. Hanack, C. Hedtmann-Rein, A. Datz, U. Keppeler and X. Münz, *Synth. Met.*, 19 (1987) 787.

150. M. Hanack and C. Hedtmann-Rein, *Z. Naturforsch.*, 40b (1985) 1087.

151. J. F. Myers, G. W. R. Canham and A. B. P. Lever, *Inorg. Chem.*, 14 (1975) 461.

152. P. A. Barrett, C. E. Dent and R. P. Linstead, *J. Chem. Soc.*, (1936) 1719.

153. M. Hanack, S. Deger, U. Keppeler, A. Lange, A. Leverenz and M. Rein, *Synth. Met.*, 19 (1983) 739.

154. H. Meier, W. Albrecht, E. Zimmerhackl, M. Hanack and J. Metz, *Synth. Met.*, 11 (1985) 333.

155. J. Metz and M. Hanack, *Chem. Ber.*, 120 (1987) 1307.

156. U. Keppeler, S. Deger, A. Lange and M. Hanack, *Angew. Chem.*, 99 (1987) 349.

157. U. Keppeler and M. Hanack, *Chem. Ber.*, 119 (1986) 3363.

158. W. Kobel and M. Hanack, *Inorg. Chem.*, 25 (1986) 103.

159. B. N. Diehl, T. Inabe, N. K. Jaggi, J. W. Lyding, O. Schneider, M. Hanack, C. R. Kannewurf, T. J. Marks and L. H. Schwartz, *J. Am. Chem. Soc.*, 106 (1984) 3207.

160. O. Schneider and M. Hanack, *Chem. Ber.*, 116 (1983) 2088.

161. O. Schneider and M. Hanack, *Angew. Chem.*, 95 (1983) 804.

162. S. Deger and M. Hanack, *Synth. Met.*, 13 (1986) 319; *Isr. J. Chem.*, 27 (1986) 347.

163. J. W. Koch and M. Hanack, *Chem. Ber.*, 120 (1987) 1853.

164. J. Metz, O. Schneider and M. Hanack, *Spectrochim. Acta*, 38 A (1982) 1265.

165. M. Hanack and A. Leverenz, *Synth. Met.*, 22 (1987) 9.

166. M. Hanack, U. Keppeler and H.-J. Schulze, *Synth. Met.*, 20 (1987) 347.

167. M. Hanack, S. Deger, U. Keppeler, A. Lange, A. Leverenz and M. Rein, Polyphthalo-cyanines, in *Conducting Polymers*, L. Alcacer, Ed., D. Reidel, Amsterdam, 1987.

168. M. Hanack, S. Deger, A. Lange and T. Zipplies, *Synth. Met.*, 15 (1986) 207.

169. D. Wöhrle and G. Krawczyk, *Makromol. Chem*, 187 (1986) 2535.

170. D. Wöhrle, G. Krawczyk and M. Paliuras, *Makromol. Chem*, 189 (1988) 1001.

171. D. Wöhrle, G. Krawczyk and M. Paliuras, *Makromol. Chem.*, 189 (1988) 1013.

172. J. Gitzel, H. Ohno, E. Tsuchida and D. Wöhrle, *Polymer*, 27 (1986) 1781.

173. J. Gitzel, H. Ohno, E. Tsuchida and D. Wöhrle, *Makromol. Chem., Rapid Commun.*, 7 (1986) 397.

174. D. Wöhrle, J. Gitzel, G. Krawczyk, E. Tsuchida, H. Ohno, I. Okura and T. Nishisaka, *J. Macromol. Sci.*, A25 (1988) 1227.

175. D. Wöhrle and J. Gitzel, *Makromol. Chem., Rapid Commun.*, 9 (1988) 229.

176. T. A. Maas, M. Kuijer and J. Zwart, *J. Chem. Soc., Chem. Commun.*, (1976) 86.

177. H. Shirai, A. Maruyama, K. Kobayashi and N. Hojo, *Makromol. Chem.*, 181 (1980) 575.

178. S. Higaki, K. Hanabusa, H. Shirai and N. Hojo, *Makromol. Chem.*, 184 (1983) 691.

179. O. Hirabaru, T. Nakase, K. Hanabusa, H. Shirai, K. Takemoto and N. Hojo, *J. Chem. Soc., Chem. Commun.*, (1983) 481; *Angew. Makromol. Chem.*, 121 (1984) 59.

180. H. Shirai, S. Higaki, K. Hanabusa, N. Hojo and O. Hirabaru, *J. Chem. Soc., Chem. Commun.*, (1983) 751.

181. H. Yamaguchi, R. Fujiwara and K. Kusuda, *Makromol. Chem., Rapid Commun.*, 7 (1986) 225.

182. J. H. Shutten and J. Zwart, *J. Mol. Catal.*, 5 (1979) 109.

183. J. H. Shutten, *Angew. Makromol. Chem.*, 89 (1980) 201.

184. M. Gebler, *J. Inorg. Nucl. Chem.*, 43 (1981) 2759.

185. K. Hanabusa, C. Kobayashi, T. Koyama, E. Masuda, H. Shirai, Y. Kondo, K. Takemoto, E. Izuka and N. Hojo, *Makromol. Chem.*, 187 (1986) 753.

186. T. W. Hall, S. Greenberg, C. R. McArthur, B. Khouw and C. C. Leznoff, *Nouv. J. Chim.*, 6 (1982) 653.

187. D. Wöhrle and G. Krawczyk, *Polym. Bull.*, 15 (1986) 193.

188. H. R. Allcock and T. X. Neeman, *Macromolecules*, 19 (1986) 1495.

189. H. Itoh, S. Kondo, E. Matsuda, K. Hanabusa and H. Shirai, N. Hojo, *Makromol. Chem., Rapid Commun.*, 7 (1986) 585.

190. A. Yamada and K. Shigehara, in preparation.

191. J. Zwart, H. C. van der Weide, N. Bröker, C. Rummens, G. C. A. Schuit and A. L. German, *J. Mol. Catal.*, 3 (1977/78) 151.

192. W. M. Brouwer, P. Piet and A. L. German, *J. Mol. Catal.*, 31 (1985) 169.

193. H. Shirai, A. Maruyama, M. Konishi and N. Hojo, *Makromol. Chem.*, 181 (1980) 1003.

194. J. H. Shutten and J. Zwart, *J. Mol. Catal.*, 5 (1979) 109.

195. W. M. Brouwer, P. Piet and A. L. German, *J. Mol. Catal.*, 29 (1985) 335.

196. J. H. Shutten, P. Piet and A. L. German, *Makromol. Chem.*, 180 (1979) 2341.

197. J. van Welzen, A. M. van Herk and A. L. German, *Makromol. Chem.*, 188 (1987) 1923.

198. A. M. van Herk, A. H. J. Tullemans, J. van Welzen and A. L. German, *J. Mol. Catal.*, 44 (1988) 269.

199. M. Hassanein and W. T. Ford, *Macromolecules*, 21 (1988) 526.

200. H. Przywarska-Boniecka, L. Trynda and E. Antonini, *Eur. J. Biochem.*, 52 (1975) 567.

201. J. D. Spikes, *Photochem. Photobiol.*, 43 (1986) 691.

202. *Ullmanns Enzycl. Tech. Chem.*, 4 Aufl. 1979, Vol. 18, p. 501.

203. F. H. Moser and A. L. Thomas, *Phthalocyanine Compounds*, Reinhold, New York; Chapman and Hall, London, 1963; *The Phthalocyanines*, CRC Press, Boca Raton, FL, 1983.

204. Several articles in the series *The Chemistry of Synthetic Dyestuffs*, K. Venkataraman, Ed., Academic Press, New York, London, 1965, 1971, 1972.

205. A. R. Gutierrez, J. Friedrich, D. Haarer and H. Wolfrum, *IBM J. Res. Dev.*, 26 (1982) 198.

206. J. Friedrich and D. Haarer, *Angew. Chem.*, 96 (1984) 96.

207. K.-Y. Law, *J. Phys. Chem.*, 89 (1985) 2652.

208. T. J. Klofta, P. C. Rieke, C. A. Linkous, W. J. Buttner, A. Nanthakumar, T. D. Mewborn and N. R. Armstrong, *J. Electrochem. Soc.*, 132 (1985) 2136.

209. T. J. Kofta, J. Danziger, P. Lee, J. Pankow, K. W. Nebesny and N. R. Armstrong, *J. Phys. Chem.*, 91 (1987) 5646.

210. H. Shirai, K. Hanabusa, M. Kitamura, E. Masuda, O. Hirabaru and N. Hojo, *Makromol. Chem.*, 185 (1984) 2537.

211. D. Wöhrle, H. Kaune, B. Schumann and N. Jaeger, *Makromol. Chem.*, 187 (1986) 2947.

212. B. Schumann, D. Wöhrle and N. Jaeger, *J. Electrochem. Soc.*, 132 (1985) 2144.

213. G. A. Chamberlain, *Solar Cells*, 8 (1983) 47.

214. M. Kaneko and A. Yamada, *Adv. Polym. Sci.*, 55 (1984) 1.

215. J. Simon and J. J. André, *Molecular Semiconductors*, Springer-Verlag, Berlin, 1985.

216. M. Kaneko and D. Wöhrle, *Adv. Polym. Sci.*, 84 (1988) 141.

217. R. O. Loutfy, J. H. Sharp, C. K. Hsiao and R. Ho, *J. Appl. Phys.*, 52 (1981) 5218.

218. N. Minami, K. Sasaki and K. Tsuda, *J. Appl. Phys.*, 54 (1983) 6764.

219. M. Shimura and H. Baba, *Denki Kagaku*, 50 (1982) 678.

220. Y. Soeda, S. Tasaki, S. Miyata, A. Yamada and H. Sasabe, *Polym. Prep. Jpn.*, 33 (1984) 475.

221. K. Yamashita, Y. Harima and H. Iwashima, *J. Phys. Chem.*, 91 (1987) 3055.

222. R. O. Loutfy and L. F. McIntyre, *Can. J. Chem.*, 61 (1983) 72.

223. M. Kaneko, D. Wöhrle, D. Schlettwein and V. Schmidt, *Makromol. Chem.*, 189 (1988) 2419.

224. E. Yeager, *Electrochim. Acta*, 29 (1984) 1527.

225. H. Jahnke, *Chimia*, 34 (1980) 58.

226. H. Jahnke, M. Schönborn and G. Zimmermann, *Top. Curr. Chem.*, 61 (1976) 133.

227. A. van der Putten, W. Visscher and E. Barendrecht, *J. Electroanal. Chem.*, 195 (1985) 63; 205 (1986) 233; 221 (1987) 95.

228. T. Hirai and J. Yamaki, *J. Electrochem. Soc.*, 132 (1985) 2125. *J. Appl. Electrochem.*, 15 (1985) 77, 441.

229. H. P. Dhar, R. Darby, V. Y. Young and R. E. White, *Electrochim. Acta*, 30 (1983) 423.

230. M. Kirschenmann, D. Wöhrle and W. Vielstich, *Ber. Bunsenges. Phys. Chem.*, 92 (1988) 403.

231. J. A. R. van Veen, H. A. Coolijn and J. F. van Baar, *Electrochim. Acta*, 33 (1988) 801.

232. M. Yamana, R. Darby and R. E. White, *Electrochim. Acta*, 29 (1984) 329.

233. G. D. Hobson, *Modern Petroleum Technology*, Applied Science Publishers, Barking, Great Britain, 1973, p. 783.

234. R. A. Meyers (Ed.), *Handbook of Petroleum Refining Processes*, McGraw-Hill, New York, 1986.

235. A. Leito, C. Costa and A. Rodrigues, *Chem. Eng. Sci.*, 42 (1987) 2291.

236. D. Wöhrle, T. Buck, U. Hündorf, G. Schulz-Ekloff and A. Andreev, *Makromol. Chem.*, 190 (1989) 961.

237. D. Wöhrle, U. Hündorf, G. Schulz-Ekloff and E. Ignatzek, *Z. Naturforsch.*, 41b (1986) 179.

238. G. Mercati, F. Morazonni, M. Barzaghi, P. Carniti and V. Ragaini, *J. Chem. Soc., Faraday Trans. I*, (1979) 1857.

239. V. Ragaini and R. Saravalle, *React. Kinet. Catal. Lett.*, 1 (1974) 271.

240. L. Prahov and A. Andreev, *React. Kinet. Catal. Lett.*, 3 (1975) 315.

241. J. R. Harbour, B. Dietelbach and J. Duff, *J. Phys. Chem.*, 87 (1983) 5456.

242. K. Uchida, M. Soma, T. Onishi and K. Tamaru, *Z. Phys. Chem. Neue Folge*, 106 (1977) 317.

243. V. Yu. Zakharov, O. M. Zakharova, B. V. Romanovskii and R. E. Mardelishvili, *React. Kinet. Catal. Lett.*, 6 (1977) 133.

244. G. Meyer, D. Wöhrle, M. Mohl and G. Schulz-Ekloff, *Zeolites*, 4 (1984) 30.

245. H. Diegruber, P. J. Plath, G. Schulz-Ekloff and M. Mohl, *J. Mol. Catal.*, 24 (1984) 115.

246. P. Gomez-Romero, Y. S. Lee, and M. Kertesz, *Inorg. Chem.*, 27 (1988) 3672.

3

Absorption and Magnetic Circular Dichroism Spectral Properties of Phthalocyanines[1]

Part 1: Complexes of the Dianion, Pc(− 2)

Martin J. Stillman and Tebello Nyokong

135

137

A. INTRODUCTION[2]

The history of the phthalocyanines begins with their remarkable spectral properties [1,2]. The purity and depth of the color of these dyes arise from the unique property of having an isolated, single band located in the far red end of the visible spectrum near 670 nm, with a molar absorptivity often exceeding 10^5 liters mol^{-1} cm^{-1} [3,4]. Unlike many other molecules, the next most energetic set of transitions is generally much less intense, lying just to the blue of the visual region near 340 nm [3–7]. We should note that although the peak absorbances near 670 nm are high, the integrated absorbance that makes up the total oscillator strength (f) of the bands between 250 and 360 nm may exceed that of the 670-nm band. (Where f is defined as $4.32 \times 10^{-9} \int \epsilon \, dv$ [8].) Figure 1 shows absorption and magnetic circular dichroism spectra for MgPc, coordinated by imidazole and dissolved in methylene chloride [9]. We choose MgPc because it exhibits a spectrum typical of many phthalocyanine complexes. It is this somewhat strange absorption spectrum that results in the purity of the beautiful blues and, by introduction of additional bands into the 500-nm "window" region (for example, from charge transfer transitions between the central metal and the π ring [10–13]), the greenish hues that are found for complexes of CrPc and MnPc [10,14]. These special spectral features of the phthalocyanines (as characterized by both λ_{max} and ϵ_{max}), more closely resemble the chlorophylls [15–17], than the protoporphyrin IX group in the hemes [18, 19], or the synthetic TPP or OEP porphyrins [5,20,21].

When a compound exhibits such intense absorption in a single band in the visible region, as that observed for the majority of metallophthalo-cyanines, it is not surprising that the variation in colors of solutions and crystals containing different metals and axial coordinating ligands (the solvent in many cases) should attract the attention of synthetic chemists and spectroscopists alike. It is of interest to read through some of Linstead's early papers on the phthalocyanines and see how the colors of the complexes were used to guide the preparations [1,22–24]. This combination of the special

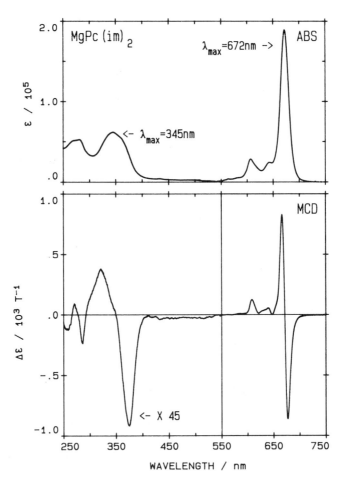

Figure 1 Absorption and magnetic circular dichroism spectra for (imid)$_2$MgPc in methylene chloride. The band at 672 nm is the Q band, while the band at 345 nm marks the B band envelope [21]. The MCD spectrum is characterized by an A term under the Q band, and an overlapping series of A terms in the B band region. (See the text for details of the MCD experiment.) Reproduced with permission from [9] (Fig. 2).

spectroscopic properties of the phthalocyanine ring itself, together with the "tuning" aspect of the central metal atom (which can include the group 1 and 2[3] metals [9,25,26], transition metals [10,12,27], lanthanides and actinides [28–30], and even main group metals and metalloids [31]), has resulted in a wealth of spectroscopic data being reported. Probably equally important are the relationship between the phthalocyanines and the parent porphyrins, the interest in model compounds that mimic the reactivity of heme proteins, and a recognition that the phthalocyanines can mimic some of the properties of the chlorophylls [32].

Perhaps the most colorful phthalocyanine solutions are obtained during redox chemistry that involves the ring or a central transition metal. Oxidation and reduction reactions of metallophthalocyanines in which both the metal and the ring can take part are dominated by the appearance of beautifully colored solutions, with dramatic changes in hue as the reactions are carried to completion [33–35].

The spectral properties of phthalocyanines are central to studies of several of their chemical and electronic properties. First, in solution chemistry, where axial ligation, charge transfer between the metal and Pc ring, and central metal oxidation state have been examined, there has been particular emphasis lately on the photochemical and electrochemical properties. Phthalocyanine complexes exhibit a quite remarkable range of redox properties [33,35]. The variety of spectra that are found for transition metal phthalocyanines also attracted interest in the search for model compounds that mimic complex biological systems, for example, like those found in photosynthesis, with the early work on MnPc [36]. Exciting electrochromic behavior has been found for dimeric $Lu(Pc)_2$ species [37,38], and the promise of molecular metals seems to be realized with the partially oxidized species formed by doping I_2 into NiPc [39], and in the monomeric and polymeric species based on chains of —SiPc—O—SiPc— units [40,41]. The well-known redox chemistry of the porphyrin family owes much of its origins to the early demonstration of the quite unique chemistry of the heme groups in the peroxidases and catalases, and to the electron transfer reactions of the cytochromes and chlorophylls. Extensive studies of the spectral changes that accompany the redox and photochemical reactions of synthetic porphyrins and phthalocyanines have addressed these properties in considerable depth [13,42–45].

There are several specific spectral properties of phthalocyanines that distinguish their observed photochemical properties from those of many porphyrins. The energies of the first singlet (the Q band) and the lowest triplet lie well to the red of the Q band in porphyrins. Because of the presence of the aza-bridges in phthalocyanines, the electronic origin of the Q band in phthalocyanines (an a_{1u} to e_g transition) is quite different when compared with porphyrins like TPP and OEP (an a_{2u} to e_g transition). As a result of a significant energy separation between the two top filled molecular orbitals (the a_{1u} and the a_{2u}), the Q state is considerably less coupled to the second singlet excited state than in the porphyrins, where extensive coupling occurs between the two lowest energy singlet states. The extinction coefficient of the Q_{00} band in phthalocyanines is up to 10 times that of the Q_{00} (or α)-band in the porphyrins. This effect is specially marked for (M)TPP complexes where Q_{00} is only observed as a shoulder on the side of the more intense, vibronically allowed Q_{01} band.

Possibly the most significant use of phthalocyanine complexes to date results directly from their special spectral properties in the 600- to 750-nm region. Peripherally substituted phthalocyanine complexes have been proposed as replacements for hematoporphyrin derivative (HPD) in photosensitized cancer treatment. This is because the Q band near 670 nm has such a high extinction coefficient when compared with protoporphyrin IX derivatives [46,47]. In photodynamic therapy (PDT), tumor destruction is greatly enhanced by the chemistry of excited state phthalocyanine molecules. Photosensitization occurs with these compounds because the 650- to 670-nm absorption band of the Q band in water-soluble sulfonate derivatives can be efficiently excited directly through tissue [46–48].

Next, spectral properties play an important part in the study of the solid state chemistry of phthalocyanines. Because of intense absorptivity, crystals are usually too optically dense or too small for study by transmission techniques, although reflection spectra have been reported [49,50]. However, many phthalocyanines sublime readily near 350°C in a high vacuum and can be laid down as thin, polymorphic films on a substrate. As a result, the optical properties of thin films on quartz disks have been extensively studied [51–54]. The band structure of such films has been exploited for use electrochemically by forming films on conducting electrodes. Phthalocyanines incorporated into a single site in a Shpol'skii matrix (a matrix formed in some organic solvents at low temperatures) exhibit highly complicated spectra that depend on the sites occupied and on solvent–solute interactions [55]. Vapor phase [56,57] and matrix-isolated [58,59] phthalocyanines have been studied as a means of obtaining solvent-free absorption and magnetic circular dichroism spectra from the Q band to the far UV (extending toward 150 nm using synchrotron radiation) [59].

Finally, the spectral properties act as an excellent guide to the accuracy of theoretical treatments of the electronic structure of these molecules. The high symmetry, and the combination of a large aromatic ring that coordinates a central metal, provides a most interesting system to work with. While there are not as many calculations for phthalocyanines as for the porphyrins and hemes, several groups have reported theoretically calculated spectral data and have discussed the theoretical origins of specific bands (for example, bands that are charge transfer in origin) [12,60–64]. However, it is clear that although the phthalocyanine ring itself can be symmetric, perturbations to the geometry of the central metal due to axial ligands or solvents, and the size effects of different central metals used in the measurement of the spectra data, have been difficult to parameterize. It is only recently that any reports of full spectral deconvolution calculations have been available that identify the location of the first four main, $\pi–\pi^*$ bands in the spectra of phthalocyanines as simple as MgPc or ZnPc, [9,65].

In this chapter, we separate the discussion of the spectral data for phthalocyanines into several parts: (1) a general description of the spectral properties, (2) the effects of charge transfer, (3) the effects of aggregation, and (4) a discussion of progress in assigning the spectra. The second half of the review includes a description of the spectral properties of solutions of a variety of MPc complexes, followed by a discussion of the spectra of MPc complexes in the solid state.

We have also compiled a data base of spectral data from MPc complexes dissolved in organic solvents, concentrated sulfuric acid, and as thin films. An ASCII-encoded copy of the data base and a listing of the references used for the whole review is available on floppy disk (IBM PC 360 K format) from the authors.

i. General Literature on the Electronic Structure of the Phthalocyanines

The last comprehensive review of the spectral properties of phthalocyanine complexes in solution was that of Lever in 1965 [4]. Many other reviews of both porphyrins and phthalocyanines have included spectral information that is helpful in assigning the spectra of phthalocyanines, for example, Taube [12]. The most recent, complete, review of the electronic structure of porphyrin-ring compounds, including the phthalocyanines, is contained in Gouterman's chapter in Volume III of *The Porphyrins* [21]. While much of Gouterman's review refers specifically to the porphyrins, the general theme of characterizing deviations from many "normal" spectral properties with "hypso" and "hyper" spectral features that arise from coupling between $\pi-\pi^*$ transitions and CT transitions, can be applied equally to the phthalocyanines [31]. Gouterman [21] includes spectral data for many different substituted porphyrins so that his review does provide an overview of the differences between the various members of the porphyrin class of molecule. Illustration of the effects on the spectra of different central metals and different peripheral substituents also applies directly to the phthalocyanines [21]. Gouterman's discussion of the theoretical aspects of the spectral properties of the porphyrin is extremely valuable in relating the predicted energy levels to those actually observed from absorption and emission spectra. Chapters by several other authors in Volume III of *The Porphyrins* [66] also relate directly to the spectral properties of phthalocyanines.

It is useful to note that the probable reason why there has not yet been a correlation between a parameter such as IR stretching frequency, with the Q band wavelength (see, as an example, the detailed work of Shelnutt and Ondrias [67] for the porphyrins), is because of the different top filled molecular orbital order in the phthalocyanines compared with the porphyrins. The Q band in phthalocyanines (equivalent to the α band in porphyrins) is fairly constant in energy, lying somewhere close to 670 nm, for a great many "normal" phthalocyanine complexes. The band that is equivalent to the porphyrin γ or Soret band is the phthalocyanine B band. Although for porphyrins it is relatively easy to identify the energy of the Soret band, this is not as straightforward to do as for the phthalocyanines, because the bands in the 300 nm region of the phthalocyanine spectrum are much broader and less well defined. It is only recently that reliable values for transition energies have become available for the B band through deconvolution of absorption and MCD spectra, primarily, from our laboratory [9,65].

The detailed review of the chemical properties of the phthalocyanines published by Moser and Thomas, first in 1963 [68], and then in a greatly revised form in 1983 [2], also includes several sections dealing with the spectral properties of phthalocyanines with a wide variety of spectroscopic techniques. The two parts of the 1983 book [2] contain invaluable references to the Russian and Japanese literature. A comprehensive review by Simon and Andre [318] relates structural, spectroscopic and electrochemical properties of the phthalocyanines to their electrical and photovoltaic properties. A recent review by Owens and O'Connor [19] describes the spectra of iron-porphyrins in considerable detail, which is of interest with respect to FePc.

With the advent of computerized data bases, searching for reference material has become much easier. Because of the enormous variety of uses the phthalocyanines are put to, some selectivity in searching is needed. The data base maintained by the Radiation Laboratory at University of Notre Dame holds a very complete listing of publications that deal with the photochemical properties of porphyrins and phthalocyanines. Many of these papers contain extensive spectral data. The data base can be accessed readily (and, at the moment, free of charge) via a modem once an account number and a password have been set up.[4]

It is appropriate to note at this point that we have not in this review tried to reference all spectra reported for phthalocyanines. In order to reduce the total length of the text, we have chosen representative examples of the spectral data. Cross-referencing with library search tools should allow for access to material not included here. While the spectra for crystals and thin films are quite old, the quality of the reported data is as good as much present day solution data, so we have included details in our description of these spectra. In preparing this chapter, it became apparent to us, that for researchers to make the fullest use of spectral traces of a single class of molecule, both the spectral traces themselves and a table of band maxima should be published. In the absence of such a table in a publication, we have had to estimate band positions from enlarged copies of figures. This procedure is clearly not very precise.

We strongly recommend that publications including detailed spectral data also include tabular data. As digital images of spectral data become more widespread, it is also now time for a common format to be adopted for UV-visible spectral data interchange, much as the JCAMP-DX format is commonly used for IR data [69]. Computer programs like Spectra Manager [70], that can manipulate spectral data in a "what you see is what you get" graphical window on the computer screen and can assemble spectra into camera ready pages of plots, are going to make the process of searching for examples of spectra very much easier over the next 5 years or so. It is clear that spectral data could be readily transferred via electronic mail between quite different computers for display by this type of program.

ii. The Phthalocyanine Molecule: Relationship to the Porphyrins

The apparent closeness between the phthalocyanines, which are known systematically as tetraazatetrabenzporphyrins (TABP), and the well-known synthetic porphyrins, like tetraphenylporphyrin (TPP), tetraazaporphyrin (TAP), and tetrabenzporphyrin (TBP) (Fig. 2), is not observed for many spectroscopic properties. Metallation of the phthalocyanine dianion, with a metal that maintains the planarity of the molecule increases the symmetry from the D_{2h} of H_2Pc to the D_{4h} of MPc. The symmetry will drop to C_{4v} for metals that do not fit inside the ring. The diagonal N—N distance is smaller [71] in phthalocyanines (396 pm) than in most porphyrins (402 pm). For example, the Fe—N distance in FePc is 192.6 pm [72], while in FeTPP it is 197.2 pm

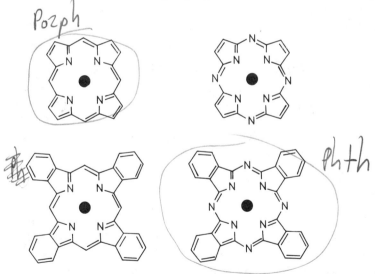

Figure 2 The molecular geometry of metalloporphyrins: phthalocyanine (lower right), porphine (upper left), tetrabenzporphyrin (lower left), and tetraazaporphyrin (upper right).

[73]. This means that in some cases metals that form planar porphyrin complexes will form square pyramids in phthalocyanine complexes. While we will return to discuss the closeness in the alignment between the theoretical calculations and the observed spectral data later, it is useful at this point to describe briefly our operational model for the molecular orbitals in the phthalocyanines.

Although there have been several detailed theoretical calculations for the molecular orbitals in the phthalocyanines [60–64,74,75,318], we will follow the scheme proposed by Gouterman's group [21,63,75], which uses a four-orbital model, based on the top two occupied molecular orbitals (a_{2u} and a_{1u}) and the degenerate, lowest unoccupied orbital (e_g), to set up the states that account quite well for the first two or three allowed transitions in the visible-UV region of the spectrum. We will consider the detailed results from these different calculations with respect to the predicted spectral properties in Section C.

Generally, the substitution of the porphine ring by either the aza groups or the benzo groups breaks the accidental degeneracy of the top filled molecular orbitals, the a_{2u} and a_{1u} orbitals in porphyrins [21,60,63] (Fig. 3). (See below for a more complete description of the molecular orbitals involved in the spectroscopy.) In Gouterman's four-orbital model [21], the first two allowed $\pi–\pi^*$ bands arise from transitions out of a_{1u} and a_{2u} into the same e_g orbital. For phthalocyanines, a_{1u} lies well above a_{2u}, whereas for many porphyrins a_{1u} and a_{2u} are close enough together that extensive configuration interaction occurs. Thus TBP, TAP, and phthalocyanine all exhibit a significant red shift in the energy, and an intensification relative to the B band, of the lowest energy, $\pi–\pi^*$, singlet transition compared with any of the normal porphyrins [4,27,60,76–78]. The spectrum of ZnTBP in pyridine [78] illustrates how the spectra of the TBP and TAP derivatives mix the intense, narrow Soret band of porphyrins like MTPP and octaethylporphyrin (MOEP), with the intense Q band of the phthalocyanine. The effect of increasing the number of pyrrole rings from four to five is demonstrated by the spectrum of the superphthalocyanine (SPc) formed on metallation with UO_2 [29], where the Q band red shifts still further to a maximum at 900 nm.

iii. An Overview of the Optical Properties of Phthalocyanines

Spectra of phthalocyanines have been reported from crystals, from thin films, from phthalocyanine complexes isolated in cryogenic matrices, widely from solution, and from vapor phase experiments. Phthalocyanine complexes exhibit essentially five types of absorption spectrum, illustrated in Figs. 1, 4, 5, 6, 7, and 8.

   1. *The free base spectrum, in which all the states are nondegenerate, and*

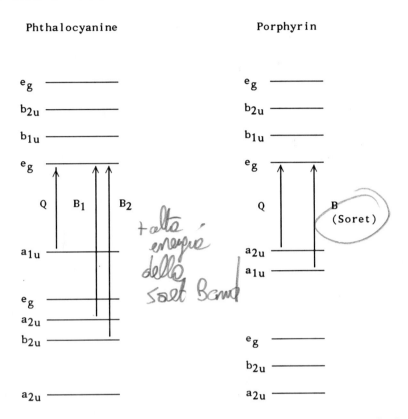

Figure 3 Origin of absorption in the region of the first two $\pi-\pi^*$ transitions, the Q and B bands, in phthalocyanines and porphyrins. For closed-shell metals, the ground state has $^1A_{1g}$ symmetry, while the allowed π^* states have either $^1A_{2u}$ or ^1E_u symmetry. For phthalocyanines the HOMOs, the a_{1u} and a_{2u}, are widely separated [21], while for porphyrins, the a_{1u} and a_{2u} HOMOs are accidently degenerate, which leads to extensive configuration interaction for transitions to the e_g LUMO [21].

so the major transitions are polarized in either the x or y direction (z-polarized transitions are also allowed by vibronic coupling in the excited states, but these states do not give rise to major intensity). For H_2Pc, the absorption and MCD spectra resemble that of chlorophyll *a* [15] (Fig. 4).

2. *The D_{4h} spectrum of phthalocyanine complexes, in which only $\pi-\pi^*$ bands are observed between 230 and 800 nm (Figs. 1 and 4).* This type of spectrum is characterized by an intense band near 670 nm between the A_{1g} (a_{1u}^2) ground state and the first excited singlet state which has E_u symmetry, from the $(a_{1u}^1 e_g^1)$ configuration [33,63] (Fig. 3). As a result of the D_{4h} molecular symmetry, allowed optical transitions must transform with either E_u

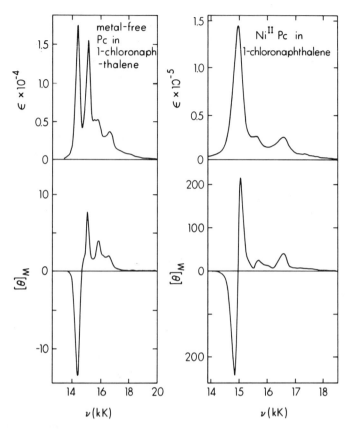

Figure 4 Absorption and MCD spectra in the Q band region of H_2Pc and NiPc dissolved in α-chloronaphthalene. Note, the chloronaphthalene absorbs strongly before any further phthalocyanine-related bands are seen, and the ordinate axes are on different scales, in particular, the NiPc MCD A term is some 20 times more intense than the pair of B terms observed for H_2Pc. The ordinate units are ϵ and $[\theta]_m$. Reproduced with permission from [26] (Fig. 1).

or A_{2u} symmetry. E_u transitions are polarized in the x/y plane of the molecule, while A_{2u} transitions are polarized in the z direction. Each of the main $\pi-\pi^*$ transitions in the 230- to 800-nm region is predicted to be x/y polarized to degenerate excited states and to exhibit a positive, Faraday A term in the MCD spectrum [59,63] (Fig. 1). The vibronic coupling in the excited state does introduce transitions with A_{2u} symmetry which connect nondegenerate states giving rise to B terms in the MCD spectrum (Fig. 10) [59,63].

3. *Spectra in which additional bands, usually assigned as metal to ligand or ligand to metal charge transfer (MLCT or LMCT, respectively), lie in the*

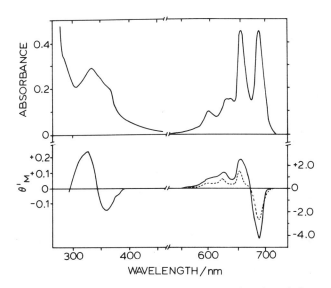

Figure 5 Absorption and MCD spectra of H_2Pc dissolved in dimethyl acetamide. The H_2Pc was prepared by acid-mediated demetallation of Li_2Pc. The ordinate scales are absorbance and angle of rotation, for absorption and MCD spectra, respectively, because of difficulties in assessing the actual concentrations. Reproduced with permission from [80] (Fig. 1).

same region as the Pc ring $\pi-\pi^*$ *set [33].* We describe the origins of charge transfer bands in more detail below (in Section C, ii). Because charge transfer transitions involve MOs of the π ring, extensive mixing can occur so that the $\pi-\pi^*$ set of transitions is blended into the CT set (these effects have been the subject of considerable discussion for the porphyrins [21]). While the most well-known charge transfer bands lie in the window region between 450 and 600 nm, additional weak bands are also found to the red edge of the Q band between 700 and 1500 nm, which may also be charge transfer in character. Figure 6 shows the absorption and MCD spectra measured for Co(I)Pc(-2), which was prepared by reducing Co(II)Pc(-2) with $NaBH_4$ [3]. The Q band intensity often diminishes when a set of CT bands is located in the 500-nm region, for CoPc the bands at 500 nm are assigned as CT [3].

4. *The spectra of dimeric complexes in solution.* The appearance of a strong band near 620 nm, which is associated with an MCD A term, results from exciton coupling between the two π systems which leads to a blue-shifted Q band. This spectral envelope is characteristic of dimerization. The spectrum of the dimeric 15-crown-5 substituted H_2Pc is typical of the spectral properties observed for dimerization in solution (Fig. 7). We describe in more detail below the origins of these spectral effects.

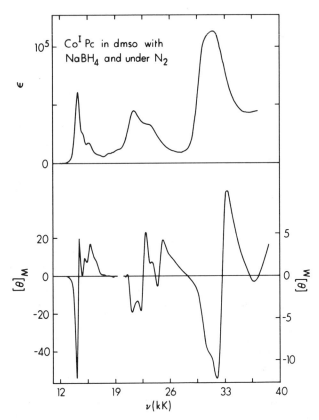

Figure 6 Absorption and MCD spectra of Co(I)Pc dissolved in DMSO, under N_2 in the presence of excess $NaBH_4$. The ordinate units are ϵ and $[\theta]_m$. The band of special interest lies in the 500-nm (20,000 cm^{-1}) region. Reproduced with permission from [3] (Fig. 6).

5. *The spectra of phthalocyanines in the solid state.* A typical example is the spectrum of a thin film of Co(II)Pc on a quartz disk. The spectrum shown in Fig. 8 is characteristic of the α polymorph, which exhibits a very broad band in the 500- to 750-nm region, that is quite unlike the narrow Q band observed for solutions. These spectral effects arise from extensive exciton coupling between adjacent rings [6].

iv. Labeling the Spectral Features

We will adopt a modification of the nomenclature proposed by Platt [79] and maintained by Gouterman's group [21] to describe the bands in the visible-UV region of the spectrum. The sequence of bands identified in the absorption

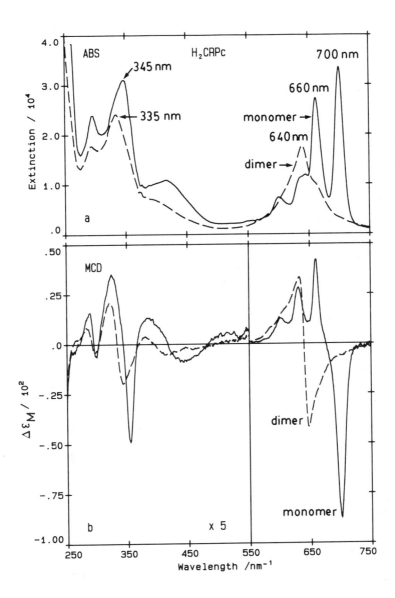

Figure 7 Absorption and MCD spectra for H_2-(15-crown-5)$_4$Pc dissolved in CHCl$_3$. The monomer (solid line) is converted into the dimer by the addition of CH$_3$COOK in CHCl$_3$ containing 0.1% v/v MeOH. Reproduced with permission from [214] (Fig. 2).

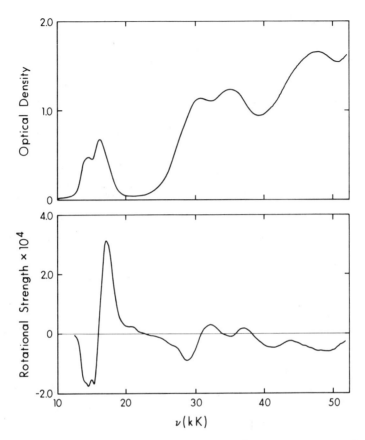

Figure 8 Absorption and MCD spectra of α-CoPc. The spectra were recorded at room temperature from CoPc sublimed to form a thin film on a quartz disk. The ordinate units are absorbance and θ_m/tesla. Reproduced with permission from [52] (Fig. 2).

spectrum of many "normal" phthalocyanines of Q, B, N, L, and C was first described systematically in a series of experiments by Gouterman's group, for example, see [57,58], in which vapor phase spectra were recorded in order to overcome the notorious solubility problems of the phthalocyanines. It has been suggested recently, through the use of UV-transparent solvents and matrix isolation, which has allowed absorption and MCD spectra to be obtained for quite a wide range of phthalocyanines, that an extra band is part of the red edge of the B band envelope in the 300- to 400-nm region [3,9,26,27,59,65,80].

We propose that the original sequence is modified to read (in order of increasing energy): Q, B_1, B_2, N, L, and C. The individual transitions are associated with orbital assignments suggested by Hollebone and Stillman [52] and Van Colt et al. [59] (Fig. 3). We describe in more detail below (Section C, iii) the evidence for the presence of B_1 and B_2.

The use of matrix isolation techniques, especially when coupled to synchrotron radiation sources [59], is clearly going to provide in the future very high-quality, wide-wavelength range spectral data that can be used to determine the complete assignment of the normal phthalocyanine spectrum. We should note though, that spectra from phthalocyanines in the vapor phase or phthalocyanine complexes that have been matrix isolated will be of the isolated phthalocyanine molecule, rather than of a molecule with axial ligation, so many of the special properties of these ring compounds that are perhaps best demonstrated by the remarkable chemistry of the heme group in the catalases [81,82], peroxidases [83–85], and cytochrome *P*-450 [86] cannot be studied.

v. Comparison between the Absorption Spectra of Porphyrins and Phthalocyanines

The spectral properties of the phthalocyanines have not received the highly detailed attention given to the absorption spectra of the porphyrins, so that the distinction between spectra that are "regular" and "irregular" is not often made for the phthalocyanines. However, the underlying electronic features that result in the large changes in the porphyrin spectra are also apparent in many phthalocyanine spectra [10]. Gouterman has reviewed the spectra of metalloid porphyrins and phthalocyanines, where "hyper" spectra are observed [31]. Many of the band shifting effects observed for porphyrins arise as a result of charge transfer bands lying in the same energy region as the main B or Soret band. As described below (Section C, ii), only CT transitions that transform as E_u or A_{2u} will be observed, these being the same transformations as the main $\pi{-}\pi^*$ transitions. When CT and $\pi{-}\pi^*$ transitions overlap, two or more strong bands are found as a result of the extensive configurational mixing between the Soret (or B) state and the CT state [21], and either a "hypso," blue-shifted, or "hyper," red-shifted, main Soret band is observed [67]. The origins of these effects have been assigned to charge transfer for complexes such as CrPc [10], MnPc [10,36], FePc [3,10,11,87], CoPc [3,88], Ge(II)Pc, and Pb(II)Pc [31], but also to solvent or axial ligation effects [3,89], and to oxidation or reduction of the π ring itself [27,65,90].

The much reduced coupling between the Q and B states, and the lower energy of the Q band in phthalocyanines, means that in most cases the Q band remains unaffected by CT bands lying to higher energy. However, MCD studies of Co(II)Pc [3] and RuPc [90], have shown that the Q band is affected when CT bands lie very near the 670-nm Q band. For these complexes, the MCD spectrum specifically monitors the mixing which is taking place between the two states in terms of enhanced B term intensity and band broadening.

Important differences in the characteristic spectral features of phthalo-cyanines and porphyrins include the well-resolved, intense Q band near 670 nm of the phthalocyanines, compared with the intense Soret band of the porphyrins, which is located near 410 nm. The Q band in phthalocyanines is reasonably constant in energy from complex to complex. Q band λ_{max} measured for a range of phthalocyanine complexes by the method of moments, provides the following set of λ_{max} (Q) values for the series Li_2Pc (651 nm), FePc (637 nm), Co(II) Pc (638 nm), $[(CN^-)_2Co(III)Pc]^-$ (656 nm), NiPc (652 nm), and ZnPc (657 nm). We should note that while extraction of λ_{max} in this manner is a very reliable method, these values are not quite the same as the λ_{max} that is simply read off the spectrum, because the whole envelope is included in the calculation of the first moment of the absorption (see Section B, ii). Values of λ_{max} that are closer to the observed value are found when the integration range is restricted to solely the main, Q_{00} band.

While the Q band region is relatively insensitive to a change in axial ligand or central metal (unless CT bands are introduced), the envelope below 450 nm comprises several overlapping bands that move considerably with changes of the axial ligand, and even more so with different metals. To some extent, this character is the reverse of the situation commonly observed with porphyrins like TPP, where the B and N bands are usually well resolved, while the Q band can frequently be observed clearly only from the MCD spectrum [91].

The distinction between the spectra of different peripherally substituted phthalocyanines is not as pronounced as between different porphyrins, in part this arises because the Q band in MPc complexes is essentially an allowed transition, which is not so dependent on the intensity-gaining mechanisms that are required by the Q band in the porphyrins.

vi. Measurement of the Absorption Spectra of Phthalocyanines in Solution

Most phthalocyanines dissolve in strongly coordinating solvents like pyridine, in concentrated sulfuric acids, and in highly aromatic solvents like α-chloronaphthalene and dichlorobenzene. Some phthalocyanines are also soluble, although the Ksp's can be far lower, in solvents which offer greater utility in UV-visible spectroscopy and in electrochemistry. Many of the spectra quoted in this review will be from phthalocyanines dissolved in one of the following solvents, either DMA, DMSO, or DMF, all more useful spectro-scopically than the aromatic solvents commonly used in the earlier, but pioneering spectroscopic studies, where pyridine and toluene were favorite solvents [76].

Recently, with the development of the redox chemistry of the phthalo-cyanines, through electrochemical and photochemical studies, acetonitrile and methylene chloride have been used [9,65], although axial ligation is often necessary to aid in the solubility. For example, ZnPc dissolves in methylene chloride as the pyridine complex [92], but is almost insoluble in the absence of a strong axial ligand. Surprisingly, in view of the consistency of many other physical properties, solubilities vary greatly with coordinated central metal as well, thus RuPc, Li_2Pc, and MgPc [6,9,26,27] are readily soluble in acetone, acetonitrile, and methylene chloride, solvents that most other phthalocyanine complexes do not dissolve in at all. Perhaps because of past problems with solubility, we find today that very few sets of spectral data are available for ranges of phthalocyanines dissolved in the same solvent. The importance of using the same weakly coordinating solvent for spectral measurements can be seen from the series of spectra reported by Dale [11]. In this study of $(L)_2FePc$, the absorption spectrum between 300 and 500 nm changes dramatically as a function of the ligand. The following papers report spectral data for a series of complexes using a single solvent [3,6,9,11,26,65,76,90,92].

Among these solvents, we find a few general trends. For axial ligand studies, the weakly coordinating DMA, DMSO, and DMF provide a consistent environment from which to compare different metals and different axial ligands. For a wide variety of phthalocyanines, the Q band is found near 667 nm; as an example see the spectral data for ZnPc in DMF (Table 4). In aqueous and alcohol-based solvents, dimerization begins to be problematical in terms of obtaining spectral data of monomeric species. Concentrated sulfuric acid, today, provides a new means by which to characterize the spectral properties of phthalocyanines. Spectra recorded from H_2SO_4 solutions exhibit a red-shifted Q band, which lies between 700 and 850 nm. Table 4 summarizes data collected together in Table 3. The wide spectral range available, from 900 to 200 nm, makes these data very useful for assignment purposes. We can see clearly the extent of the red shift, by comparing the data for Zn-octabutoxy-Pc in DMF (visible region λ_{max} = 678 nm) and H_2SO_4 (visible region λ_{max} = 827 and 729). The Q band maximum is considerably more sensitive to the metal than in, for example, DMA. Compare Q λ_{max} measured in H_2SO_4 for Co^{2+} (738 nm), with Cu^{2+} (821/779 nm), with Fe^{3+} (765/685 nm), Ni^{2+} (738 nm), and Zn^{2+} (783/696 nm).

vii. Related Spectral Data

While this review is primarily concerned with the absorption spectra of phthalocyanine complexes, related spectra for porphyrins reviewed by Gouterman [21] are extremely helpful. Phosphorescence and fluorescence spectra of phthalocyanines have been reported by Vincett et al. [93] and Eastwood et al. [56]. Fluorescence excitation spectra measured near 15,000 cm^{-1} for H_2Pc and MgPc, using supersonic molecular jets [94], provide a remarkable insight into the complexity of the vibronic interactions observed more generally as a broad envelope of transitions in the Q band region, and labeled in this paper as Q_{vib}. Similarly, the fluorescence and excitation spectra from phthalocyanines in Shpol'skii matrices [55] also provide very detailed information on the vibronic structure in the Q band region of the phthalocyanine ring. Huang et al. [55] have reported fluorescence and fluorescence excitation spectra for H_2Pc, MgPc, ClAlPc, ZnPc, and CuPc from a single site in a mixed α-chloronaphthalene/octane solvent system measured at 4.2 K. Shpol'skii matrices are formed at low temperatures using mixed organic solvents (in this case [55], a mixture of α-chloronaphthalene and octane), and very sharp line spectra are observed from molecules at specific sites in the matrix.

viii. Some Experimental Comments

Impurities in phthalocyanine complexes prepared in the laboratory or purchased from major suppliers frequently include various carbonaceous decomposition products, assorted complexes of the type $(L)_nMPc$, where the ligands might be variable, free base H_2Pc, and unreacted phthalonitrile. We have found that at a minimum, phthalocyanine from commercial suppliers should be sublimed once under high vacuum, or repeatedly extracted in a Soxhlet apparatus using a variety of solvents. H_2Pc, sometimes detected from mass spectra if the fractions can be quite high. In absorption spectra, H_2Pc can usually be detected by the presence of a shoulder on the red edge of the Q band, somewhere near 690 nm. Phthalonitrile can be observed from its characteristic aromatic absorption and MCD spectrum in the 250- to 290-nm region. H_2Pc will bind metals at elevated temperatures so that matrix isolation and high temperature monolayer formation must take place from inert crucibles. Low temperature MCD spectral measurements are particularly sensitive to the presence of paramagnetic impurities.

Solvents like DMF, DMA, and DMSO can decompose to form new material with quite different axial ligation properties. This can result in multiple Q bands being observed as shoulders in the spectrum of the phthalocyanine. We have specifically found that even careful distillation of DMA can result in an increase in impurities that quite significantly change the absorption spectrum of ZnPc. A similar comment concerning the blue and green solutions obtained when $Lu(Pc)_2$ is dissolved in DMF has been made by Corker et al. [95]. Traces of dimethylamine in DMF are suspected of reducing the green $Lu(Pc)_2$ to the blue $[Lu(Pc)_2]^-$ [96]. Traces of acid will often demetallate phthalocyanine complexes.

ix. Comparison between Spectra Recorded from Samples in Solution, in the Vapor Phase and in the Solid State

After the early measurements of the spectra of phthalocyanines in solution, generally, in either pyridine [76] or concentrated acids, the first systematic studies of the phthalocyanine spectral properties were reported by

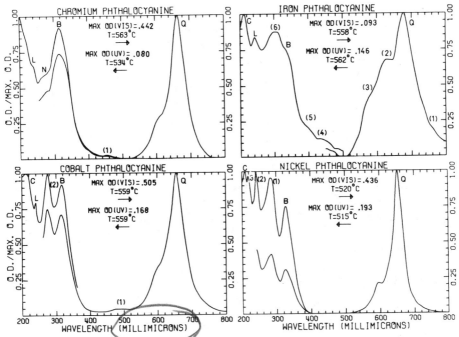

Figure 9 Vapor-phase absorption spectra of CrPc, CoPc, FePc, and NiPc. Temperature and maximum OD (corrected to 1-cm pathlength) for two runs are indicated. Reproduced with permission from [57] (Fig. 2).

Gouterman's group during the 1960s and 1970s [56,57]. The most resolved data were from the vapor phase at temperatures in the region of 500°C, where the absence of solvent allowed bands to be measured to the cut-off of the spectrometer, near 200 nm. A typical set of these spectra is shown as Fig. 9 for CrPc, CoPc, FePc, and NiPc [57]. Notice the broadness in each of the Q bands near 670 nm when compared with MgPc in CH_2Cl_2 (Fig. 1), the single resolved B band near 315 nm, and the series of bands running into the rising background absorption due to the benzene rings. The broadness in all of these bands appears to be a general, but not unexpected, feature of the vapor phase experiment (see also [78]). Comparison between the spectra of MgPc (Fig. 1) and FePc (Fig. 9) shows that the metal plays an important part in determining the number of optically accessible states in the visible-UV region. New bands are clearly present for FePc, and the assignment to Q, B_1, B_2, N, L, and C is not as easy. Solid state spectra, especially from thin films, are complicated by the broadening that arises from the Davydov effect, so that while spectra have been recorded to 200 nm, the spectral envelopes are not at all similar to the envelopes observed for either vapor or solution phases [51,52]. Figure 8 illustrates this with the absorption and MCD spectra of α-CoPc as a thin film on quartz.

x. Effects of Aggregation

As with the porphyrins [77,97], the phthalocyanines dimerize and polymerize readily in solution. Dimeric (or polymeric) species can be formed (1) as a result of direct linkages, or bridges, between two or more phthalocyanine rings, that place the rings close enough in space that intramolecular association can occur (see, for example, the spectral properties of the peripherally linked phthalocyanines and "clam-shell" phthalocyanines [98–101]), (2) through covalent bonding involving the metal, frequently as μ oxo links, especially for Fe and Si containing phthalocyanines (see, for example [102]), (3) sandwich-type complexes, in which two rings share one central metal, for example, $Sn(Pc)_2$ [103] and $Lu(Pc)_2$ [104], or (4) through weak association where peripheral substitution holds two rings adjacent in space. See, for example, the effect of substitution with crown ether rings on the Q band in the spectrum of CuPc in polar solvents [105]. Further discussion of Section C.vii includes the effects of aggregation on the spectral properties of MPc complexes.

B. MAGNETIC CIRCULAR DICHROISM SPECTROSCOPY

MCD spectra have played an important role in the understanding of the absorption spectra of heme groups in proteins [81,83,84,86,106–116], chlorophylls [15,17,108,117–124], synthetic porphyrins both neutral [20,122,125–131], and following one or two electron oxidation or reduction at the porphyrin π ring [131–133], and, finally, of neutral [3,51,90,92,127,134–136], [6,26,27,52,65,78,80,87,137–141], and ring reduced or oxidized [9,27,65,78,90,92] phthalocyanine complexes. A brief introduction is essential for the purposes of this review. While the Faraday arrangement of the magnetic field in the MCD technique has dominated the spectral characterization of the porphyrins and phthalocyanines, studies have also been reported that involve the Zeeman effect. For example, Chen et al. [142] have examined the effect of high magnetic fields on the excited states of PtPc using emission spectroscopy.

More complete descriptions of the theoretical and practical aspects of the technique may be found in the original literature of Buckingham and Stephens [143], Stephens [127,144], Briat [145], and Schatz and McCaffery [140], as well as in discussions of the specific MCD spectroscopy of porphyrins [20,127,146,147] and heme proteins [106,148–150]. A recent, and very comprehensive, discussion of the theory of MCD is contained in the book by Piepho and Schatz [151].

i. Description of the Technique

The MCD signal can be observed when a magnetic field is aligned parallel with the optical path, and circularly polarized light is passed through a sample located at the center of the magnetic field. In this orientation of the magnetic field, the so-called Faraday arrangement, only transitions between states with $\Delta M_J = \pm 1$ can absorb light. Transitions with $\Delta M_J = +1$ absorb selectively left, circularly polarized light (lcp), while transitions with $\Delta M_J = -1$, absorb right circularly polarized light (rcp). In the more traditional Zeeman arrangement, that is with the magnetic field perpendicular to the optical path, transitions with $\Delta M_J = \pm 1$ and 0 give rise to absorption with σ ($\Delta M_J = \pm 1$) and π ($= 0$) polarization.

Figure 10 shows how each of the three special spectral features that characterize the MCD spectrum, namely the Faraday A, B and C terms, arises. The observation of the distinctive A term (a derivative-shaped band), of either positive sign (positive lobe to high energy of the crossover point) or negative sign (positive lobe to low energy of the crossover point) initially pointed to the power of this technique in providing assignment criteria for complicated absorption spectra. This is because the A term is observed only when the excited state is orbitally degenerate. Usually, and this is the case for many complexes of the porphyrins and phthalocyanines, the ground state is orbitally nondegenerate. The converse of this, when the ground state is orbitally degenerate and the excited state is orbitally nondegenerate, results in C terms, which are identified by a temperature-dependent spectrum. Therefore, observation of A and C terms is dependent on the symmetry of the molecular orbitals concerned with the transition.

For example, by simply comparing the Q band region MCD spectra of H_2Pc with that of NiPc, both recorded for dilute solutions (Fig. 4), we see that H_2Pc must be a species with a symmetry lower than the D_{4h} of NiPc. The A term centered exactly on the absorption band maximum in NiPc unambiguously identifies the degeneracy of the Q band excited state [26]. In the same way, the 2 B terms (Gaussian-shaped bands or either negative or positive sign), lying exactly under the two absorption components of the H_2Pc, indicate that these are the symmetry-split x and y polarized transitions of the D_{4h} degenerate excited state. Derivative-shaped signals can be observed when oppositely signed Gaussian envelopes lie within a few cm^{-1} of each other, and when the band widths are large. Under these conditions, it is very difficult to distinguish between an A term and a B term.

The utility of the MCD technique arises because the MCD spectrum can be recorded from solutions of the type used for UV-visible measurements, at room temperature. Aligned crystals are not required. For phthalocyanines and porphyrins, where the extinction coefficients are often greater than 10^4, this is an important feature.

The Faraday C term is observed when the ground state is orbitally degenerate. Temperature dependence in the spectral intensities arises from the change in Boltzmann distribution across the split orbital components of the degenerate ground state. The C term will usually be larger than all other MCD spectral effects at temperatures below 20 K, because the population of the lowest energy component far exceeds that of the upper component. The shape of the C term is the same as a B term, that is, simply a Gaussian distribution of

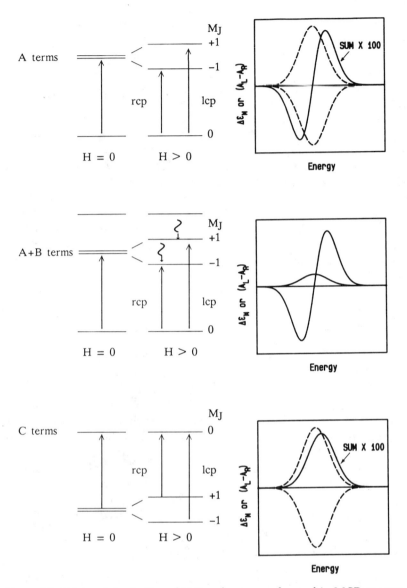

Figure 10 Origin of the A, B, and C Faraday terms observed in MCD spectra.

either positive or negative sign. Because the CD spectrometer is tuned to measure differences in $A_l - A_r$ of the order 10^{-4}, when one polarization is absorbed selectively with $A_l - A_r$ of 10^{-1}, this signal completely dominates the spectrum. Because the ground state splits by several cm^{-1} under the influence of the magnetic field, complete depopulation of the upper state can occur. Saturation effects become noticeable at very low temperatures (< 10 K) and high magnetic fields (> 4 T). Analysis of the change in MCD signal as a function of temperature and magnetic field can yield the ground state g factor; see examples of the application of the theory by Schatz [151], Vickery [150], Hatano and Nozawa [149], and Thomson [152].

Normally spin degeneracy in the ground state does not result in C terms being observed, unless this is accompanied by orbital degeneracy. However, when spin orbit coupling can operate in the excited state, a special spectral feature is observed as a function of temperature. In effect selective access to different spin–orbit components in the excited state from a ground state that has split spin components occurs. This selectivity operates because the $\Delta m_s = 0$ selection rule applies. As a result it appears that A terms are growing in intensity as the temperature drops; these new bands are sometimes known as "temperature-dependent A terms," even though the change in intensity has truly a ground state origin and therefore can arise only from a C term. A detailed discussion is contained in the book by Piepho and Schatz [151], and a nicely worked example for hemoglobin has been given by Treu and Hopfield [153]. Some examples include the low spin ferric heme proteins [83,130], and recently the low-temperature spectra of $[Co(III)OEP(-1)]^{2+}[ClO_4^-]_2$ [154].

No temperature-dependent MCD spectra of phthalocyanines have been fully described to date, although it is to be expected that the ground state of the ring can become orbitally degenerate as a result of coupling to paramagnetic central metals. In our laboratory, we have been particularly interested in the spectral effects created by the $^2A_{1u}$ ground state in π cation radical species; to date we have found that while $[Co(III)OEP(-1)]^{2+}$ exhibits temperature-dependent MCD spectra [154], the MCD spectra of $[MgPc(-1)]^+$ [155] and $[ZnPc(-1)]^+$ [65] are essentially temperature independent to 4.2 K. The most widely studied examples of the porphyrins involve iron, where oxidation and spin effects are known to introduce degeneracy into the ground state of the ring [106,149,150].

A major complication in assigning MCD spectra arises from the presence of B terms (Fig. 10). B terms gain signal intensity from magnetic-field-induced mixing between all states. Because there is a $1/\Delta E$ term in the intensity-gaining mechanism of B terms, the states adjacent to the one of interest usually contribute the greatest contribution to the B term signal. B terms will sum to zero throughout the spectrum. For those metallophthalocyanines for which there is no charge transfer in the Q band region, we observe, typically, very symmetric A terms, with little contribution from B terms [9,26,65]. However, the complexes like Co(II)Pc, where charge transfer bands lie near either the Q band [3], or the B band [65], B term intensity increases quite significantly, so that assignments become very much more difficult. Spectral deconvolution that places both an A term and a B term function under the spectral envelope, each with the same band energy and band width, is the only way of analyzing such data [9,65,156].

ii. Theoretical Description of the MCD Parameters

Stephens [144] and Piepho and Schatz [151] have provided a very good description of the theoretical basis of the MCD technique. For our purposes, we need to state only the relationship between the observed signal and the three "terms" described above, and describe how we may extract parameters from a spectrum.

First, rewriting equation (A.4.16) from Piepho and Schatz [151], we find that for an MCD spectral lineshape described in terms of ΔA_m (where a nonzero value for $A_l - A_r$ is induced by a magnetic field), that $\Delta \epsilon_m cl/E = 152.5 Bcl[A_1(-df/dE) + (B_0 + C_0/kT)f]$, where A_1, B_0, and C_0 are the Faraday terms, B is the magnetic induction in tesla, c is the molar concentration, l is the cell pathlength in cm, E is the spectral energy, and f is the lineshape. This equation illustrates how only the C_0 term is temperature dependent. The values of the individual parameters are best determined by integration over the band. This is the "method of moments" technique, described in detail by Henry et al. [157], and modified for MCD solution work by Stephens [144,158]. We will follow the equations of Piepho and Schatz [151].

Numeric integration is carried out over a band with an absorption coefficient $k(E)$, at an energy E, which has a line shape $f(E)$. Rearranging the general equation for any moment of a lineshape, $\langle f \rangle_n = \int^{\text{whole band}} f(E)(\bar{E}-E)^n dE$, for absorption and MCD spectra, we choose f to have units of A/E and $(\Delta A/E)_m$, respectively, and $\bar{E}$ is usually chosen to be the average energy of the band, which will be close to, but not identical to, the traditional band center. This equation can be rearranged [151] to separate intensity that originates from a lineshape f, with a single sign, for example, a simple Gaussian shape, from intensity that originates from a lineshape with a derivative shape, df/dE. In the MCD spectrum, this extracts the A term intensity from the B and C terms. We use the absorption spectrum to obtain the dipole strength and the average band center.

Thus, the dipole strength is obtained from, $\langle A \rangle_0 = 326.6 D_0 cl$, while setting $\langle A \rangle_1 = 0$, provides E. Integration of the MCD spectrum gives the B and C terms, $\langle \Delta A_m/E \rangle_0 = 152.5[B_0 + C_0/kT]Bcl$, while $\langle \Delta A_m/E \rangle_1 = 152.5 A_1 Bcl$ provides the A term where, B is the field strength in tesla, E is the spectral energy in cm^{-1}, and the electric dipole intensity, D, is in Debye units. The magnetic moment information is then best extracted by dividing out the electric dipole intensity. This is achieved by dividing each term by D_0, to give A_1/D_0, B_0/D_0, and C_0/D_0. We also believe that it is important for the analysis of MCD data to follow a common path, and suggest that the equations described by Piepho and Schatz as (A.4.16–A.4.22) [151] should be adopted. In order to introduce some uniformity in the use of numerical methods to obtain A_1, B_0, and C_0, we would recommend that the actual values of the zeroth and first moments of both the absorption and MCD spectra be published (that is, $\langle\Delta\epsilon_m\rangle_1$, $\langle\Delta\epsilon_m\rangle_0$, and $\langle\epsilon_m\rangle_0$) as well as the final A_1/D_0, B_0/D_0, and C_0/D_0 terms.

Clearly, temperature dependence experiments allow the values of B_0 and C_0 to be separated out. The MCD experiment can provide excited state values readily at room temperature when the ground state is nondegenerate. Variable temperature measurements, down to liquid helium temperatures, are necessary when the ground state exhibits either spin or orbital degeneracy. When calculations of the moments of a band can be carried out over a complete transition, excited state angular momentum values may be obtained with much greater precision and accuracy than is possible by the deconvolution route; however, when bands overlap, deconvolution is the only technique available [158–161]. We have described the implementation of the method of moments in the analysis of the absorption and MCD spectra of a range of s^2 ions doped into alkali halide crystals [162,163].

iii. Description of Current Experimental Practice for MCD Spectroscopy

Because circularly polarized light is required for the Faraday experiment, the MCD spectrum can be conveniently measured in a CD spectrometer, as long as the sample compartment is large enough to take the magnet, and the electronic and photoacoustic equipment is far enough away, or adequately screened from, the 5- to 7-T magnetic field commonly used, so that the spectrometer continues to operate normally. While the traditional light source is a Xe lamp, Schatz and co-workers have recently used a beam line on the synchrotron at Madison to provide light of energies well below 180 nm [59]. For complexes that exhibit strong MCD intensities, electromagnets and even permanent magnets, with fields of 1 T, can be used.

Variable temperature studies can be performed readily in the superconducting magnet between 1.7 K and room temperature. The greatest technical problem to overcome for low-temperature studies is to obtain strain-free, frozen glasses at these low temperatures, so that the circularly polarized light is not depolarized (even slight depolarization will nullify the CD polarization and either reduce or greatly increase the observed intensity). As for all spectroscopic techniques, the role of the computer in the acquisition of MCD spectral data, and the subsequent data analysis, has been significant since the first reports of MCD spectra were made [112,127]. The availability of digital data is specially important in the case of C terms, because once the qualitative analysis of the MCD spectrum has been carried out, quantitative analysis requires the calculation of magnetic moments using either the method of moments [26,127,158] or spectral deconvolution [9,17,52,65,127,164].

iv. Contribution of the MCD Spectral Data to the Understanding of MPc Spectra

While magnetic optical rotatory dispersion (MORD) spectra were the first to be obtained using the Faraday arrangement of the magnet, the presence of rotation from all of the components in the optical path, including those that were optically transparent, meant that the spectra were very difficult to analyze [165]. The use of CD instrumentation to measure MCD spectra eliminated these problems; see, for example, the early paper on the charge transfer bands by Schatz et al. [166]. MORD data for the phthalocyanines [165] were first analyzed by Stephens et al. [127]. The major contribution of the MCD technique to the understanding of absorption spectrum of the pthalocyanines to date has been in the assignment of each band. By comparing the spectral envelope measured in the absorption spectrum, with envelope measured in the MCD spectrum, it is often relatively straightforward to determine whether the transition responsible is to a degenerate or nondegenerate state. Thus the MCD spectrum can provide polarization information directly from a dilute solution at room temperature.

Calculations of the magnitudes of the magnetic moments of excited states have been reported by several authors using both spectral deconvolution and method of moments techniques [9,26,65,127,158,159]. The values of the excited and ground state magnetic moments represent a quantitative measure of the orbital contribution to any state, and as such allow models to be tested quantitatively. The angular momentum of the excited states is directly dependent on the electronic configuration.

The "Four Orbital" theory described by Gouterman's group [21] predicts an angular momentum for the Q band in phthalocyanines of 3.1 Bohr magnetons (BM) and approximately 1 BM in the B band. Although many more MCD spectra of porphyrins have been analyzed than phthalocyanines, it is not possible to compare the porphyrin data directly with the phthalocyanines data. This is because the "accidental" degeneracy of the top two filled MOs in most porphyrins, the a_{1u} and a_{2u} (Fig. 3) [21], results in such significant configuration interaction between the Q and B (or Soret) states that there is little similarity between the electronic structures of the porphyrins and phthalocyanines [63,75,167]. Compare, for example, the MCD spectrum of ZnTPP [91] with that of ZnPc [65,92].

Qualitatively, the angular momentum of the phthalocyanine complexes measured to date compares favorably with the value predicted theoretically for the Q band [63]. In the absence of interaction between the metal and the ring, a value for the magnetic moment near 3 BM seems to be normal [26,63]. The values calculated for a range of complexes of $(L)_2MgPc$ average to 2.60, with little of the variability with axial ligand observed for ZnPc [65]. The data of Stillman and Thomson [26] suggest that there is fine balance in the angular momentum of the Q band excited state so that variations of a factor of two can occur when only the axial ligand is changed. With so few data available, it is not yet possible to determine a general trend, although it appears that axial ligands that result in the central metal being strongly decoupled from the ring, result in the highest angular momentum values, for example $(CN^-)_2Co(III)Pc$. When CT bands are close to the Q band in energy, the angular momentum drops, and, coincidentally, we also observed a broadening in the Q band. MCD spectra from matrix isolated H_2Pc and CuPc [141], and ZnPc [59], have been reported. As expected, the spectra are very well-resolved, although the effects of multiple sites are an added complication. The low-temperature MCD spectra of Mn(II)Pc show significant temperature dependence [168]. Because the MCD spectrum is so sensitive to degeneracy in the ground state, measurement of the temperature dependence of the spectrum provides unambiguous data on coupling between a paramagnetic central metal and the diamagnetic π ring.

Although theoretical calculations for the porphyrins are quite advanced in their predictive powers, the precision for the phthalocyanines is not as good. MCD spectra of neutral, diamagnetic complexes, such as ZnPc and MgPc, provide far more detail about the complexity of the spectral envelope between 240 and 400 nm than is available from the absorption spectra alone [9,65]. Unlike the parent porphyrins [19,21], several bands overlap in this region making it difficult to measure precise band centers and oscillator strengths. Deconvolution by our group, of both absorption and MCD spectra [9,65,156], has proven to be a useful technique for obtaining reliable band parameter data for several phthalocyanine complexes. It is difficult to calculate fits that are clearly unambiguous for spectral data with so many overlapping bands. In our work we have tended to accept more readily fits that put bands where we see inflection points in the absorption spectrum, rather than those sets of parameters that simply fill the envelope most efficiently.

C. ANALYSIS OF THE SPECTRAL DATA

i. Molecular Orbital Calculations

Because the spectral properties of hemoglobin have been known for such a long time, many theories to account for the color changes observed for the protoporphyrin IX heme group have been published. It is useful for the phthalocyanine spectroscopist to examine these theories put forward for porphyrins, because the electronic structure of a phthalocyanine can be considered to be that of a highly perturbed porphyrin. Among the earliest to identify orbital assignments with the α and Soret bands was Simpson [169].

The straightforward molecular orbital picture described by Simpson [169] for the π ring in porphyrins remarkably applies quite well today as an introduction to the spectral properties of both porphyrins and phthalocyanines. The essence of his scheme lay in placing 18 π electrons into a stack of the 18 π MOs, in a 1:2:2:2:2:2:2:2:2:1 arrangement, in the manner used for any cyclic polyene, in which the angular momentum in the top filled molecular orbital is ± 4, and in the lowest unoccupied molecular orbital it is ± 5. In this scheme, transitions involved either $\Delta M_1 = \pm 1$ or ± 9, where the ± 1 transition resulted in the allowed Soret band, found near 400 nm, and the ± 9 transition was the forbidden, α or Q, band found in porphyrins near 580 nm. The theory did not predict the energies of these bands, but the identification of forbidden and allowed transitions was itself a considerable advance. Since that time, there have been many discussions of systematic features observed in the spectra of porphyrins and heme proteins [19,60,170]. Measurement of MCD spectra led the way to obtaining magnetic moment values for the two lowest energy π^* states; these values were surprisingly close to those predicted by Simpson [26].

Molecular orbital calculations for porphyrins were reported in 1950 from Platt's group [79], where bands observed in the absorption spectra of porphine were compared directly with predicted band energies, and x/y degeneracies were associated with the two main transitions, which are located near 625 nm (16,000 cm^{-1}) and 435 nm (23,000 cm^{-1}). Basu [171], carried out a similar calculation for the phthalocyanines. From Platt's work, Gouterman's group contributed the next major step in calculating optical data for a wide range of porphyrin complexes, including, various hemes (see Gouterman's review in *The Porphyrins* for a summary of these papers [21]). In 1972 and 1973, this group introduced calculations that were specific for the phthalocyanines [63,75,172], using a similar set of labels for the first five bands that were used in the porphyrin analyses: for phthalocyanines these are Q, B, N, L, and C. Schaffer et al. [75] reported extended Hückel calculations for NiPc and CoPc, comparing the energy levels with those of NiTAP and NiTAP (Fig. 11). Although no state energies were calculated, the energies of the π, σ, and d orbitals were calculated. From the energies of the metal d orbitals, it was suggested that MLCT bands would overlay the π–π^* spectrum. (See also the paper by Lever et al. [10], for a detailed discussion of the spectral effects of CT.)

In the SCMO–PPP–CI model used in 1972 [63], the energies, oscillator strengths, and angular momenta for each of these five states were calculated. With configuration interaction included up to 50 nm (Fig. 12) [63], the Q band is clearly $a_{1u} \rightarrow e_g$ in character. Unlike many porphyrins, for example, ZnTPP [91], the $A_{1g} \rightarrow E_u$ transition for Q_{00} in phthalocyanines is much more intense than the vibrational components, Q_{vib}. Indeed, it is a feature of phthalocyanines that the Q_{vib} progression has an absorbance of less than 10% of the main Q_{00} band, unless CT bands overlay the envelope. Compare the spectrum of $(CN^-)ZnPc$ [65], with, for example, the spectra of ZnTPP [91]

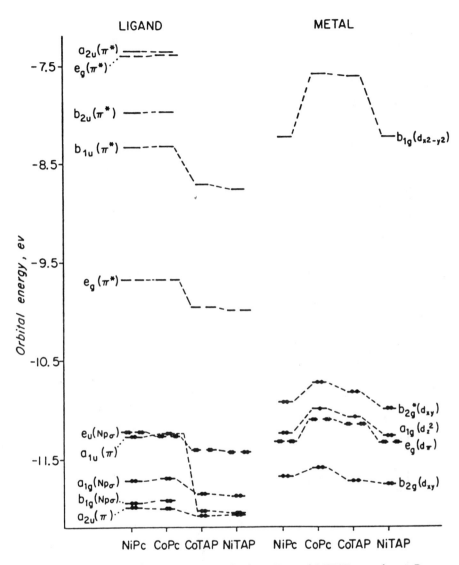

Figure 11 Top filled and lowest empty MOs for MPc and MTAP complexes. Reproduced with permission from [75] (Fig. 4).

and oxygenated ferrous myoglobin [173], where the Q_{01} band, the β band, has comparable or even greater intensity than the Q_{00} or α band. We should note that the vibrational progression in porphyrins is complicated and the vibronic interaction can lead to the reversal of the usual positive sign of MCD A terms predicted for the A_{1g}–E_u transition [59,174]. Between 330 and 250 nm (30,000–40,000 cm^{-1}), the individual transitions were predicted to blend

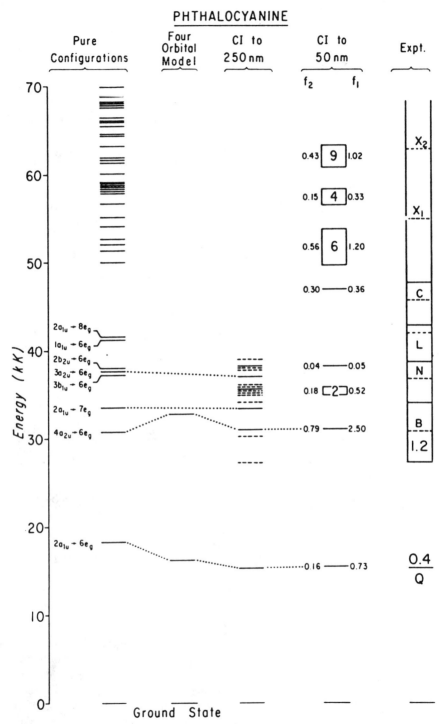

Figure 12 The effect of increasing extents of configuration interaction on the state composition for ZnPc. Reproduced with permission from [63] (Fig. 4).

together as a result of configurational interaction, to form bands of multiple transitions. At that time (1970s), there was no experimental evidence to support the existence of two or more individual bands under the B envelope, as is now considered to be the case for the B_1 and B_2 bands that have been observed in the spectra of ZnPc, MgPc, and Li_2Pc [9,26,59,65].

McHugh et al. [63] also provided the first detailed calculation of the angular momentum in the degenerate states. Figure 12 shows the effect of configuration interaction on both the energies of the states and the orbital composition. For the ground and first excited state, they predicted orbital Zeeman terms of 2.12 and 1.82 in units of Bohr magnetons, respectively. In addition they calculated M_{jj} values, where M_{jj} is the magnetic moment of the excited, degenerate state j, state for Q, B, N, and L, comparing phthalocyanine with porphine tetraazaporphyrin and tetrabenzporphyrin. M_{jj} can be obtained readily from MCD measurements of A/D values, using the relationship $M_{jj} = (2/\beta) \times A/D$, and noting that the splitting of the degenerate state as a function of the applied magnetic field, H, is given by $2M_{jj}BH_z$ [9,26,65,127,143,151]. For the Q, B, and N band regions, McHugh et al. [63], calculated magnetic moments of -4.3, $+0.03$, and $+0.30$ BM, respectively. Our measured values for a range of complexes indicate that the sign of M_{jj} is negative for the each of the first four bands (which is seen from the positive A terms observed under each band). The Q band magnetic moment, calculated by the method of moments over the whole Q band envelope [9,26,151], is found to be dependent on both the metal and the axial ligand. In our recent work [9], we have found it useful to report the $\langle\Delta\epsilon_m\rangle_1$, $\langle\Delta\epsilon_m\rangle_0$, and $\langle\epsilon_m\rangle_0$ experimental values, as well as the computed A_1/D_0 and μ_B parameters. These data allow for direct comparison by other workers in the field. Because we see B_1 and B_2 under the "B" band, it is not clear to us whether the values predicted for this region can be used directly at present.

In a separate study, Henriksson et al., using the SCF–MO method, calculated state energies, oscillator strengths, and state symmetry for H_2Pc (Fig. 13) [62] and CuPc [61]. For CuPc, valence electrons on the metal atom were included using the Peel method, and metal-related transitions were predicted in the 9,800 (1,020 nm) to 55,100 cm^{-1} (181 nm) spectral region. Spin doublet, orbitally degenerate, $\pi-\pi^*$ states, with ^2E_u symmetry, were calculated to lie (listing only transitions with $f > 0.01$) at 18,400 ($f = 0.44$; 543 nm), 34,500 (2.1; 290 nm), 35,300 (1.4; 283 nm), 47,700 (0.11; 210 nm), 50,000 (0.79; 200 nm), 52,600 (0.08; 190 nm), 53,900 (0.05; 186 nm), and 55,100 cm^{-1} (0.38; 181 nm). Significantly, it was suggested [61] that the states at 34,500 (290 nm) and 35,300 cm^{-1} (283 nm) involved considerable configuration interaction such that several orbital transitions contribute to the observed oscillator strengths. Because it is in the B band region that we have found the spectral overlap to be most complicated, it is useful to examine the predictions made by Henriksson et al. for the ^2E_u state at 34,500 cm^{-1} (290 nm). This state is comprised contributions from a number of transitions: $0.57(4a_{2u}-7e_g) - 0.55(3b_{2u}-7e_g) + 0.31(2a_{1u}-7e_g) - 0.20(3a_{2u}-7e_g) + \cdots$,

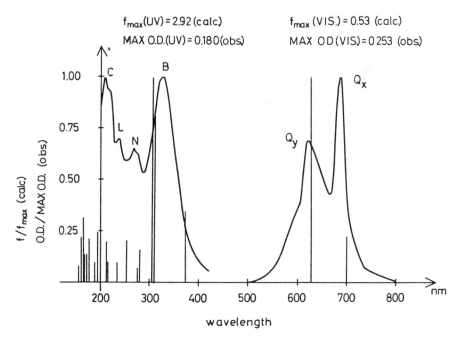

Figure 13 Observed vapor phase absorption spectra for H_2Pc and calculated $\pi-\pi^*$ singlets with $f > 0.2$. Reproduced with permission from [62] (Fig. 3).

while the close lying state at 35,300 cm^{-1} involves even more transitions: $0.61(4a_{2u}-7e_g) + 0.50(3b_{2u}-7e_g) - 0.28(6e_g-3b_{1u}) - 0.25(3a_{2u}-7e_g) + 0.23(2a_{1u}-7e_g) - 0.21(2b_{1u}-7e_g) \cdots$. There seems to be no reason why this state should not be observed separately from B_2 at 290 nm.

PPP LCAO–SCF–CI calculations by Marks and Stojakovic [30] that modeled both ZnPc and $SPcUO_2$ were carried out using Gouterman's techniques. From this calculation (Table 4), bands were reported at 698.9 and 695.0 nm (the Q band), and 331.3 and 330.7 nm (the B band). Details of the components of the major excited states were not reported. Khatib et al. [352] have very recently reported the results of an MSX_α calculation on ZnPc. Their calculations place a_{1u} as the highest occupied MO and e_g as the lowest unoccupied MO. We will discuss the spectral predictions below (in Section E.xxviii).

More recent calculations have been reported by Lee et al. [60], using CNDO/S methods for the phthalocyanine, TAP, TBP, and P rings. The results of these calculations compared well with the gas phase spectral data for MgPc [57], predicting that four bands would be present with significant oscillator strength. The authors do not include the compositions of the states so it is not possible to compare these data with those of Henriksson et al. [61]. As with the four-orbital model first used by Gouterman's group, there is no suggestion that more than one band exists in the 330-nm region, the "B" band region. It would be interesting to see how the transitions that are available could be combined in a way to generate two degenerate states of energies close to those of B_1 and B_2.

ii. Charge Transfer Spectra

With the obvious exceptions of d^0 and d^{10} metals, charge transfer (CT) transitions, either metal to ligand (MLCT) or ligand to metal (LMCT), can be expected in the spectral region 200 nm (50,000 cm^{-1}) to 1,000 nm (10,000 cm^{-1}) [3,10,12,21,36,57,88]. The direction and the resultant energies of these CT bands are dependent on both the spin and oxidation state of the central metal.

In addition to simply putting new bands into the spectrum, the effect of new, metal-related states can be to mix ring and metal orbitals together and hence further increase the complexity of the orbitals involved in each state observed as an absorption band in the spectrum.

Lever's group [10,13] has attempted to correlate the energies of bands assigned as CT in a very wide range of phthalocyanine spectra, with the known electrode potentials of the ring and the central metal atom. By using this method, similar to that suggested by Jorgensen for "optical electronegativities" [176], it is possible to calculate quite detailed assignments that include both the direction of the CT transition and a prediction of its energy. For LMCT, ring oxidation is coupled with reduction of the metal. For MLCT, ring reduction is coupled with metal oxidation. Thus, the redox potentials provide a very good starting point in the calculations. Transitions from the metal d orbitals into the empty, π^* molecular orbitals will be separated by the energy differences between the π^* states, the lowest being e_g and b_{1u} and b_{2u} (see Fig. 3). There are assumptions made that will contribute some inaccuracy to the final values for the energies of the transitions [10]. For example, this type of treatment assumes that the same orbitals are used for electrochemical processes as for the spectroscopic processes. The system also assumes that reorganizational effects when the electron moves from one part of the molecule to another in charge transfer are not a significant part of the total energy of the transition so that data can be used that are obtained from half reactions, for example, metal→oxidized metal, without the coupled ring reduction being included.

As an example, consider the spectrum of Na$_3$Cr(III)TSPc (Fig. 14). The possible redox directions are

$$\text{LMCT Cr(III) } d^3 + \text{Pc}(-2) \rightarrow \text{Cr(II) } d^4 + \text{Pc}(-1) \tag{1}$$

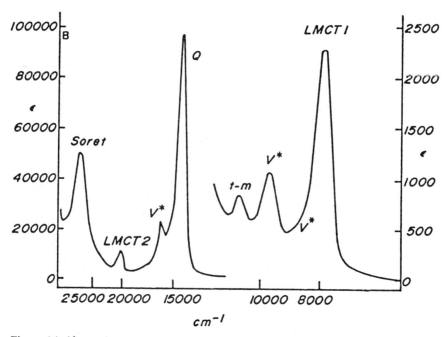

Figure 14 Absorption spectrum of $Na_2Cr(III)TSPc$ in DMF. LMCT denotes a ligand-to-metal charge transfer transition, V^* denotes excitation to a vibrationally excited state, and t-m denotes a trip–multiplet transition. Reproduced with permission from [10] (Fig. 3b).

and

$$MLCT\ Cr(III)\ d^3 + Pc(-2) \rightarrow Cr(IV)\ d^2 + Pc(-3). \tag{2}$$

The number and energies of the CT transitions observed will depend on the energies of the initial and final states for both the Cr(III) and the Pc(−2) ring. Transitions in both directions are possible for a single ion, but the energies will be quite different.

First, we can construct a diagram to show the possible, symmetry-allowed transitions. Thus, Fig. 15 illustrates possible transitions for a low spin d^6 metal complex. The selection rules applied in these diagrams allow only those transitions that transform as either E_u or A_{2u}, which for the lowest energy ligand to metal direction involve the top two π HOMOs. Because transitions for MLCT end on the b_{1u} and b_{2u} orbitals, we expect that these CI bands will lie above the Q band in energy, whereas for LMCT the energies of the transitions can straddle the Q band. Similar diagrams can be constructed for metals with other configurations.

From a consideration of the redox data of d^3 Cr(III), Lever et al. [10]

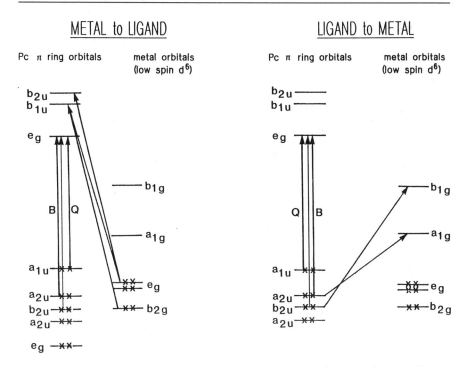

Figure 15 Possible directions for charge transfer transitions between the central metal and the Pc ring.

predicted that bands at 1,266 nm (7,900 cm^{-1}), 481 nm (20,080 cm^{-1}), and one other with a calculated wavelength of 422 nm (23,720 cm^{-1}) were LMCT in nature. Absorption to the blue of the 1,266-nm band (marked v*) was assigned as vibronic, like the Q_{vib} bands to the blue of the Q band. Finally, absorption at 912 nm (10,965 cm^{-1}) was assigned to a trip–multiplet origin, much like similar bands observed in the absorption spectra of Mn(II)Pc, at 937 nm, Mn(III)Pc at 937 nm, and Cu(II)Pc at 1,081 nm. Clearly, the assignment problem for complexes where there are many bands additional to the set normally observed for a complex like MgPc [9] requires the combination of many different pieces of information, including electrochemical data, as well as the absorption and MCD spectra.

Figure 16 summarizes the origins of the CT and π–π^* transitions observed for complexes such as Cr(III)Pc [10]. Well-resolved bands, with significant intensity, have been assigned to a CT origin for $[Co(I)Pc(-2)]^{-1}$ (with bands at 500 nm) (Fig. 6) [3]. Transitions in the 500-nm region of Pc(-2) species are typically assigned in terms of charge transfer, for example, Mn(II)Pc and Fe(II)Pc. Although some of the transitions to the red of the Q band may be charge transfer in character, both *d–d* and trip–multiplet assignments are also possible. However, the major problem in these assign-

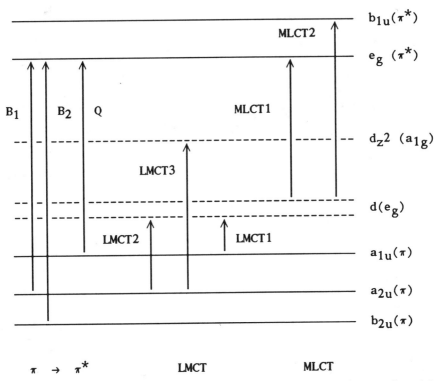

Figure 16 Scheme of the energy levels in a typical phthalocyanine showing the origin of the various LMCT, MLCT, Q, and B bands. Reproduced with permission from [10] (Fig. 2).

ments may well be in identifying the individual transitions accurately and reliably; for this task low-temperature MCD data would help.

iii. Deconvolution of the Spectral Data

The absorption spectra of transition metal porphyrins and phthalocyanines have long been regarded as difficult to fully assign. In much of the spectroscopic work, interest has centered on charge transfer transitions that lie within the $\pi-\pi^*$ envelope. For the phthalocyanines, the problem of counting the number of bands present between 800 and 200 nm is difficult because at energies greater than the window region near 500 nm, the bands overlap continuously toward the solvent cutoff.

The vapor phase work, for example [57], has resulted in significant advances in the general understanding of how many bands to expect in phthalocyanine spectra, but the high-temperature broadening effects were serious and sometimes specific features in the spectra could not always be reproduced from solutions. The thin film data, for example [54], suffer from Davydov effect band broadening and doubling, and, therefore, the resultant spectra are difficult to compare with solution phase data. Measurement of matrix isolated species has not unfortunately, to date, yielded spectral data of the same quality as that routinely obtained from solutions, largely because of the presence of multiple sites. Data from complexes cast into polymer films, such as polyvinyl alcohol and polymethylmethacrylate, are of high quality, and do allow for the measurement of low-temperature spectra, but very few phthalocyanine complexes can be dissolved completely into these polymeric materials without undergoing significant chemical reactions.

We, and others, have approached the problem of assigning the absorption spectra of phthalocyanines by measuring both absorption and MCD spectral data in UV-transparent solvents. The limitations here are the lack of solubility of many phthalocyanine complexes in solvents like DCM or ACN. In the past, the solvents DMA, DMSO and DMF have played important roles as the weak axial coordination enhances solubility yet at the same time allows for substitution with stoichiometric amounts of stronger field ligands. In this manner, absorption spectra of complexes of the type $(L)_2MPc$, where L = pyridine, imidazole, cyanide, CO, etc., have been studied between 260 and 400 nm, a region often obscured or distorted, by absorption from the axial ligand if it was the solvent as well.

The Q band in phthalocyanine complexes can be unambiguously assigned in most species. Measurement of MCD data is useful to confirm cases where the Q band is heavily overlaid by CT bands, for example (py)RuPc [6]. It is in the region below 450 nm that the assignment is most complicated. Unlike porphyrin complexes, no clear band can be identified as resulting from the $a_{2u} \rightarrow e_g$ transition, the "B" band.

The only way in which this type of overlap can be properly resolved is through deconvolution calculations, "band fitting" [9,52,65,156]. However, this type of analysis is fraught with problems. The problems increase with the degree of overlap and with the number of bands present. The deconvolution is also very much more difficult when there is no baseline on either side of the spectral envelope. The UV spectral region of phthalocyanine complexes falls under a "worst case" scenario for fitting: overlap of bands is significant at room temperature, there are several bands present, the envelope starts near zero absorbance at 500 nm, but in the 220- to 250-nm region, there is an intense absorbance related to the fused benzo groups.

In order to improve the quality of the results obtained from the deconvolution technique, we have attempted to fit absorption and MCD data that were measured in DCM, a solvent which extends the usable UV cutoff to 250 nm. We use the same number of bands for both absorption and MCD spectra, which are described by the same band energy and band halfwidth parameters. Despite the inclusion of the additional MCD spectral data, the deconvolution calculations are very time consuming to carry out. We have

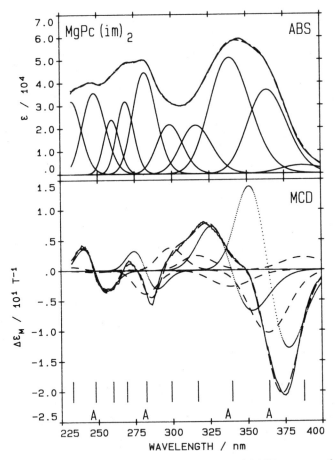

Figure 17 Deconvolution results for the absorption and MCD spectra of (imid)$_2$MgPc recorded in methylene chloride. The vertical bars marked with A indicate the positions of A terms that arise from degenerate transitions. Reproduced with permission from [9] (Fig. 4).

made the choice to limit the band shape to a simple Gaussian lineshape (and derivative, for MCD A terms) [156]. While a variety of band shapes could be tested, and skewed halfwidths could be used, we felt that in view of the complexity of the envelope itself, it would be safer to accept understandable mismatches in the fitting, rather than have the computer program alter, for example, the skewness of the band to fill a hole. We also set up the parameters for the initial best guess so that the program puts bands where we see inflections in the spectra, in other words, where we would instinctively place bands by eye.

    Do the deconvolution calculations provide a reliable means of obtaining the number of bands? We think that use of the MCD spectrum does help considerably, but up to now we have concentrated only on finding transitions

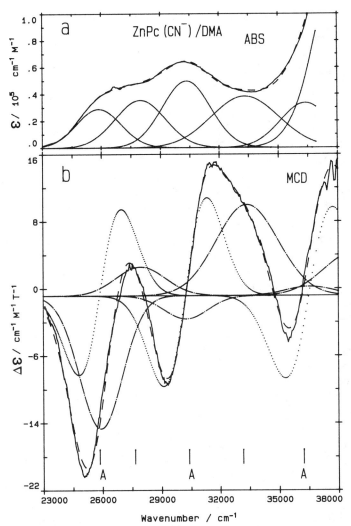

Figure 18 Band analysis of the UV region of the absorption and MCD spectra of (CN⁻)ZnPc in DMA. The recorded spectra are indicated by the solid lines, fitted spectra by the dashed lines, and individual A term contributions by the dotted lines. The vertical bars show the positions of the band centers used for both absorption and MCD components; the A indicates the use of an A term required to fit a degenerate transition. Reproduced with permission from [65] (Fig. 4).

to degenerate states, which are identified by MCD A terms. Do the deconvolution calculations provide good estimates for band centers? Spectral data for ZnPc [65] and MgPc [9], independently analyzed in our laboratory, agree remarkably closely (Figs. 17 and 18). Recently, Schatz' group at the University of Virginia in Charlottesville [59] has obtained absorption and MCD data for ZnPc matrix isolated in an argon matrix. Their fitting, which was carried out with completely different software, identified a similar set of bands in the B region at much the same energies that we have found.

iv. Alignment between Theory and Experiment

The overall conclusions from these spectral deconvolution calculations are (1) that there are two bands in the "B" region and (2) the MCD spectrum is almost essential in limiting the freedom of the program in placing bands throughout the envelope.

It is clear that the theoretical calculations reported to date need revising to take into account the presence of a second band in the B region. From the theoretical papers reviewed above, we think that the problem lies in fully accounting for configuration interaction. However, when the state is predicted to arise from several transitions, as is the case for the CuPc calculations of Henriksson et al. [61], then the characterization of a band as a transition between just two orbitals becomes meaningless. We support the proposal, based on the recent spectral data for ZnPc [59,65] and MgPc [9], that the two bands be assigned as B_1 and B_2, so that the traditional order adopted by Gouterman's group be maintained.

In summary, for an "unperturbed" metallophthalocyanine, five degenerate excited states are accessible between 250 and 800 nm, named (using λ_{max} values for (imid)ZnPc [65]) L (246 nm), N (282 nm), B_2 (336 nm), B_1 (368 nm), and Q (671 nm). The energies of the B_1 and B_2 bands appear to be quite variable, whereas the N band is invariably located close to 280 nm. Additional bands lying near 500 nm in neutral complexes are usually identified as charge transfer, although ring oxidation to the cation radical also results in absorption at 500 nm [65]. Weak bands to the red of the Q band near 900 nm may be trip–multiplet absorption [10], while other weak bands are probably also charge transfer in origin.

v. Peripheral Substitution

Substitution at the benzo group periphery has been a popular method by which to impart water or alcohol solubility. In addition, substitution with strongly electron-withdrawing (for example, as octacyano-Pc [177–179,348]), or strongly electron-donating (for example, octabutoxy- [180]), perturbs the excited state energies, as well as the redox properties. Examples of spectra reported from a range of peripheral substituents include tetraaminophthalocyanines [181], tetracarboxyphthalocyanines (MTcPc) [138], tetraalkyphthalocyanines [182–184], tetrasulfonated (MTSPc) [46,102,185], tetraquinone- [186], tetra(15-crown-5-ether)phthalocyanine, MTCRPc [187], and, perhaps the most exotic, tetra(tetraphenylporphyrin)phthalocyanine [186]. Peripheral substitution can also lead to modified electrochemical properties, for example Cu(CN)$_8$Pc [177], or new chemistry, for example, with halogenated peripheral positions [188]. Photochemical sensitization of the oxidation of L-tryptophan with MTSPc complexes using (Cl)AlTSPc and (Cl)GaTSPc is efficient when compared with hematoporphyrin [46,47]. Sensitization of the oxidation of EDTA in DMF in the presence of MV^{2+}, involving polymer-bound, peripherally substituted ZnPc and ZnTPP, has been reported following the reaction of tetrakis(4-hydroxyphenoxy)Pc and Zn(tetrakis(4-aminophenyl)porphyrin), with a linear styrene [189]. The absorption spectrum of the substituted polymer (compound 6) resembled the simple addition of the spectra of the MPc and MTPP [189]. Similar spectra were obtained when H$_2$-tetratolylporphyrin (TTP) was linked to Zn tetra-*t*-butylphthalocyanine through a bridging mesophenoxy group on the H$_2$TTP, which indicated that there is a lack of strong intramolecular coupling between the two rings [190].

For many peripherally substituted complexes, aggregation occurs readily at the low concentrations typically used to record absorption spectra, making the spectral data more difficult to interpret. In these cases, quite different absorption spectra are often observed in water, when compared with a solvent like DMF, which usually indicates the presence of extensive aggregation [185,191,192,193,194]. Dimers can often be dissociated by the addition of 3–10% DMF, for example, see the spectra reported by Harriman and Richoux for Zn-tetrasulfophthalocyanine in water at pH 7 [194]. The dimerization reaction, 2 MTSPc ↔ [MTSPc]$_2$, behaves as expected with respect to temperature: as the temperature is raised so monomers predominate, for example, between 20 and 85°C, the Q band of Co(II)TSPc red shifts to the λ_{max} of 664 nm of the monomer in water [191]. Monahan, et al. [183] also demonstrate clearly how the Q band broadens with association for the water soluble Cu, Zn, and VO tetraoctadecylsulfonamidophthalocyanines, when dissolved in benzene.

Because derivatives of the tetrasulfonated phthalocyanines (TSPc) are generally soluble in aqueous solutions, a considerable body of spectral data has been reported for a wide range of molecules; examples include H$_2$ [185,192], Mn [195,196], Fe [197–199], Co [102,191–193,197–199], Cu [59,192,200,201], Zn [192,194], and Pd [202]. Photochemical (Co and Fe [198], Cu and Co [201], Zn [194], Zn [194], Zn and Pd [202]) binding toward albumin (Co and Fe [199]), and electrochemical (Co [102,197,203]), properties have also been reported. In many cases, the spectra in the Q band region show the effects of aggregation, typically a blue shift of the maximum absorbance from 670 toward 620 nm and an overall broadening of the Q envelope with loss of resolution of the vibrational components (see the spectral data in [185,200]).

Kobayashi and co-workers [186] have reported the absorption and emission spectra of H$_2$(anthraquinone)$_4$Pc and H$_2$(tetraphenyporphyrin)$_4$Pc. The spectral data of the anthraquinone-substituted complex do not exhibit the usually split Q$_x$/Q$_y$ spectrum; in its place a very broad envelope is found that spans the 580- to 780-nm region [186]. It is probable that this spectrum results from dimerization, as a result of the large anthraquinone groups stacking in solution.

Sielcken et al. [105] (Fig. 19) and Kobayashi and Lever [187] (Fig. 20) have reported spectral data for phthalocyanines which have crown ether subunits attached to the peripheral benzo groups. For H_2CR_4Pc and $CuCR_4Pc$, the Q band spectra resemble monomeric species in chloroform, but association clearly takes place in methanol [105]. As is often observed for H_2Pc species, association leads to a loss in the geometrically induced splitting between the Q_x and Q_y components [187]. For $CuCR_4Pc$ [105], and also the Co, Ni, and Zn derivatives [187], the Q band near 675 nm is replaced by a weaker and broader band centered near 630 nm. Kobayashi and Lever [187] also report on the cation-induced association of MTCRPc in chloroform; addition of K^+ leads to dimerization [214].

While the lack of solubility of phthalocyanines has been noted many times for the phthalocyanine complexes in general, several phthalocyanines readily dissolve in many spectroscopically and electrochemically useful solvents; the list includes Li_2Pc and $MgPc$ [9,26]. Attempts to increase solubility by adding peripheral alkyl groups to MnPc [204] resulted in a compound with reduced solubility of the monomer. Octacyano phthalocyanines tend to aggregate, so that spectral data [177] do not resemble the monomeric parent in dilute solution.

vi. Polynuclear, Covalently Linked Polymers

In these complexes, covalent linkages between groups attached to the rings of two or more phthalocyanine rings are made. When the rings interact electronically, the absorption, MCD, and emission spectra show distinct evidence of dimerization. In the absorption spectrum, the effects observed for the MTSPc derivatives are also found. Aggregation leads, in the visible region, to blue shifts and broadening for the Q band, that results from the exciton split excited states [102,205]. Lever and Leznoff and co-workers have reported preparation of "clamshell-like" polynuclear phthalocyanines which are built from units of [tetra(neopentoxy)phthalocyanine]cobalt(II) (CoN_4Pc) [99,101,206]. CoN_4Pc aggregates above 10^{-4} M concentrations in DCB solution [206]. From analysis of the Q band spectral data, Nevin et al. [206] calculate K_D for the reaction $2MPc \leftrightarrow (MPc)_2$ as 2.57×10^3 M^{-1}. Absorption spectra that were calculated from spectral data obtained for complexes similar to the CoN_4Pc complex illustrate how the dimer spectrum is dominated by a splitting in the Q band maxima. We should note, however, that the spectral data for the monomers reported in this work [206] are not in themselves typical of monomeric, unsubstituted CoPc complexes, so some care has to be taken in extracting purely monomeric and dimeric spectral envelopes itera-

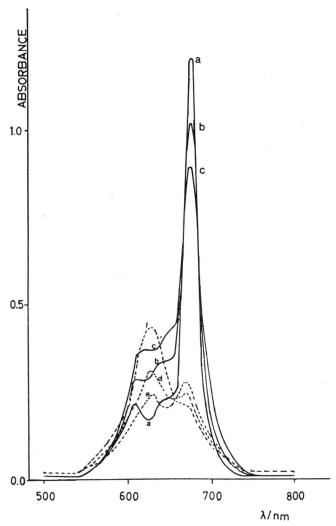

Figure 19 Absorption spectra of CuPc (with peripheral substitution by four 18-crown-6 rings). Dimerization of the complex occurs to different extents depending on solvent, ranging from monomeric in (a) CHCl$_3$, to (b) CH$_2$Cl$_2$, (c) pyridine, (d) ethanol, (e) 1-butanol, and (f) methanol, where the complex is increasingly aggregated. Reproduced with permission from [105] (Fig. 3).

tively in this fashion. For example, while the absorption spectrum of the mononuclear Co(II) N$_4$Pc prepared by Lever's group [102,207] resembles that of monomeric Co(II)Pc's in solution [3], the binuclear- [207] and tetranuclear-phthalocyanines [99] (peripherally substituted, with chains that link two phthalocyanine rings together, see Chapter 1, this volume), exhibit significant dimerization even at concentrations as low as 3.8×10^{-6} M. For the

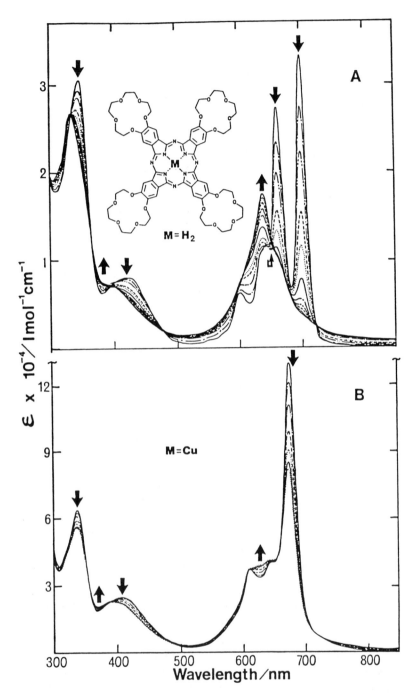

Figure 20 Absorption spectra of H_2Pc (A) and CuPc (B) (each with peripheral substitution by four 15-crown-5 rings) in $CHCl_3$. Dimerization of the complex occurs to different extents depending on the cation concentration. The arrows indicate the effect of the addition of aliquots of (A) CH_3COOK or (B) CH_3COONa, to the cuvettes. Reproduced with permission from [187] (Fig. 1).

tetranuclear-phthalocyanine species, the spectral envelope of the dimer (λ_{max} 625 and 676 nm) is replaced at lower concentrations by a red-shifted maximum at 676 nm. (In these complexes, exciton splitting to give a band at 625 nm can take place both intermolecularly for dimeric and intramolecularly for the monomer in which attached rings couple together [99].) Figure 21 shows the spectral data reported for a series of mononuclear and binuclear neopentoxyphthalocyanines [208].

Wöhrle et al. [349,350] have reported the preparation of a wide range of polymeric phthalocyanines, where the polymeric chain links phthalocyanine rings through substitution at the benzo groups. The spectra of thin films of these polymers resemble the β polymorphic form rather than simple dimers. Water soluble phthalocyanines bound to a linear polymer, which is based on poly(methacrylic acid), exhibit dimer spectra in water, but monomeric spectral properties at high DMF:water ratios [351].

Finally among these examples of polynuclear complexes, Gaspard et al. [190] have prepared a polynuclear molecule comprising ZnPc covalently linked through the benzo groups to ZnTPP. The spectrum of this molecule resembled an admixture of the individual spectra of the ZnPc and ZnTPP, which suggest little electronic mixing, although quenching of the porphyrin emission indicates that electron transfer does take place between the two rings [190].

vii. Interpretation of the Spectral Effects of Dimerization

Dimerization of symmetrical porphyrins and phthalocyanines results in spectral effects that extend from band broadening, to a blue shift of the Q and B bands by some 500–2,000 cm^{-1}, to an observed splitting of the Q band (and sometimes also observed for the B band) by between 500 and 3,000 cm^{-1}. The extent of the spectral effects on association depends on the closeness of approach of the rings, the overlap position, the tilt angle that the rings adopt, the bulkiness of the peripheral groups, and the extinction coefficients of the electronic bands involved. Gouterman et al. [209] and Zgierski [210,211] have dealt in some detail with the case of dimeric porphyrins, and Hush and Woolsey [8] have described calculations for dimeric SiPc, while Dodsworth et al. [208] have described the case of dimeric free base phthalocyanines in solution.

The strongest perturbations in the optical spectra are found in the solid state, where the α, β, and χ polymorphic forms obtained for thin films on quartz disks (discussed below in detail) exhibit spectra between 250 and 800 nm, which resemble only very slightly their solution analogues. The band broadening that accompanies the exciton (or Davydov coupling) effects changes the visible region spectrum quite dramatically. These effects, and the different experimental techniques used in characterizing the aggregation, have been discussed by many authors (see, as examples [52,53,212,213]).

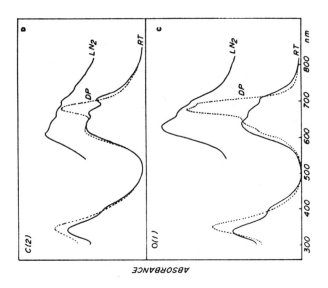

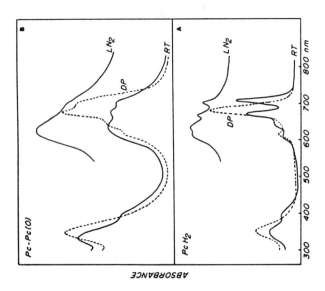

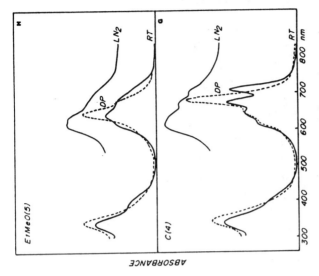

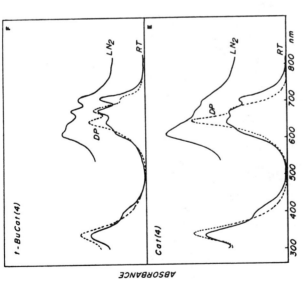

Figure 21 Absorption spectra of a range of mononuclear and binuclear neopentoxyphthalocyanine species at room temperature (RT) in toluene/ethanol (3:2 v:v) solutions, or glass at 77 K (LN$_2$) in the same solvent mixture. DP indicates a deprotonated species obtained by addition of 0.1 ml tetrabutylammonium hydroxide in methanol. In these compounds, two phthalocyanine rings are linked via a benzene ring in position 3 or 4; the number of parentheses is the number of atoms in the bridge link. (A) Free base tetraneopentoxyphthalocyanine, PcH$_2$; (B) Pc—Pc(0), phthalocyanines rings directly linked; (C) O(1), phthalocyanine rings linked via an oxygen atom; (D) C(2), phthalocyanine rings linked via —CH$_2$—CH$_2$—; (E) Cat(4), phthalocyanine rings linked via 4-t-butyl-o-catecholate; (F) t-BuCat(4), phthalocyanine rings linked via 4-t-butyl-o-catecholate; (G) C(4), phthalocyanine rings linked via —(CH$_2$)$_4$—; (H) EtMeO(5), phthalocyanine rings linked via —OCH$_2$C(Me)(Et)CH$_2$O—. Reproduced with permission from [208] (Fig. 1).

It is of interest to include discussions of both porphyrins and phthalo-cyanines, as the results for the two systems are rather different. Gouterman et al. [209] discuss in detail the effects of dimerization on the absorption and fluorescence data of [Sc(III)OEP$_2$]O and [Sc(III)TPP]$_2$O, in terms of an exciton coupling between pairs of degenerate $\pi-\pi^*$ excited states of the porphyrin rings. They show that the nature of the coupling depends on the geometry of the dimer, in particular on the angle between the two rings. For an overall D_{4h} symmetry, where the two rings lie parallel, they find that the degenerate pairs of states (the Q and B) each split into two further pairs of degenerate states, the upper state is of E_u and the lower state of E_g symmetry. They report the presence of a strong band, assigned as B$^+$, blue shifted from the monomer B position for the OEP dimer, and very weak absorption on the red edge of the B band, tentatively assigned as B$^-$. The exciton splitting [half the energy of $E(B^+) - E(B^-)$] was estimated from both absorption and fluorescence data as between 1,400 and 2,090 cm^{-1} [209]. They observe that the exciton splitting is dependent on the extinction coefficient of the monomeric band. Thus, the exciton splitting of the Q band in both the OEP and TPP complexes was almost insignificant [209], mainly because ϵ_Q for OEP and TPP complexes is very low, when compared with the splittings reported for many phthalo-cyanine dimers, where ϵ_Q for MPc is high.

Dodsworth et al. [208] have similarly described the absorption and emission spectra observed for dimeric binuclear metal-free neopentoxy-phthalocyanine species. Unlike the situation for the porphyrins, dimeric phthalocyanines usually exhibit a significant blue shift in the Q band region, together with some band broadening, but much less splitting in the B band region. Four points should be noted when one considers the phthalocyanines in comparison with the porphyrins. First, the geometrically induced splitting of the Q and B bands in H$_2$Pc is much less pronounced in the B band than in the Q band. Second, the extinction coefficients of the bands under the B envelope are very much less than those of typical Q bands. Third, the Q band is well isolated from other electronic transitions so that any splittings are readily visible. In the B band envelope, at least two electronic transitions are overlapped in the monomer, and the band widths are significantly greater than the Q band so that band duplication is harder to quantify. Finally, fourth, the Q band arises from an $a_{1u} \rightarrow e_g$ transition, which is the same as the B band in porphyrins.

Figure 22, which is based on the analysis by Gouterman et al. [209], Hush and Woolsey [8], and Dodsworth et al. [208], shows how two coplanar D_{4h} metallophthalocyanine rings interact. If additional vibronic interactions are ignored, it can be predicted [208,209] that four degenerate states will be formed. Transitions are formally allowed only to the upper pairs of states. Transitions to the lower energy pairs of states are forbidden. The spectral properties expected following dimerization will include a blue shift of the Q band and B bands. It is predicted from this picture that the exciton split

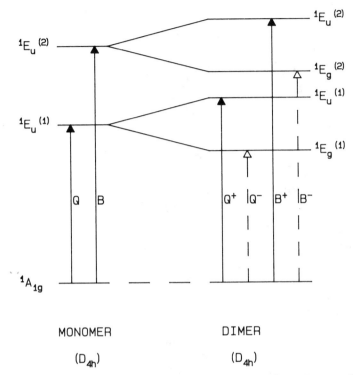

MONOMER   DIMER

(D_{4h}) (D_{4h})

Figure 22 Interaction between molecular orbitals on two coplanar D_{4h} metallophthalocyanine rings. The diagram is taken from [214], which was based on the analysis by Gouterman et al. [209], Hush and Woolsey [8], and Dodsworth et al. [208].
The full arrows indicate allowed transitions. The dashed lines indicate forbidden
transitions. Reproduced with permission from [214] (Fig. 10).

components Q^+/Q^-, and B^+/B^-, will exhibit A terms in the MCD spectrum
(and this is observed [214]). For free base phthalocyanines [208], the
degeneracy is lifted and the Q_x, Q_y, B_x, and B_y components will couple to form
pairs of Q_x^+/Q_x^-, etc., with the upper states being accessible optically.

From the large number of absorption, emission, and MCD spectra
reported for dimeric phthalocyanine complexes, we find that dimerization in
phthalocyanines is indeed characterized by a blue shift of both the Q and B
bands [8,187,215]. Considerable band broadening is often observed
[177,216], which is especially noticeable in the Q band of solid state spectra
[52,53,101,212,217]. The extent of the broadening is also related to the
degree of coupling between the two rings; see, for example, the Q band half
width data for binuclear CoPc complexes [101]. Enhanced absorption
observed on the red side of the Q band suggests that the forbidden Q^- band
may still exhibit some (perhaps 5–10%) of the extinction of the allowed Q^+
band [8]. The exciton splittings in the Q band region of phthalocyanines

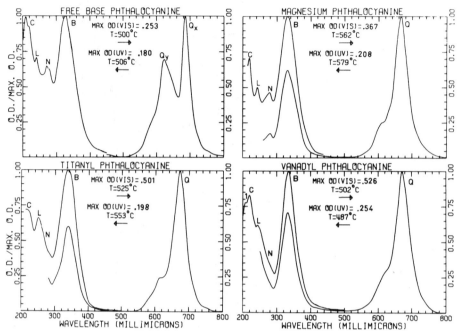

Figure 23 Vapor-phase absorption spectra of H_2Pc, TiOPc, MgPc, and VOPc. Temperature and maximum OD (corrected to 1-cm pathlength) for two runs are indicated. Reproduced with permission from [57] (Fig. 1).

[8,183] are comparable to those reported for the OEP dimers by Gouterman et al. [209]. Exciton splittings of 1,100–1,500 cm^{-1} have been reported by Monahan et al. [183] for copper, zinc, and vanadyl tetraoctadecylsulfon-amidophthalocyanines in benzene. Further spectroscopic evidence for association is found in emission quenching that also occurs with aggregation [102,186,194].

D. SPECTRAL PROPERTIES OF SOLUTIONS (MAIN GROUPS)

i. Absorption Spectra of H_2Pc

Spectra of H_2Pc have been reported from the vapor phase (Fig. 23) [56,57], from the solid phase for single crystals [218], and thin α-, β-, and χ-polymorphic phase films on quartz (see Fig. 24 for the spectrum of α-H_2Pc and Fig. 25 for χ-H_2Pc [26,53]), from matrix isolated H_2Pc [58,59,141,219], from solutions in aromatic solvents like α-chloronaphthalene (Fig. 4) [76], and pyridine [76], as well as from solutions in UV-transmitting solvents like DMA

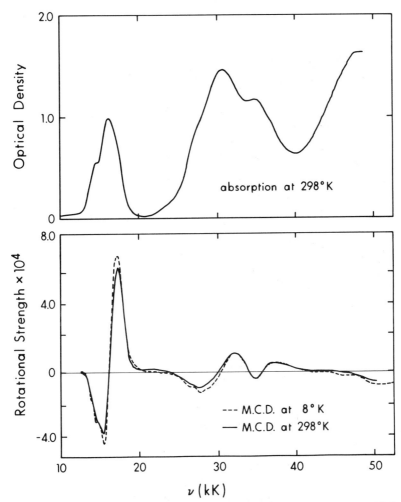

Figure 24 Absorption and MCD spectra of α-H$_2$Pc. The spectra were recorded at room temperature (solid line) and 8 K (dashed line for MCD) from α-H$_2$Pc sublimed to form a thin film on a quartz disk. The inset shows the Q band region absorption spectrum for H$_2$Pc in α-chloronaphthalene at room temperature. The ordinate units are absorbance and θ_m/tesla. Reproduced with permission from [52] (Fig. 1).

(Fig. 5) [80]. In strongly basic solvents, the pyrrole protons are acidic enough to dissociate, thus the spectra are of the dianionic, symmetric, Pc(-2), species, in which a single Q band is observed.

Spectral data have also been reported from the peripherally substituted H$_2$Pc's, for example H$_2$-tetra-t-butyl-Pc in CH$_2$Cl$_2$ [182], and in solution with surfactant vesicles [182], and H$_2$(CN)$_8$Pc [348], where the eight electron-withdrawing groups quite significantly alter the redox potentials of the ring [177].

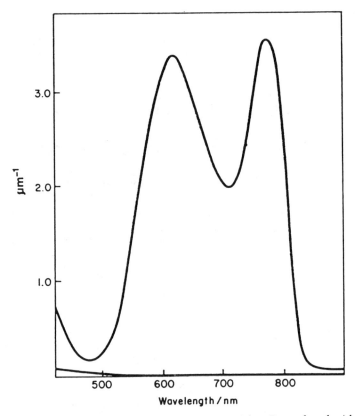

Figure 25 Absorption spectrum of 60-nm χ-H₂Pc particles. Reproduced with permission from [331] (Fig. 4).

With the exception of H_2Pc dissolved in basic solvents, or laid down as thin films [26,53], a split Q band is always clearly observed (Figs. 4, 5, and 23). The two Q components of the split band are polarized in the x and y directions, the MCD spectra clearly identifying the absence of degenerate transitions.

Full-wavelength range absorption spectra of H_2Pc in solution have proven difficult to obtain because of insolubility in UV-transparent solvents. In chlorinated aromatic solvents, for example, α-chloronaphthalene, the Q band splits into two components [220], each oppositely polarized. Low-temperature measurements on matrix isolated H_2Pc [141,219] reveal the presence of many vibrational components on the high-energy edges of Q_x and Q_y. These spectra quite closely resemble those of chlorin and chlorophyll a [15,17,118]. H_2Pc can be formed by demetallation of Li_2Pc in DMA and DMSO [80]; in this case spectra that extend from 300 to 800 nm can be obtained, which exhibit well-resolved split Q_x and Q_y components. While the absorption spectrum in the 350-nm region looks like that of the Pc(-2) anion

[80], the MCD spectrum shows that the D_{2h} symmetry breaks the degeneracy of the states responsible for the B bands, and only weak, poorly resolved MCD bands are observed.

The Q band spectrum of H_2Pc exhibits a strong pH dependence [220]. Spectra of the anion, Pc(-2), measured in DMSO, and stabilized by NH_4^+, exhibits a regular D_{4h} spectrum, with λ_{max} at 669 nm associated with a strong A term, and two well-resolved bands in the B region are also associated with A terms. Similarly, H_2Pc in pyridine also exhibits the D_{4h} spectrum, with its single Q band, that is characteristic of the dianion. The presence of the two bands in the 370-nm region, both x/y polarized, has been used by our group to suggest that unlike the porphyrins, there are two main π^* states, both degenerate, under the B or Soret band envelope. As can be seen below, the state to lower energy is very sensitive to the axial ligand.

ii. H_2Pc: Peripheral Substitution

Bernauer and Fallab have described the preparation of the tetrasulfonated species, H_2TSPc (and also the Co, Cu and Zn derivatives) [192], and also the equilibrium and kinetic properties of the dimerization to give $[H_2TSPc]_2$ [185]. Whereas in alcohol, each of these complexes exhibited typical monomeric spectra in the Q region, in water, the Q region of each species broadens, and the maximum absorbance is found blue-shifted from the Q band of the monomeric species measured in alcohol. Log K_D is reported as 7 for H_2TSPc, 5.3 for CoTSPc, 7.2 for CuTSPc, and 6 for ZnTSPc, when measured in water at temperatures near 60°C. These authors also reported on the deprotonation reaction of H_2TSPc to form the $HTSPc^-$ monomer, in the presence of base. $HTSPc^-$ is characterized by a single Q band, with λ_{max} at 680 nm, and $\epsilon = 2.4 \times 10^5$. For H_2TSPc, the dimerization coefficient, $K_D = 9.3 \times 10^7$ at 20°C, and $pK_a = 9.6$. Rollmann and Iwamoto [221] report from the absorption spectra that monomeric H_2TSPc exists in DMSO, DMF, and MeOH, but dimers exist in water.

Kobayashi and co-workers [187,214], have studied the dimerization reaction of the 15-crown-5 (R) peripherally substituted H_2R_4Pc. The absorption and MCD spectra of the monomeric species in $CHCl_3$ [187,214] provide the clearest, and widest-available wavelength range, spectral data to date for room temperature solutions of the H_2Pc molecule (Fig. 7). In the Q band, the symmetry-split Q_x and Q_y bands at 700 and 660 nm, respectively, are associated with the expected B terms of opposite sign. In the B band region, we find broad absorption between 350 and 450 nm, followed by two well-resolved bands at 345 and 300 nm. The MCD spectrum under the 345-nm band provides the first indication that the B band is split into the B_x and B_y components; the negative B term near 350 nm is skewed away from the crossover point at 345 nm. The origin of the 350–450 nm absorption is unclear, because we find a similar envelope of bands for CoTCRPc,

CuTCRPc, NiTCRPc, and ZnTCRPc [214]. Either the peripheral groups perturb the π system such that transitions which normally exhibit a very low extinction coefficient become allowed, for example $n \to \pi^*$, or this absorption is related to the 15-crown-5/benzo attachment itself. Similar broad absorption is not observed for unsubstituted phthalocyanines. The deconvolution of these spectra is very complicated, requiring 21 bands to fill both absorption and MCD envelopes [214].

Addition of CH_3COOK to the solution of H_2TCRPc in $CHCl_3$ promotes dimerization [187]. As with other free base dimers, we note the lack of the symmetry splitting in the blue-shifted, Q band (now at 640 nm) and the similarity between monomeric and dimeric spectral envelopes in the B region [214].

The octasubstituted species, H_2—$(CN)_8$—Pc, is aggregated in DMF solutions [177], so does not exhibit the normal two-banded Q band either, the spectrum resembling more closely the β-polymorphic phase.

Binuclear H_2Pcs described by Marcuccio et al. [222], and by Dodsworth et al. [208], in which two rings are covalently linked through bridges attached to benzo groups, exhibit highly red-shifted absorption spectra in the visible region (e.g., compound 7a [222], shows bands at 708 (log ϵ = 4.88), 676 (4.92), 642 (4.84), 620 (4.75), 388 (4.75), and 336 nm (4.98), while a substituted monomer (compound 19) shows bands at 698 (log ϵ = 5.11), 664 (5.05), 604 (4.45), 342 (4.88), and 292 nm (4.76)). Dodsworth et al. [208] interpret these spectral effects in terms of coupling, as described in Section C, vii.

iii. Group 1: LiPc, NaPc, and KPc

These alkali metals are considered to form M_2Pc complexes. The Li_2Pc and Na_2Pc complexes are very soluble and can act as precursors in metallation reactions [223] from which a variety of phthalocyanine complexes can be made quite easily. Both Li_2Pc, and several derivatives, for example, LiHPc and [(tri-n-dodecyl-butyl)ammonium]LiPc (TDBA)LiPc, have been prepared and studied spectroscopically [26,223–226]. Figure 26 shows the spectral data for LiHPc [226]. The absorption spectrum of LiHPc in pyridine indicates high symmetry, because the Q band at 667 nm is well resolved and symmetric in nature, which underscores the unusual nature of the alkali metal phthalocyanines. Presumably this is a spectrum of the dianion. The spectrum measured at 10 K from LiHPc in a KBr pellet resembles spectra measured from thin films, rather than a diluted solution. Homborg and Katz [225] have reported the absorption spectrum of (tri-n-dodecyl-butyl)ammonium lithium phthalocyanine, (TDBA)LiPc, in CH_2Cl_2, as a thin film at 10 K, and in a KBr pellet also at 10 K [225]. In the CH_2Cl_2 solution, the spectrum gave λ_{max} = 240, 267, 330, 380, 629, 636, and 667 nm. The presence of both 330- and

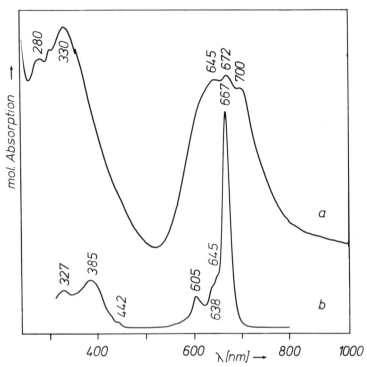

Figure 26 Absorption spectrum of LiHPc(-2). (a) In a KBr pellet at 10 K; (b) dissolved in pyridine. Reproduced with permission from [226] (Fig. 1).

380-nm bands suggests that the B_1/B_2 splitting may be observed for this complex.

Homborg and Kalz have also reported absorption spectra for AgLiPc(-2) and CuLiPc(-2) [227], with Q band maxima at 672 and 678 nm, respectively. The Ag(I)LiPc(-2) complex exhibits additional bands at 435, 482, and 812 nm, which are typical of an oxidized ring, see for example the spectrum of [ZnPc(-1)]$^+$ [27,92].

X-ray data for K_2Pc show that the K atoms are located above and below the phthalocyanine plane [25,228], with a surprising degree of flexibility demonstrated by the ring in the solid state. The absorption and MCD spectra of Li_2Pc [26] with CN$^-$ axial ligands are good examples of unperturbed Pc(-2) $\pi-\pi^*$ spectra (Fig. 27) and provide strong evidence for a D_{4h} symmetry in DMSO and DMA solutions. As has been found for several other phthalocyanines [3,65], the B band splits apart under the influence of strong σ donors to reveal two bands lying between 245 and 470 nm. The appearance of similar, positive A terms under each of these bands indicates [65] that there are two degenerate states that are sufficiently close in energy that they are not generally observed with the unligated M_2Pc species (or in the vapor phase).

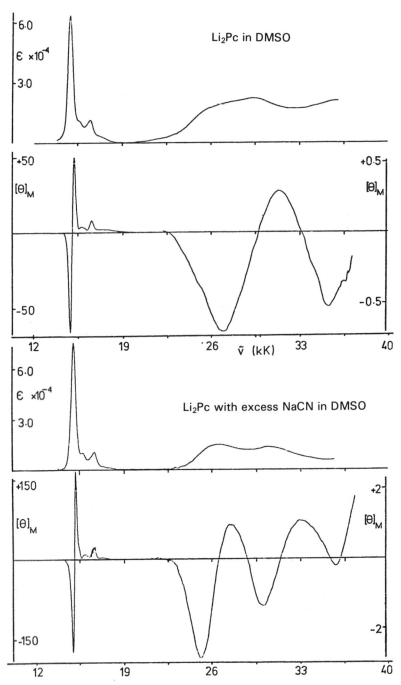

Figure 27 Absorption and MCD spectra of $Li_2Pc(-2)$ dissolved in DMSO; (top) without additional ligands; (bottom) in the presence of excess NaCN. The ordinate units are ϵ and $[\theta]_m$. Reproduced with permission from [26] (Fig. 2).

iv. Group 2: MgPc

MgPc is very soluble in a wide range of solvents (for example, CH_2Cl_2 [9,229] and THF [230]). The absorption spectrum of $(H_2O)_2MgPc(-2)$ in THF [230] exhibits a series of bands typical [9,26,65] of an unperturbed phthalocyanine π system: a narrow, highly resolved Q band at 670 nm, with a broad, poorly resolved B band (345 nm), and then at higher energies, two weak bands, at 282 and 266 nm. Vapor phase absorption spectra (Fig. 23) [56,57] are broad, but show [57] four resolved bands below 400 nm, assigned by Gouterman and co-workers at B, N, L, and C.

The first MCD spectra of MgPc were reported by Linder et al. [231] in a study of the spectral properties of the reduced $[MgPc(-3)]^-$. More extensive studies of the absorption and MCD spectra for MgPc in the presence of several axial ligands (Figs. 1 and 28) [6,9] illustrate a behavior similar to that of Li_2Pc [26] and ZnPc [6,65,92]. The sequence of bands for $(imid)_2MgPc$ in CH_2Cl_2 (measured from the spectra rather than with fitted bands) is 281, 345, 607, 643, and 672 nm. Figure 17 shows the results of deconvolution calculations that fitted both the absorption and MCD spectra with the same set of bands for the imidazole complex [9,156]. These calculations identified a number of degenerate states that we now associate with the L, N, B_2, B_1, and Q states [21,63], with band centers of 248, 282, 339, 364, and 672 nm. Degenerate transitions are found at 339 and 364 nm, bands which are now assigned as B_1 and B_2; see also the analysis of ZnPc discussed above [65].

X-Ray data for monohydrated dipyridinated MgPc [71] indicated that the Mg atom was up to 49.6 pm out of the plane of the ring, with Mg—N bond lengths of ca. 204 pm. This loss of planarity should be observed in the MCD spectrum in the Q band region by a broadening of the Q band width, and a reduction in the magnetic moment of the excited state, as measured by the A/D value of the A term as the degeneracy of Q_x/Q_y components is reduced. These values for the excited state magnetic moment [9] are comparable with those reported for ZnPc [65].

v. MgR$_n$Pc: Peripheral Substitution

Absorption and fluorescence data have been reported for Mg-tetra-*t*-butylphthalocyanine dissolved in a wide range of solvents, including, pyridine, *n*-butanol, isopropanol, and ethanol [184]; Q λ_{max} is located near 672 nm.

vi. Group 13: AlPc

Homborg and Murray [229] report the preparation and spectra of $[(X)_2AlPc(-2)]^-$ species (where X = F^-, Br^-, Cl^-, and I^-) in CH_2Cl_2 and as thin films on quartz (Fig. 29). The $[(F)_2AlPc]^-$ spectrum in CH_2Cl_2 closely resembles those of ZnPc and MgPc, except that the Q band maximum is blue shifted from the more usual 670 to 662 nm.

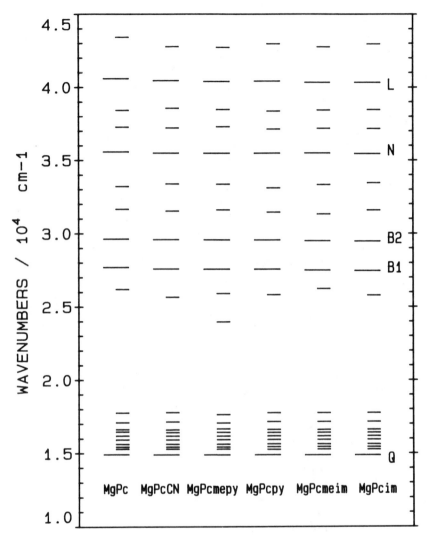

Figure 28 Comparison of the band energies determined by deconvolution calculations for a range of MgPc complexes illustrating the presence of two degenerate transitions in the 320-nm region. Reproduced with permission from [9] (Fig. 10).

vii. AlR*n*Pc: Peripheral Substitution

Freyer et al. [184] report absorption and fluorescence spectra for (Cl)Al-tetra-*t*-butylphthalocyanine in a range of solvents, for isopropanol, Q λ_{max} = 674 nm. Homborg and Murray [229] have reported the spectrum in CH_2Cl_2, of the μ-oxo dimer, $[((F)AlPc(-2))_2O]^{2-}$, formed by the hydrolysis of suspensions of $[(F)AlPc(-2)]$ (Fig. 29). Compared with the spectrum of the

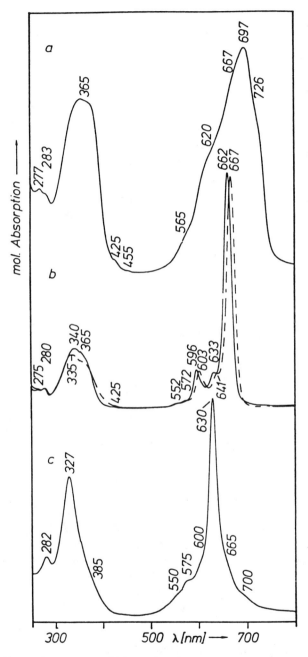

Figure 29 Absorption spectra of (a) tetrabutylammonium[(F$^-$)$_2$AlPc($-$2)]$^-$ as a film on a quartz slide at 10 K; (b) a solution of [(F$^-$)$_2$AlPc($-$2)]$^-$ (solid line) and [(F$^-$)ZnPc($-$2)]$^-$ (dashed line) in CH$_2$Cl$_2$; (c) a solution of the dimeric species [((F$^-$)AlPc($-$2))$_2$O]$^{2-}$ in CH$_2$Cl$_2$. Reproduced with permission from [229] (Fig. 1).

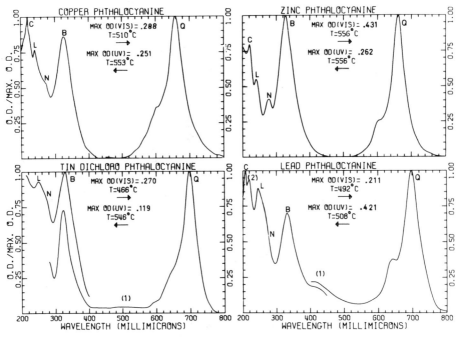

Figure 30 Vapor-phase absorption spectra of CuPc, (Cl)$_2$SnPc, ZnPc, and PbPc. Temperature and maximum OD (corrected to 1-cm pathlength) for two runs are indicated. Reproduced with permission from [57] (Fig. 3).

monomer [229], we find three significant new features. (1) The Q band is blue shifted by 30 nm (662–630 nm), and (2) a very weak shoulder at 700 nm, that could be due to the low energy, forbidden component of the exciton coupling between the two rings, is seen. (Alternatively this 700-nm band can arise from demetallation [224], a very common and characteristic feature of phthalocyanine chemistry.) Finally, (3) there is a single, resolved band at 327 nm.

Thin films of surfactant–(C$_{16}$H$_{33}$)AlPc exhibit photoactivity toward a range of redox agents. The action spectrum followed the absorption spectra in the Q band region, which is characteristic of an aggregated species [232]. The photoactive efficiencies of (Cl)AlPc, and the water-soluble derivative, chloro-aluminum phthalocyanine sulfonate, (Cl)AlTSPc, have been compared in promoting cultured cell destruction following illumination with white light [48]. The Q bands of both (Cl)AlTSPc(Cl) and (Cl)AlPc lie near 677 nm [48].

viii. Group 14: SiPc, GePc, SnPc, PbPc

The spectral properties of Ge, Sn, and Pb phthalocyanines have been reviewed by Sayer et al. [31]. The spectra of monomeric forms of these complexes are essentially the same as for ZnPc, with the exception of new bands being found in the 450-nm region. Figure 30 shows vapor phase spectra

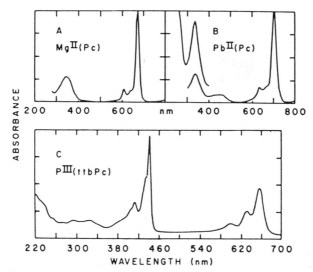

Figure 31 Absorption spectra of (A) MgPc in DMSO, (B) Pb(II)Pc in DMSO, and (C) P(III)tetra-*t*-butyl-Pc in ethanol. Reproduced with permission from [31] (Fig. 5).

for $(Cl)_2SnPc$ and PbPc, while Fig. 31 shows absorption spectra for DMSO solutions of Pb(II)Pc. The Q band in the M(II)Pc complexes red shifts for Ge(II) (655 nm) to Sn(II) (682 nm) to Pb(II) (ca. 702 nm, in DMSO). Sayer et al. [31] describe the unusual B band envelope observed for Ge(II) as "hyper," to emphasize the extent of the red shift and the multiple band character of the envelope. For Ge(II) and Pb(II), bands are also observed in the 440-nm region.

A much more extensive literature exists for dimeric and polymeric SiPc and GePc complexes. The spectra of monomeric SiPc, for example, $(Cl)_2SiPc$ in pyridine [175], exhibit bands typical of ZnPc, with λ_{max} = 314, 367, and 699 nm, or for $(OH)_2SiPc$ in THF, λ_{max} = 318, 363, 377, and 667 nm [175]. Axial substitution by two $-OSi[C(CH_3)_3](CH_3)_2$ groups (R) also yields a monomeric spectrum in benzene for $(R)_2SiPc$, with λ_{max} = 330, 353, and 668 nm [175].

Ciliberto et al. [175] and Hale et al. [64] have described results from Hartree–Fock–Slater calculations on the energy levels involved in the spectral properties of a range of derivatives of SiPc. Hale et al. [64] also report specific calculations for $Si—(CN)_8Pc$ and $Si—(F)_8Pc$. They find that octasubstitution with these strongly electron-withdrawing groups shifts both occupied and unoccupied orbitals to lower energy. The a_{1u} remains the HOMO and the e_g remains the LUMO [64].

ix. SiPc: Axial and Peripheral Substitution

The absorption spectra data for cofacially oriented complexes based on SiPc rings covalently linked through Si—O—Si repeat units, with capping

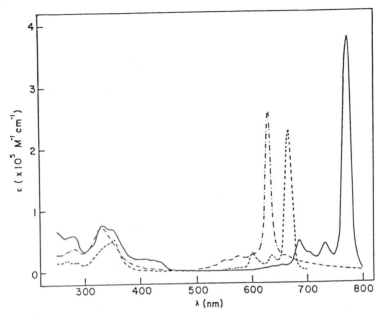

Figure 32 Absorption spectra of SiPc[OSi(n-C$_6$H$_{13}$)$_3$]$_2$ (----), (n-C$_6$H$_{13}$)$_3$SiO(SiPcO)$_2$-Si(n-C$_6$H$_{13}$)$_3$ (–·–·–), and SiNc[OSi(n-C$_6$H$_{13}$)$_3$]$_2$ (——) (----) in CH$_2$Cl$_2$. SiNc is silicon 2,3-naphthalocyanine. Reproduced with permission from [236] (Fig. 7).

groups involving alkyl groups (R), such as C$_6$H$_{13}$ [41], are complicated by the effects of dimerization and more extensive polymerization as the number of repeat units increases. Only a brief selection of the very large literature that describes spectral properties of the group 14 phthalocyanines is included in this review, for Si [8,40,41,175,215,233–237] and for Ge [40,215]. Citations in these papers are extensive, and a full search is not difficult to carry out.

Figure 32 shows spectral data for monomeric and dimeric units of SiPc. The dimers are connected through Si—O— links to the other SiPc molecule [236]. These data are particularly useful because the spectral range extends below 300 nm. These linked species exhibit the characteristic blue shift of the Q band of the SiPc following dimerization; see, as examples, the spectra in [175,236]. As the number of stacked rings increases, for example to three and four, so the Q band loses the typical MPc Q band character, blue shifting to near 600 nm, and a long tail to the red side grows in that could contain the "forbidden" Q$^-$ exciton component; see the spectra of [(n-C$_6$H$_{13}$)$_3$SiO]-(SiPcO)$_n$[Si(n-C$_6$H$_{13}$)$_3$], where n = 3 and 4 [41].

Absorption spectra of ((((CH$_3$)$_3$SiO)$_2$(CH$_3$)SiO)(PcSiO)$_{1-4}$ (Si(CH$_3$)(OSi)-(CH$_3$)$_3$)$_2$), and ((((CH$_3$)$_3$SiO)$_2$(CH$_3$)SiO)(PcGeO)$_{1-2}$ (Si(CH$_3$)(OSi)(CH$_3$)$_3$)$_2$), in cyclohexane (Si) or n-hexane (Ge) [215], also show that as the degree of oligomerization increased, so the Q and B bands blue shifted. For $[(R)_2SiPc]_n$, where R is a connecting Si—O—Si group, Q λ_{max} blue shifted from 665, to 630, to 618, to 615 nm, for oligomers with one to four stacked rings, respectively. The spectrum of the three-ring $[SiPcO]_3$ complex exhibited extensive broadening in the Q band region [215], that was later assigned as arising from exciton coupling between the rings [8]. Monomeric $(Cl)_2SiPc$ exhibits band maxima, in a Nujol mull, at 229, 276, 388, 670, and 742 nm [235]. Polymerization to form a poly-yne-polymer, using *p*-diethynylbenzo bridging groups, shifts the Q band from 670 to 648 nm [235]. The 648-nm band is presumably the allowed component of a pair of exciton split bands where the forbidden component lies to the red of the Q band observed for the monomer.

Similar spectral data are observed for other monomer/dimer systems, for example, Q λ_{max} for (OH)SiPc in THF = 667 nm, while for $(HO)[SiPcO]_2H$, Q λ_{max} = 630 nm [175]. Partial oxidation of chains of SiPc—O—SiPc units is thought to be a route towards metal behavior in these molecular system [175]. Dopants have been added to $[SiPc(O)]_n$ and $[GePcO]_n$, in the solid state, to form π cation radical species [40,41]. Absorption spectra taken from Nujol mulls [40] are reminiscent of the thin film spectra described below, rather than solution spectra. Oxidation of these samples is accompanied by the appearance of several new bands, the most important from the characterization viewpoint being the presence of a red-shifted Q band and the 490-nm band. Curiously, although $[(SiPcO)(I_3)_{0.37}]_n$ exhibits the 490-nm band that has been shown to be characteristic of many π cation radical species [27,65,90], SiPcO species doped with either SbF_6^-, PF_6^-, or BF_4^- do not show any absorption near 500 nm [40].

x. Group 15: PPc, AsPc, SbPc

Gouterman et al. [238] have reported the spectra of both P(III)Pc and P(V)Pc in pyridine. The spectrum of P(III)Pc is very unusual in the B band region; band centers are found at 413, 435, 442, 597, and 655 nm. Not only is the intensity of the 442 nm band extremely high for a phthalocyanine complex, but the band is also split into at least three components. A similar spectrum is reported for the P(III)(tetra-*t*-butyl)Pc species (Fig. 31), where no bands are found between 460 and 580 nm, but the intensity and sharpness of the 446-nm band is unique among phthalocyanine spectra. Sayer et al. [31] compared this spectrum with the "hyper" spectrum observed for many porphyrins. The spectrum of P(V)Pc in pyridine [238] was reported to be much more normal, with λ_{max} = 413, 593, and 653 nm, resembling more the spectra of $(imid)_2MgPc$ [9]. In a later review Sayer et al. [31] include spectral data for P(III)Pc (λ_{max} in EtOH = 325, 438, and 651 nm) and (Cl)As(III)Pc (λ_{max} in DMF = 340 and 580 nm). These authors also report that the spectra from AsPc and SbPc complexes have not been characterized [31].

E. TRANSITION METAL PHTHALOCYANINES

i. VOPc

Vanadyl phthalocyanines are used for Q switching lasers, so a considerable body of data exists. Huang and Sharp reported solution and solid state absorption and fluorescence spectra for VOPc [239] (Q band spectra only were shown in 1-ClN, THF, and CH_2Cl_2; in each solvent λ_{max} (Q) was near 690 nm. From fluorescence data, they suggested that aggregation occurred even in dilute solutions (see Sections A.x and C.vii).

ii. CrPc

The first spectral characterization of chromium phthalocyanines was reported by Elvidge and Lever [240]. The coordination and redox chemistries of the CrPcs are rich as is to be expected from the availability of both Cr(II) and Cr(III). Cr(II)Pc is reported [240] to be air sensitive. The Q band is found between 669 and 689 nm, with $\epsilon_Q = 10^{4.4}$ liters mol^{-1} cm^{-1}, for each of the CrPc species studied. Bands near 475 nm for the Cr(III)Pc complexes result in the green color of the solutions and most probably arise from LMCT (see Fig. 14) [10]. Bands are also observed in methanol solutions near 270, 308, 321, and 345 nm.

iii. MnPc

MnPc was one of the first phthalocyanines for which extensive spectroscopic and photochemical data were reported [36]. A major interest was the involvement of manganese-containing complexes in reversible oxygen binding [241], and in photosynthetic reactions [36,242]. Because of the variety of metal oxidation chemistry, together with the various spin states, available to the d^5 Mn(II), a colorful chemistry has been reported, in particular, with the reactions of axial ligands, especially with O_2 [36,241,242–246]. A number of MnPc complexes have been shown to be photoactive [36,246]. MCD data have been reported by Gall and Simkin [139].

The spectral properties of monomeric MnPc complexes with central metal oxidation states of Mn(II) or Mn(III), are distinctive [36,244,246], and the Q band region can be used diagnostically (see below). μ-Oxo dimer formation is prevalent in the presence of oxygen, water, and mild acids. The (L)Pc(−2)Mn(III)—O—Mn(III)Pc(−2)(L) complex is uniquely characterized by a "Q" band near 616 nm in DMA [244,246]; compare, for example, Fig. 33, showing the absorption spectrum of monomeric (N-Me-imid)$_2$Mn(II)Pc, with Fig. 34, the μ-oxo dimer. It has been suggested [35] that rather than the Mn(I) oxidation state being formed when Mn(II)Pc(−2) is electrochemically

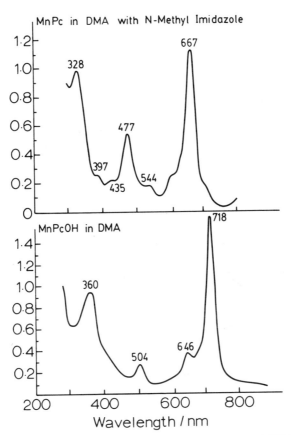

Figure 33 (Top) (N-Methyl imid)$_2$Mn(II)Pc in DMA. (Bottom) (OH)Mn(III)Pc in DMA (J. Wilshire and A. B. P. Lever, unpublished data).

reduced, the ring is reduced to form [Mn(II)Pc(-3)]$^-$. This conclusion was suggested by Clack and Yandle [35] because the spectral properties of the reduced Mn(II)Pc(-2) resembled those of [MgPc(-3)]$^-$ and [ZnPc(-3)]$^-$. The Mn(IV)Pc(-2) species inferred by Engelsma et al. [36] from their spectrochemical studies, is probably the μ-oxo dimer [246].

Single crystal X-ray diffraction studies of β-Mn(II)Pc at room temperature [72], and also at 116 K [247], indicate that the Mn—N bond length is about 193.7 pm, compared with about 192.7 pm for FePc [72] and about 208 pm for Mn(III)TPP [248], with the Mn lying in the coordination plane. The spin state of Mn(II)Pc in the solid state is $S = 3/2$, i.e., an intermediate spin. In the d^4 Mn(III) μ-oxo dimer species, the Mn—N distance expands to about 197 pm [249], with a nearly linear Mn—O—Mn bond. The Mn—O distance was reported to be 171 pm [249].

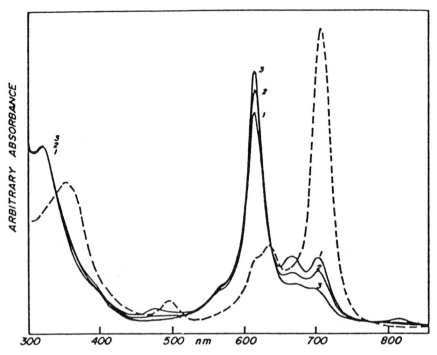

Figure 34 Reaction of N-methylimidazole with (O₂)PcMn in DMA under O₂; (O₂)MnPc (dashed line) and PcMn(III)—O—Mn(III)Pc (line 3). Reproduced with permission from [246] (Fig. 1).

iv. Mn(II)Pc

The spectral data reported by Engelsma et al. [36], Yamamoto et al. [242], and Lever et al. [244], for Mn(II)Pc in pyridine, α-chloronaphthalene, and ethanol, show that (L)₂Mn(II)Pc(−2) is identified with deep blue solutions which exhibit an absorption spectrum with bands extending throughout the visible and near IR region. The band center of the single, well-resolved Q band, λ_{max} = 674 nm, in DMA [244], or 660 nm in pyridine [36,242], is sensitive to the axial ligand. This effect often arises because of spin changes induced by strongly coordinating axial ligands, where coordination by pyridine results in low spin Mn(II), but in DMA solutions the high spin Mn(II) is found [10]. In pyridine, Engelsma et al. [36] report the spectrum between 320 and 1000 nm, as 323 (B), 390, 467, 520, 557, 596, 643, 660 (Q), 835, and 880 nm. However, the broadness of the Q band suggests that there may have been a mixture of species present. Wilshire and Lever (unpublished data), shown here as Fig. 33, report λ_{max} for (N-methyl-imid)₂Mn(II)Pc in DMA as 328 (B), 392, 435, 477, 544, and 667 (Q) nm, with further bands to the red of the Q band. This spectrum is qualitatively the same as for (imid)$_2$Mn(II)Pc in DMA (Q λ_{max} = 663 nm, with excess imidazole).

There are many bands in the absorption spectra of (L)$_2$Mn(II)Pc complexes in addition to those observed for nontransition metal phthalocyanines, for example, MgPc. Lever et al. [10] have characterized the origins of the additional bands in the spectrum of the low spin Mn(II)Pc as LMCT (497 nm, and 1310, with vibronic components at 1200 and 1074 nm), and trip–multiplet absorption for the 937 nm band, as with Cr(III)Pc and Cu(II)Pc [10].

v. Mn(III)Pc and LPcMn(III)—O—Mn(III)L

(OH$^-$)Mn(III)Pc can be prepared indirectly from Mn(II)Pc [246]. Oxidation of Mn(II)Pc dissolved in pyridine with air forms the μ-oxo dimer, (py)Mn(III)—O—Mn(III)(py). Stirring this solution with water, under nitrogen, results in the main band in visible region shifting from the 620 nm of the dimer to the 717 nm of the (OH)Mn(III)Pc [246]. Addition of acid to the μ-oxo dimer, also forms the bright green monomeric (L)Mn(III)Pc species which is characterized by a band near 710 nm. For (HO)Mn(III)Pc in DMA, the spectrum (Fig. 33) shows band maxima at λ_{max} = 360, 504, 646, and 718 nm (Q) (Lever and Wilshire, unpublished). Absorption in the 710 to 720-nm region for the Q band has been suggested [246] as diagnostic for monomeric (L)Mn(III)Pc.

Bubbling O$_2$ into greenish-blue solutions of Mn(II)Pc in very dry DMA forms a new monomeric complex, referred to as the oxygen adduct. This complex is characterized by Q λ_{max} = 705 nm, with bands at 295, 355, 417, and 634 nm [244,246]. The adduct is very stable, although it will revert to (L)$_2$Mn(II)Pc if nitrogen bases are present and the solution is held under vacuum [244]. Similarly, (py)$_2$Mn(II)Pc in pyridine is reported to react only very slowly with O$_2$, even in the presence of excess O$_2$, if the pyridine is very dry [246]. The (O$_2$)MnPc adduct is also photosensitive, releasing O$_2$ following irradiation with sunlight [246] to reform the monomeric (L)$_2$Mn(II)Pc. From analysis of the adduct's EPR and magnetic properties, Lever et al. concluded that the electronic distribution was better described by the formalism, (O$_2^-$) Mn(III)Pc($-$2), which requires superoxide to coordinate the Mn(III) [246]. The MCD spectrum of (O$_2$)MnPc (Fig. 35) clearly shows the presence of the additional band in the 500-nm window region. A similar band has been the subject of considerable study in Fe(II)Pc complexes, see below. The main B band envelope centered on 350 nm comprises at least two bands.

In aerated solutions, (O$_2$)MnPc can be converted to the deep blue μ-oxo dimer. The spectrum is characteristic of aggregated phthalocyanines, with a blue-shifted Q band observed near 616 nm, in DMA. Figure 34 shows a series

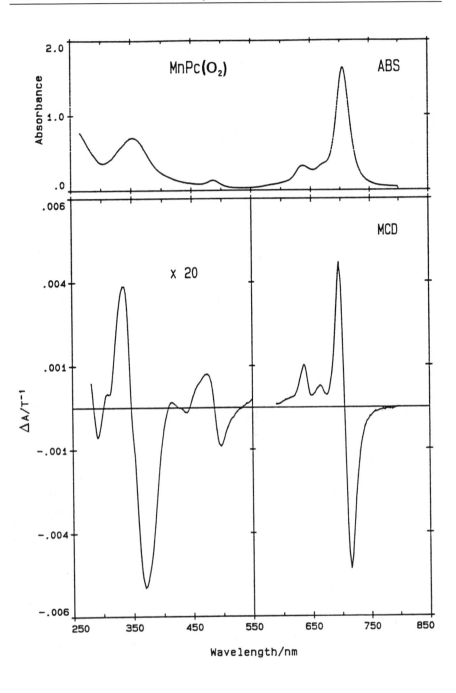

Figure 35 Absorption and MCD spectra of $(O_2)PcMn$ in DMA under O_2. Reproduced with permission from [6] (Fig. 2-25).

of spectra recorded following the addition of N-methylimidazole to a solution of $(O_2)MnPc$ in aerated DMA [246]. Initially, a spectrum of $(N\text{-Me-imid})_2$-$Mn(II)Pc$ forms (λ_{max} at 666 and 816 nm), then the μ-oxo dimer appears. The lack of any 470 nm absorption, and the major band at 616 nm, are characteristic of the dimer. Additional weak bands appear between 630 and 710 nm, and may be associated with the forbidden Q^- transition. The MCD spectrum reveals an A term located at 616 nm, identifying the degeneracy of this state and confirming the assignment as the allowed, Q^+, exciton split component of the Q band (unpublished data of M. J. Stillman and E. Ough).

vi. MnR_nPc: Peripheral Substitution

The solvent dependence of the monomer/dimer equilibrium for $Mn(II)TSPc$ is similar to that of other tetrasulfonated species. See, as an example, the absorption and EPR spectral data of Cookson et al. [195], in which the $Mn(II)TSPc$ dimer in aqueous solutions (with Q λ_{max} = 636 and 718 nm) is dissociated by 5% DMF to give a spectrum characterized by a much more intense peak at 718 nm, and a weaker peak at 636 nm. [Presumably, Q λ_{max} = 636 nm is the dimer, and Q λ_{max} = 718 nm is monomeric $(O_2^-)Mn(III)TSPc$.] As the oxygen adduct forms in the 5% DMF/H_2O solution, resulting in further intensification at 718 nm, there is complete loss of absorbance at 636 nm [195]. Octaalkyl substitution yields the poorly soluble $Mn(II)(Et)_8Pc$ [204], which has an absorption spectrum from a DMA solution in the presence of air, of λ_{max} = 725 (Q), 664 (Q_{vib}), 508 (CT), and 360 nm; again this species is probably the oxygen adduct, (O_2^-)-$Mn(III)(Et)_8Pc$. Lever et al. [10] report a similar spectrum for $(OAc^-)Mn(III)(t\text{-Bu})_4Pc$ in DMF, λ_{max} = 368, 497, 645, 716 (Q), 937, 1074, 1200, and 1310 nm.

vii. Fe(I)Pc, Fe(II)Pc, Fe(III)Pc, and Fe(IV)Pc

The intense blue of an FePc impurity found during the manufacture of phthalimide in Scotland [1] was the starting point for the study of the chemistry of the phthalocyanines in 1928 by Linstead's group [1,2]. Because of the close relationship between the phthalocyanines and the porphyrins, and the variety of the coordination and redox chemistry exhibited by the ferrous and ferric central metal, there are a great many reports that include spectral data for $(L)_2FePc$ complexes.

FePc exhibits a complicated axial ligand chemistry that is related to both the glyoximes and porphyrins. $(L)_2Fe(II)Pc$ complexes are always diamagnetic in strongly coordinating solvents [250–252], contrasting the chemistry of ferrous porphyrins where both intermediate and high spin configurations have been observed. X-Ray studies of β-$Fe(II)Pc$, which has an intermediate spin of $S = 1$, show the Fe(II)—N distance as 192.6 pm [72,253], compared with

Fe(II)TPP, also $S = 1$, where the Fe—N distance is 197.2 pm [73]. Quite different chemistry has been reported from complexes dissolved in noncoordinating and aromatic solvents. Additional bands are observed, and the absorption spectrum extends past the normal Q range toward 1500 nm; see, for example, [3,87,254,255]. As with the ferrous and ferric porphyrins [19,256], the spectra of Fe(II)Pc and Fe(III)Pc complexes are expected to exhibit charge transfer transitions [10,11,257]. (L)$_2$FePc complexes have been studied by many authors, those using absorption spectra include [3,57,135,250,255,257–264], [11,87,89,139,251,252,265–270], and [138,197,262,271–277]. MCD spectral data have been reported for Fe(II)Pc and Fe(III)Pc [87,134,135,138,139].

viii. Fe(I)Pc

Lever and Wilshire have reported the spectrum of $[(py)Fe(I)Pc(-2)]^-$ [265]. The purple complex was formed electrochemically in pyridine in the presence of excess LiCl. The EPR spectrum was reported to be characteristic of the d^7 Fe(I) metal. The absorption spectrum showed peaks at 327, 515, 595, 661, and 801 nm. Clearly, the presence of the Fe(I) central metal greatly disturbs the normal $\pi \to \pi^*$ spectrum, as there is no sign of the usually intense Q band in the 650 nm region (the bands at 661 and 801 nm each having ϵ values of about 12,000 liters mol^{-1} cm^{-1}). Clack and Yandle [35] also report spectral data for what they describe as a pink mono anion and a purple dianion, following electrochemical reduction of Fe(II)Pc in DMF. Bands are reported at 326, 515, 596, 665, and 800 nm for the pink solution and 340, 395, 625, and 740 nm for the purple solution. Although, there was some uncertainty expressed about the oxidation state assignment to Fe(I) for the monion species [35], the agreement with the data of Lever and Wilshire [265] suggests that the pink solution is the $[Fe(I)Pc(-2)]^-$ species.

ix. Fe(II)Pc

The vapor absorption spectrum of Fe(II)Pc measured at 558°C exhibits both extensive broadening of the $\pi–\pi^*$ transitions, and the appearance of bands assigned as CT throughout the 400- to 1,000 nm region. The range of solvents that will dissolve (L)$_2$FePc, includes the weak, axially coordinating DMSO and DMA, and the noncoordinating α-chloronaphthalene, chloroform, toluene, acetone, and dichlorobenzene. In the presence of strongly coordinating solvents or ligands (for example, py), λ_{max} for the Q band is observed in the range 650–670 nm [269], with Q_{vib} components on the blue edge that change quite noticeably as a function of axial ligand. These vibronic bands are most resolved with cyanide, and least resolved with DMSO.

Tahiri et al. [277] report spectral and X-ray data for a series of $[(\sigma\text{-alkyl})\text{-}Fe(II)Pc(-2)]^-$ derivatives, in which CH_3^-, $C_2H_5^-$, and $CH(CH_3)_2^-$ are coordinated by the Fe(II). The spectrum of each was reported to be similar in chlorobenzene solution; for L = CH_3^-, λ_{max} (log ϵ) = 479 (4.94), 540 (4.75), 639 (4.75), 675 (4.76), and 707 nm (4.96). It is perhaps useful to comment that the very high absorbances observed throughout this spectrum are rather unusual.

x. Spectral Properties That Depend on Axial Coordination of Fe(II)Pc

As the σ donor strength of the axial ligand increases, for example, from DMSO to pyridine, so a band at 420 nm becomes more prominent [3,11,257]. At the same time a second band red shifts away from the 300 nm "B" band envelope toward 380 nm, so that for $(CN)_2Fe(II)Pc$, three resolved bands are observed in the 300- to 400-nm region. Because of the changes in absorbance in the 300- to 400-nm region, the colors of $(L)_2FePc$ solutions vary between royal blue for L = DMSO, blue-green for L = tri-*n*-butyl phosphite [252], and green for L = CN^-, NH_3 and $P(Bu)_3$ in toluene [252].

Dale [11], Ouedraogo et al. [257], and Stillman and Thomson [3] have studied the change in the 250- to 500-nm region of the absorption spectrum of $(L)_2Fe(II)Pc$ dissolved in DMSO as a function of axial ligand. The absorption and MCD spectra of a series of $(L)_2FE(II)Pc$ complexes is shown in Figs. 36, 37, 38, and 39, for L = DMSO, py, NH_3, CN^-, CO, and imid. These spectra illustrate how variable the spectral properties for $(L)_2Fe(II)Pc$ complexes can be. It is possible that the transitions that give rise to the band maxima and minima that are observed throughout the 300- to 450-nm region in the MCD spectrum of $[(CN^-)_2Fe(II)Pc(-2)]^{2-}$ (Fig. 37) are always present, and, depending on the axial ligand, simply diminish or intensify. The MCD spectrum for $[(CN^-)_2Fe(II)Pc]^{2-}$ (Fig 37) is one of the most complicated that we have measured for an $M(II)Pc(-2)$ species.

Dale [11] identified four bands, which he labeled I to IV in order of increasing energy, in the 750- to 285-nm region of the absorption spectrum of $(L)_2Fe(II)Pc$, where L = py, imid, NH_3, butylamine, pip, and CN^- [11]. For $(NH_3)_2Fe(II)Pc$, these four bands are located at 320 (IV), 352 (III), 440 (II), and 671 (I) nm [3,11]. He found that two new bands to the red of the B band intensified, or at least became more resolved, as a function of the axial ligands. The wavelength of the 420-nm band (II) increasing for L = DMSO (not observed) < py < imid < pip < NH_3 < *n*-butylamine. Dale tentatively assigned band II at 420 nm as Fe(II) to Pc ligand CT, and quantified the ligand dependence of the band energies in terms of the base strength of each ligand toward Zn^{2+} [11]. Lever et al. [10] later also calculated that an MLCT band should lie in the 350- to 450-nm region. The other bands in the B band envelope region were assigned by Dale as the B (at 379 nm) and the N (at 304 nm), after Gouterman's model [278].

Ouedraogo et al. [257] report a similar analysis of the absorption

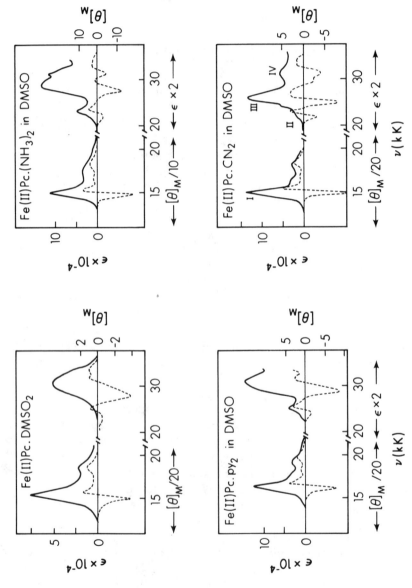

Figure 36 Absorption and MCD spectra of $(DMSO)_2FePc$, $(py)_2FePc$, $(NH_3)_2FePc$, and $[(CN^-)_2FePc]^{2-}$ in DMSO. Reproduced with permission from [3] (Fig. 2).

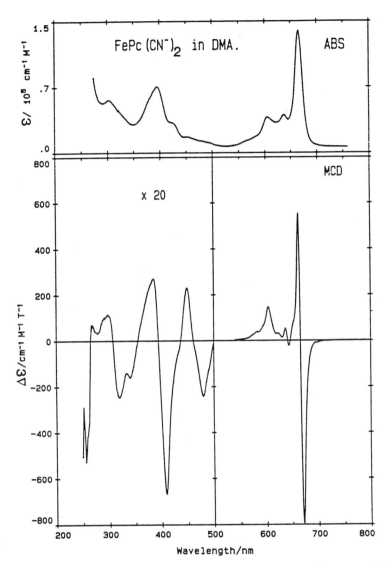

Figure 37 Absorption and MCD spectra of $[(CN^-)_2Fe(II)Pc(-2)]^{2-}$ in DMA. Reproduced with permission from [6] (Fig. 2-20).

spectrum, using a very wide range of nitrogen bases, complemented by Mossbauer data. However, these workers do not support the assignment of MLCT for the "420-nm" band on a number of grounds. Rather they assign this band as axial ligand-to-Pc CT. Stillman and Thomson [3] employed MCD spectroscopy to examine the polarization of these same bands in a similar series of $(L)_2Fe(II)Pc$ complexes (Fig. 36). The band near 420 nm was observed

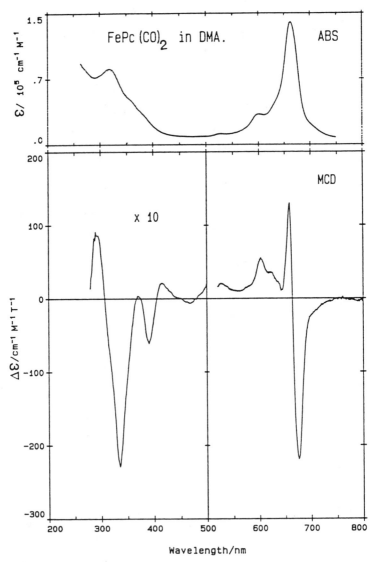

Figure 38 Absorption and MCD spectra of $(CO)_2Fe(II)Pc(-2)$ in DMA. Reproduced with permission from [6] (Fig. 2-19).

in the MCD spectra for all complexes of $(L)_2FePc$, including FePc coordinated to the solvent alone, $(DMSO)_2FePc$. In this case the band was observed under the red edge of the "B" band envelope. This casts some doubt as to whether the band near 420 nm is simply metal-related charge transfer in origin.

A band at 440 nm is clearly observable in the spectra of Fe(II) tetracarboxyphthalocyanine (FeTCPc) in 0.1 N bicarbonate solution at pH 9.0 [138]. However, the situation is more complicated here because of the

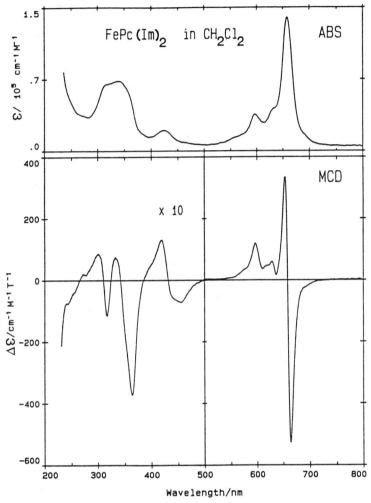

Figure 39 Absorption and MCD spectra of (imid)$_2$Fe(II)Pc(-2) in CH$_2$Cl$_2$. Reproduced with permission from [6] (Fig. 2-21).

tetracarboxy-peripheral substituent groups. This band may be analogous to the 440- to 530-nm band found in the "Group II" complexes (those with electron-withdrawing ligands, such as 4-CNpy) of Ouedraogo et al. [257], which was assigned as Fe-to-axial ligand. The MCD spectrum in the Q band region of the Fe(II)TCPc is not as simple as in MCD spectra of Fe(II)Pc in DMSO [3] (Fig. 36). There are probably two overlapping A terms in FeTCPc, which strongly suggests that there is a mixture of species. It is possible that some dimer is present, which would result in the blue shifted A term being observed somewhere in the 615 nm region, compared with a monomer Q λ_{max} of 678 nm [the normal λ_{max} for monomeric, low spin Fe(II)Pc].

It is interesting to note that this same type of "420-nm" band, appears in the spectra of complexes of both Mn(II)Pc (near 4% nm for (O_2)MnPc) and Ru(II)Pc (near 400 nm for $(pip)_2$RuPc) (see Figs. 34 and 49 respectively).

More recent studies in the MCD spectra of Li_2Pc, MgPc, and ZnPc in the 320- to 450-nm region [9,26,65] reveal that there is a general trend for a second $\pi \rightarrow \pi^*$ band to red shift out of the B band envelope as a function of the axial ligand, specifically, coordination by CN^- results in a B band envelope that spans the greatest energy range even in the absence of charge transfer transitions (see for example the spectra of $[(CN^-)_2Li_2Pc]^{2-}$ [26] and $[(CN^-)ZnPc]^-$ [26,65]). This leaves the 420-nm band as a transition metal marker band. Stynes and James [268] used the 420-nm band observed for $(py)_2$Fe(II)Pc, to monitor coordination by carbon monoxide in the presence of various pyridines. When CO is bubbled into $(py)_2$Fe(II)Pc solutions, the blue (py)(CO)FePc forms from the green $(py)_2$FePc. The intensity of the 420-nm band is ligand sensitive; specifically, in the presence of CO and a trans pyridine, the band almost disappears, while it intensifies with bis coordination by pyridines. The pyridine blocks part of the UV region so that finding the 420 nm band in the presence of the CO is difficult. The experiment might be easier with NH_3 or imidazole, ligands that coordinate to FePc strongly but contribute little extra absorbance. Stynes [266] has also reported kinetic data for ligand displacement from absorption changes in the 420 nm region for the reaction

$(py)_2$FePc (high 420 nm absorption) + dibenzyl isocyanide

$\rightarrow$(py)(RNC)FePc (low absorption at 394 nm and 420 nm)

$\rightarrow(RNC)_2$FePc (high 394 nm absorption, no 420 nm absorption).

Stymne et al. [276] report equilibrium constants (both $2 \times 10^2 \ M^{-1}$) for the formation of (py)FePc and $(py)_2$FePc in CCl_4 solutions. Several other studies on the coordination chemistry of $(L)_2$FePc have reported kinetic and formation constant data for axial ligand competition reactions [250], including $P(Bu)_3$ and $P(OBu)_3$ [252], and a range of nitrogen bases [263,266,269,279–281].

Kalz et al. [261] have reported spectral data for both Fe(II)Pc and Fe(III)Pc (Fig. 40). From this figure, we can see the strong similarities between the absorption spectra of $[(CN^-)_2Fe(II)Pc(-2)]^{2-}$ (B) and $[(CN^-)_2Co(III)\text{-}Pc(-2)]^-$ (C) in CH_2Cl_2. Bands observed in the normally empty spectral window between 350 and 500 nm have been assigned as CT for both complexes [3,261,282]. The data shown as Fig. 40 for CH_2Cl_2 solutions of Fe(II)Pc closely resemble those measured in the more coordinating DMA solvent (compare Fig. 40(B), L = CN^-, with Fig. 37).

xi. The Spectra of Unligated FePc in Chlorinated Solvents

Figure 41 shows the absorption and MCD spectra of FePc dissolved in dichlorobenzene [3]. A new band appears at 850 nm and the rest of the spectrum changes quite significantly, with loss of the resolution of the Q_{vib}

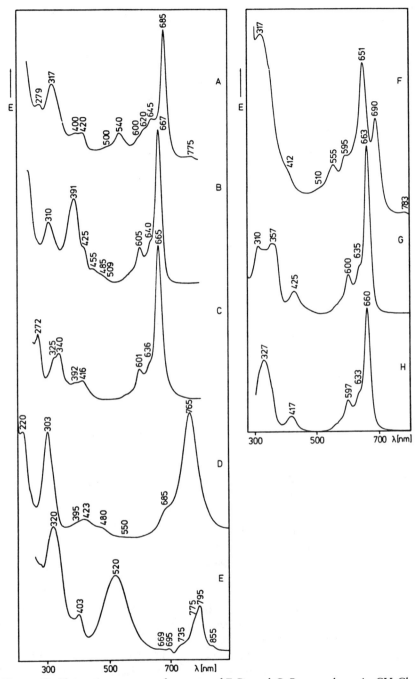

Figure 40 Absorption spectra of a range of FePc and CoPc complexes in CH_2Cl_2: (A) $[(CN^-)_2Fe(III)Pc(-2)]^-$, (B) $[(CN^-)_2Fe(II)Pc(-2)]^{2-}$, (C) $[(CN^-)_2Co(III)Pc(-2)]^-$, (D) $[(CN^-)(HSO_4^-)Fe(III)Pc(-2)]^-$ in concentrated H_2SO_4, (E) $[(CN^-)(CF_3COO^-)Fe(III)Pc(-1)]$ in CF_3COOH/CH_2Cl_2, (F) $[(CN^-)(py)Fe(III)Pc(-2)]$ in py/CH_2Cl_2, (G) $[(CN^-)(N_2H_4)Fe(III)Pc(-2)]^-$ in N_2H_4/CH_2Cl_2, and (H) $[(N_2H_4)_2Fe(II)Pc(-2)]$ in N_2H_4/CH_2Cl_2. Reproduced with permission from [261] (Fig. 4).

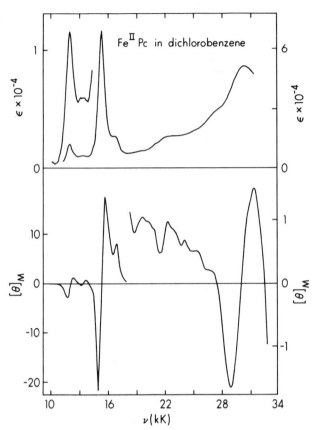

Figure 41 Absorption and MCD spectra of FePc in dichlorobenzene. Reproduced with permission from [3] (Fig. 3).

bands. Stillman and Thomson [3] characterized these effects as arising from a ferrous species with an intermediate spin; thus the 850-nm band was assigned as MLCT. Later, Gouterman [21] suggested that the 850-nm band arose from trip–doublet absorption, much like the low-energy bands observed for CuPc and Cr(III)Pc. Lever et al. [10] suggested, on the basis of a consideration of redox potentials, that the band arose from ligand to metal charge transfer. In each of these arguments, the Fe(II) group would still have to be of intermediate spin rather than the low spin common to $(L)_2Fe(II)Pc$ species. A band in the same 850-nm region has also been observed for high spin Fe(III)Pc complexes [87,134,254]. The spectra of these complexes are discussed below. Doeff and Sweigart, in a report on the ligand chemistry of RuPc [260], comment on the effect of chlorination on Fe(II)Pc. In particular, they discuss the various possible products, metal oxidation, ring chlorination, and protonation of the aza nitrogens.

xii. Fe(II)R$_n$Pc: Peripheral Substitution

From absorption and MCD spectra recorded in CH$_2$Cl$_2$, Kobayashi et al. [87] characterized the presence of low spin Fe(II)-(4,4′,4″,4‴-tetradecyloxy-carbonyl)Pc with bisimidazole coordination (λ_{max}: 350, 440, and 670 nm). This assignment was based on the absence of an EPR signal and on the appearance of a new band in the 420- to 440-nm region, which is characteristic of (L)$_2$Fe(II)Pc in general, as discussed above.

xiii. Fe(III)Pc

Kalz et al. [261] have reported the spectral properties of [(CN)$_2$Fe(III)-Pc(-2)]$^-$, using the bulky cations bistriphenylphosphinimminium and *n*-tetrabutylammonium to stabilize the ferrate group (Fig. 40). The Q band region is shown to be very sensitive to metal oxidation, Q λ_{max} for Fe(III) = 685 nm, whereas Q λ_{max} for Fe(II) = 667 nm. Bands are observed at 279, 317, 400, 420, 540, 600, 620, 645, 685, and 775 nm in the spectrum of [(CN$^-$)$_2$Fe(III)Pc(-2)]$^-$ in CH$_2$Cl$_2$. The Fe was characterized as low spin d^5, with absorption at 540, 610, and 775 nm being assigned as LMCT. (We should note that Kalz et al. [261] use the porphyrin nomenclature for labeling the bands, so that they label the 300- to 400-nm region bands as Q, and the 650 nm as B. This is slightly confusing.)

Kennedy et al. [254], reported spectra from a series of five-coordinate, (L$^-$)Fe(III)Pc(-2) complexes, in α-chloronaphthalene, where L = Cl$^-$, Br$^-$, and I$^-$, in which a band near 830 nm is prominent; for (Cl)Fe(III)Pc(-2), λ_{max}: 450, 485, 595, 655, 758, and 832 nm. The Q λ_{max} at 655 nm suggests that these complexes might be high-spin Fe(III)Pc species. Low-spin six-coordinate, (L)$_2$Fe(III)Pc(-2) spectra exhibit Q λ_{max} in the 670- to 690-nm region (for example, 685 nm for L = CN$^-$ [254]). Kennedy et al. [283] have also reported the magnetic and Mossbauer properties for (Cl)Fe(III)Pc(-2); their analysis requires the presence of a mixed spin Fe(III), with $S = 3/2$ and $5/2$, in the five-coordinate complexes, much like configuration of the heme group in horseradish peroxidase [84].

Lever and Wilshire [265] report that when Fe(II)Pc, dissolved in DMA with excess LiCl, is exposed to oxygen, the Q band splits, and red shifts from 650 nm to near 680 nm, a value also reported by Homborg et al. for Fe(III)Pc [261]. Jones and Twigg [267] report the formation of a species with a similar split Q band spectrum, following the addition of HCl to solid FePc. The lack of a typical π cation radical spectrum (specifically, the lack of a strong 500 nm band [65,92]), and an EPR spectrum which was characteristic of low-spin Fe(III), suggested [265] the presence of Fe(III)Pc(-2). Collamati [258] reports similar spectra following addition of nitrogenous bases to solutions of PcFe—O—FePc in chloroform. She finds Q λ_{max} = 636 and 682 nm, and concludes that the product is probably a Fe(III)Pc species [258].

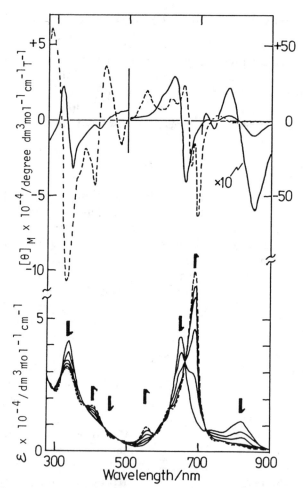

Figure 42 Absorption and MCD spectra of 1,2,8,9,15,16,22,23-octadecyloxycar-bonyl-Fe(III)Pc recorded as the concentration of Bu$_4$NCl increases. The initial high-spin Fe(III)Pc (solid line) is converted into the low spin species (dashed line). Reproduced with permission from [134] (Fig. 1).

xiv. Fe(III)R$_n$Pc: Peripheral Substitution

Kobayashi and Nishiyama report absorption and MCD spectra for Fe(III)TCPc from solutions in 0.1 N bicarbonate at pH 9.0 [138]. They interpret the appearance of a split Q band as arising from a mixture of low- and high-spin Fe(III). With the exception of the lack of the 440-nm band, the B band region for Fe(III)TCPc resembled that of Fe(II)TCPc.

Absorption and MCD data have been reported by Kobayashi et al. [134], for Fe(III)-octadecyloxycarbonyl-Pc (ODCPc) in CH$_2$Cl$_2$. Figure 42 shows MCD and absorption spectra recorded as the spin state of the Fe(III) changes from high (λ_{max} = 337, 650, 738, and 817 nm) to low (329, 410, 552, 617, and 690 nm), following the addition of tetrabutylammonium chloride. A characteristic band of the high-spin species at 817 nm (assigned as LMCT of Fe(III)Pc), diminishes when the Bu_4NCl is added. The Q band red shifts from 650 nm to the 690 nm of the low-spin Fe(III)ODCPc. The MCD spectrum of the low-spin Fe(III)ODCPc complex is quite complicated, with at least two A terms lying near 680 nm. Additional bands lying between 340 and 630 nm are not normally associated with MCD spectra of MPc complexes.

Kobayashi et al. have also reported absorption and MCD spectral properties for Fe(III)-(4,4′,4″,4‴-tetradecyloxycarbonyl)-Pc (TDCPc) [87] and Fe(III)R_nPc complexes, where R_n are the peripheral substituents carboxy- (n = 4 and 8), and bis(3′,4-dicarboxybenzoyl)- [135]. Coordination of Fe(III)–TDCPc by a single imidazole resulted in a low-spin (imid)Fe(III)TDCPc species, which exhibited a more complicated spectral envelope that extended from 280 to 1400 nm. EPR and magnetic moment results, together with the absorption and MCD spectral data for the unliganded, Fe(III)TDCPc complex, that were similar to those found for FePc dissolved in CH_2Cl_2 [3], suggested [87] the presence of a high-spin Fe(III)Pc.

The Fe(III)(octacarboxy-Pc) complex in water in pH 9 exhibits [135] bands at λ_{max} = 370, 693, and 820 nm. An A term, which is located under the 693-nm band in the MCD spectrum [135], identifies the Q band. These spectral features are interpreted as arising from a high-spin Fe(III) species.

xv. Fe(IV)Pc

The Fe(IV) oxidation state has been proposed recently for the [Fe(IV)Pc]$_2$(μ-carbido) [284] and [Fe(IV)Pc](μ-nitrido) [259,284] species. Five coordinate Pc(-2) derivatives and six-coordinate π cation radical, Pc(-1), derivatives were characterized by Mossbauer spectroscopy. No absorption data were given for the μ-carbido derivative. Spectra for the μ-nitrido complex are discussed below.

xvi. μ-Bridged Dimers: PcFe—R—FePc

A variety of μ-R dimers have been prepared, with R = oxo, carbido, and nitrido. Spectral data are most available for μ-oxo complexes of Fe(III)—O—Fe(III), with peripherally substituted rings. [Fe(III)Pc(-2)]$_2$O can be readily formed by aeration of FePc [285]. The two Fe(III) atoms are high spin and strongly coupled in the solid state. Two isomeric forms are reported to exist in the solid state [285], but the blue solutions of these isomers have identical spectral properties, with the major visible region maximum at 620 nm (ϵ = 1.54 × 10^5), for example, [Fe(III)Pc]$_2$O in pyridine [255,285]. The μ-oxo dimer derivatives are reduced back to the (L)$_2$Fe(II)Pc species following addition of coordinating electron donors, such as imid [272]. Similarly, irradiation into the Q band of [Fe(III)-tetra(dodecylsulfonamido)Pc]—O, Q

λ_{max} = 633 nm, forms the monomer, characterized by a Q band near 650 nm and the appearance of a band near 450 nm [272].

Bottomley et al. report spectroelectrochemical results for $(FePc)_2O$ [255], in which oxidation at $+0.55$ V formed a species (Q λ_{max} = 652 nm) assigned as $[Fe(III)Pc]O[Fe(IV)Pc]$, from the $[Fe(III,III)Pc]_2O$ (Q λ_{max} = 622 nm) starting compound. One-electron reduction at -0.59 V formed the Fe(II)—O—Fe(III) species, which is associated with a very broad, rather poorly resolved Q region envelope. Reduction at -0.70 V formed, in a nonisosbestic fashion, the blue-green $(py)_2Fe(II)Pc$ (Q λ_{max} = 652 nm), more extensive reduction at -1.1 V formed a pink complex assigned as Fe(I)Pc(-2) with λ_{max} = 652 nm. Similar spectroelectrochemistry involving redox reactions of the μ-nitrido complex, $(FePc)_2N$ [259] suggested the presence of a mixed valence compound, Fe(IV)–N–Fe(III), with a Q-region λ_{max} of 626 nm, as well the Fe(IV)—N—Fe(IV) and the Fe(III)—N—Fe(III). The three complexes were characterized spectroscopically in pyridine, in the presence of 0.2 M tetrabutylammonium perchlorate, by λ_{max} ($\epsilon \times 10^4$): $[Fe(IV)—N—Fe(IV)]^+$, 547 (1.18), 634 (5.76), and 648 nm (3.0), Fe(III)—N—Fe(IV), 537 (1.24), 626 (5.25), and 658 nm (1.82), and $[Fe(III)—N—Fe(III)]^-$, 594 (1.77), 627 (2.98), and 655 nm (4.26) [259].

xvii. A Comment on the Oxidation State Assignments

While it is relatively easy to characterize the oxidation and spin states of Fe(III)Pc complexes formed from the water-soluble, tetrasulfonated derivatives [286], it is clear that additional experimental evidence is required before the nature of the FePc species that are formed in chlorinated solvents, or in the presence of high concentrations of halide ion, are fully understood.

xviii. Co(I), Co(II), and Co(III)Pc

As with MnPc and FePc, the spectroscopy of CoPc is dominated by reactions taking place at the metal center, for example [14,197]. Like CoTPP and CoOEP [133], Co(II)Pc readily oxidizes to form $[Co(III)Pc(-2)]^+$, especially in the presence of strong σ donors [3,88]. Spectral data have been reported for Co(I)Pc [88,197,221,282], Co(II)Pc [3,10,14,88,181, 187,197,221,262,282], and Co(III)Pc [88,91,197,262,282,287]. MCD spectra have been reported for Co(I)Pc, Co(II)Pc, and Co(III)Pc complexes [3,138,139].

xix. Co(I)Pc

The absorption spectrum of the $[Co(I)Pc]^-$ species is specially marked by a broad and intense envelope of bands near 450 nm and a rather weak Q band near 700 nm [3], which is red shifted from the energy of the Q band in Co(II)Pc in DMSO, λ_{max} 657 nm [3].

Day et al. [88] have reported spectral data for Co(I)Pc in pyridine (λ_{max} = 311, 426, 467, 641, and 704 nm). New bands at 426 and 467 nm, which lie in the characteristic window between the B and Q bands, have been assigned as MLCT [3,88]. Stillman and Thomson [3] reported absorption and MCD spectra for Co(I)Pc in DMSO. The 450-nm band envelope comprised several transitions, at least one of which was degenerate (Fig. 6). Kobayashi and Nishiyama [138] reported similar spectra for Co(I)TCPc, which exhibits band maxima at 308, 467, and 649 nm, although the MCD resolution in the 400- to 500-nm region was not sufficient to determine the band composition. The tetrasulfonated Co(I)TSPc, generated electrochemically in DMSO with 0.1 M TEAP [221] or in aqueous solution at pH 2 [197], exhibits the strong band near 460 nm, with a Q band near 710 nm, which is red shifted from the 663 nm of the neutral $Na_4Co(II)TSPc$ [221]. Unlike many other sulfonated derivatives, the [Co(I)TSPc]$^-$ did not appear to dimerize in DMSO (this being perhaps a function of the negative molecular charge).

xx. Co(II)Pc

The characteristic spectrum of $(L)_2Co(II)Pc(-2)$ complexes exhibits a well-resolved Q band near 658 nm, a clear 450-nm window, and a broad B band envelope starting at 332 nm, taking the data of Metz et al. [262] for Co(II)Pc in pyridine, as an example. Similar data are described by Stillman and Thomson [3] for Co(II)Pc in DMSO (Fig. 43) and Day et al. [88] for Co(II) and Co(III)Pc complexes in pyridine.

In general, the absorption and MCD spectra of the low-spin, $(L)_2Co(II)Pc$ complexes in DMSO and pyridine are similar to those of $(L)_2Fe(II)Pc$. In the near-IR, several weak bands observed for $(py)_2Co(II)Pc$ have been assigned as $d–d$ [14]. However, as in the spectra of many phthalocyanines, there is some ambiguity in this assignment as the bands could also be charge transfer in character [10]. As with Cr(III), Mn(II), and Mn(III), a sharper band near 1,000 nm could arise from trip–multiplet absorption. The broadness in the B band envelope might also arise from underlying charge transfer bands.

xxi. Co(II)R$_n$Pc: Peripheral Substitution

Spectra of $Na_4Co(II)TSPc$ [191,193,221] have been measured in DMSO, DMF, MeOH, and H_2O. Rollmann and Iwamoto [221] also report electrochemical and visible region spectral data for Co(II) redox derivatives. They report that association of these tetrasulfonated species occurred above 10^{-5} M concentrations in water. In MeOH, the spectral data of the monomers give λ_{max} = 328, 593, and 656 nm, and the Q band log ϵ = 5.11 [221]. Seiders and Ward [191] report the dissociation of Co(II)TSPc dimers at 85°C in water, where the monomeric Q band at 640 nm red shifts out of the Q band envelope at room temperature. Gruen and Blagrove [193] show the effects of

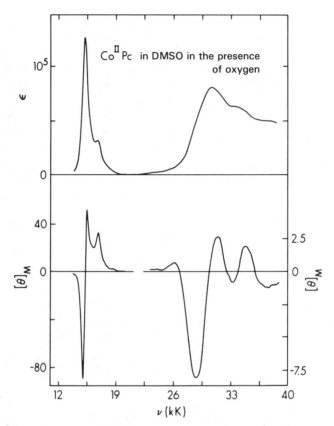

Figure 43 Absorption and MCD spectra of Co(II)Pc(−2) in DMSO. Reproduced with permission from [3] (Fig. 4).

temperature on the Q band region spectrum: the monomer Q band at 663 nm intensifies between 25 and 70°C for Co(II)TSPc in water. They calculate ϵ for the pure monomer at 10^{-7} M concentration in 20% ethanol as 1.2×10^5 at 663 nm. Addition of KCl to an aqueous solution increases for dimer fraction. In alkaline aqueous solutions, Co(II)TSPc binds O_2 to form an adduct with Q $\lambda_{max} = 670$ nm, and $\epsilon = 1.7 \times 10^5$. The appearance of splitting in the MCD envelope in the Q band region of Co(II)TCPc recorded in 0.1 N bicarbonate solution at pH 9.0 [138] suggest that there is a mixture of species present, probably dimers. In concentrated H_2SO_4, the Q band red shifts to 738 nm in the tetraamine-substituted Co(II)Pc [181], as is often reported for highly acidic solutions. Ferraudi [198] has described the photochemistry of monomeric and dimeric Co-sulfophthalocyanines.

The absorption and MCD spectra of the monomeric and dimeric Co(II)TCRPc [214] give Q λ_{max} (monomer) = 670 nm, and (dimer) = 630 nm. In the B region, there appears to be a series of overlapping bands that reach a maximum near 290 nm, which is similar to the spectrum of NiTCRPc. This part of the spectrum is unusual and CT bands may be playing a significant role.

Binuclear $CoR'_3R''_1$ Pcs have been described by Marcuccio et al. [222], Leznoff et al. [205], and Nevin et al. [207], where R' = a neopentoxy group, and R'' is a bridging group. In these compounds the two rings are covalently linked through bridges attached to the phthalocyanine benzo groups. These complexes exhibit two-banded absorption spectra in the visible region, for example, compound 28 in ref. [222] shows two bands of comparable extinction, log ϵ = 4.9, at 626 and 680 nm. This doubling of Q bands was interpreted as arising from interaction between the rings in an intramolecular fashion [207,222]. A highly substituted CoPc complex (compound 20 in ref [222]) exhibited a typical monomeric phthalocyanine spectrum with bands at 290 (log ϵ = 4.72), 330 (4.75), 608, and 672 nm (log ϵ = 5.04), which illustrates how the attached groups can be used to restrict associative mechanisms. Nevin et al. [207] report spectroelectrochemical studies on Co(II)N$_4$Pc, which has a spectrum with λ_{max} = 330, 380, 612, 645, and 678 nm (Q).

xxii. Co(III)Pc

Addition of NaCN to Co(II)Pc in DMSO results in the blue solution turning green. Absorption and MCD spectra of this species [3] are quite unusual for a phthalocyanine complex, because a continuous series of bands is observed from 500 nm to the solvent cutoff at 260 nm (in DMSO). A typical Co(III) spectrum, taking $[(CN^-)_2Co(III)Pc]^-$ in DMSO as the example (Fig. 44) gives λ_{max} = 284, 352, 427, 607, and 673 nm (Q) [3]. The absorption and MCD spectral properties are maintained even at 4.2 K with the phthalocyanine dissolved in a polyvinyl alcohol film [3,7], which suggests that the system is either totally diamagnetic or, if paramagnetic spins are involved, the spins would have to be strongly coupled. It remains for a complete spectral analysis to be carried out to determine which bands could be CT in nature. The spectrum of (py)(Cl)Co(III)Pc(-2) in pyridine solution exhibits [287] bands at 352, 429, 569, 640, and 669 nm. Homborg and Kalz [282] report the preparation and absorption spectra for a range of anionic complexes, $[(L)_2Co(III)Pc(-2)]^-$. Figure 45 [282] shows the spectra for the tetrabutylammonium salt of $[(L)_2Co(III)Pc(-2)]$ dissolved in CH$_2$Cl$_2$, with L = OH$^-$, F$^-$, Cl$^-$, and Br$^-$. For $[(Br)_2Co(III)Pc]^-$, the values of the band centers are 285, 330, 400, 430, 520, 596, 630, and 663 nm [282], which is quite similar to the spectrum of the species with cyano coordination. Although in the sequence of L = OH$^-$, F$^-$, Cl$^-$, to Br$^-$, the Q band remains unchanged, new bands begin to intensify at 400 and 430 nm.

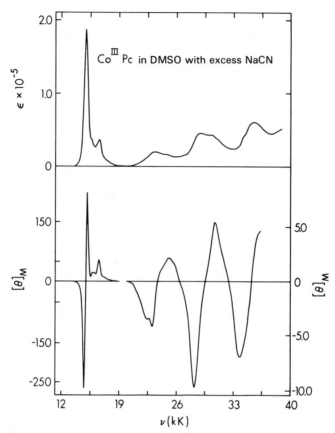

Figure 44 Absorption and MCD spectra of $[(CN^-)_2Co(III)Pc(-2)]^-$ in DMSO. Reproduced with permission from [3] (Fig. 4).

xxiii. Co(III)Pc: Peripheral Substitution

Nevin et al. [207] have reported spectral data for Co(III)–N$_4$Pc. The spectrum of $[Co(III)N_4Pc)]^+$ in DMF with 0.3 M TBAP, with $\lambda_{max} = 340$, 355, 610, and 676 nm (Q), is quite similar to the six-coordinate cyano derivative in CH$_3$CN.

Reflection spectral data for the peripherally substituted Co—$(R)_m$—Pc, with R = CH$_3$ and —OCH$_3$ and $m = 8$, and R = t-Bu and NO$_2$ with $m = 4$, as well as cyano-linked polymeric chains $[Co(III)R_mPc(CN^-)]_n$, have been reported [288]. The spectrum of Co(II)-octamethyl-Pc [288] shows bands at $\lambda_{max} = 280$, 365, 575, and 690 nm. Hanack and Fay, using EPR data ($g = 2.0028$) measured for powders at room temperature, suggest that these compounds can be written as Co(II)R$_8$Pc(-1). However, the absence of the spectral data in the published report makes direct comparison with other data for Pc(-1) species difficult.

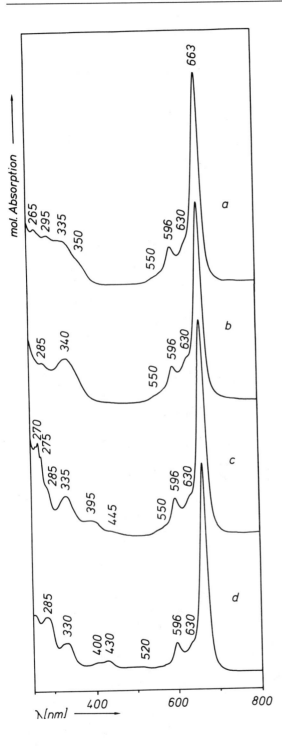

Figure 45 Absorption spectra of (tetrabutylammonium)$^+$ [(L)$_2$Co(III)Pc(-2)]$^-$ in CH$_2$Cl$_2$, for L = OH$^-$ (a), F$^-$ (b), Cl$^-$ (c), and Br$^-$ (d). Reproduced with permission from [282] (Fig. 1).

xxiv. NiPc

Nonperipherally substituted NiPc is particularly insoluble, so that only Q region absorption and MCD spectra [26] have been reported in any detail (Fig. 4).

xxv. NiR$_n$Pc: Peripheral Substitution

Tetrasulfonated NiPc, Na$_4$NiTSPc, dissolves in DMSO, DMF, and MeOH to give similar spectra [76], λ_{max} for solution in DMF = 348, 604, and 671, with Q log ϵ = 5.26. In MeOH the Q band blue shifts, λ_{max} = 662, while in water association occurs so that shoulders appear in the Q region [76]. A spectrum of Ni(II)-tetraamine-Pc solutions in concentrated H$_2$SO$_4$ has been reported, with λ_{max} = 208, 302, 380, and 738 nm [181].

NiPc peripherally substituted with 15-crown-5 (NiTCRPc) is freely soluble in CHCl$_3$. Addition of CH$_3$COOK to such solutions results in dimerization [187], which is associated with the development of a blue-shifted Q band, Q λ_{max} (ϵ) for the monomeric NiTCRPc = 669 nm (70,000), with bands at 401, 603, and 638. Q λ_{max} for the dimer is located at 632 nm [187,214]. Absorption and MCD spectra of the monomeric and dimeric species [214] extend to 250 nm, revealing a complicated B band envelope. Deconvolution calculations [214] put degenerate states at 669 nm (Q band), and 410, 318, and 276 nm. The separation of 410 and 318 nm is unusually large if these are the B$_1$ and B$_2$ bands. However, the spectrum of NiTCRPc in CHCl$_3$ is not like those of ZnPc or MgPc, so an additional perturbation must be acting on the π system.

xxvi. CuPc

CuPc is particularly insoluble in most solvents, making its optical spectra rather hard to measure, especially below 350 nm. Reports using α-CIN [49] and pyridine usually include only the Q band region. For Cu(II)Pc in α-CIN, λ_{max} = 350, 510, 526, 567, 588, 611, 648, and 678 nm, which can be compared with the red-shifted spectrum of CuPc in conc. H$_2$SO$_4$, λ_{max} = 210, 220, 254, 303, 438, 634, 698, 746, and 790 nm [289]. The only dilute "solution" MCD spectrum available for CuPc is from CuPc in an Ar matrix at 20 K; only the Q band was measured [141]. The spectrum of this matrix isolated species appears to be similar to those of other phthalocyanine complexes. Eastwood et al. [56] report a vapor phase absorption spectrum (Fig. 30) which resembles ZnPc (vapor) with λ_{max} near 320 and 660 nm. No splitting is observed in the B region.

xxvii. CuR$_n$Pc: Peripheral Substitution

The addition of four, peripherally fused, 15-crown-5 groups, leads to superior solubility in CHCl$_3$, acetic acid, CH$_3$OH, and py [187,290]. The absorption spectrum of the monomer in CHCl$_3$ [187,214,290] extends from

250 to 800 nm, with λ_{max} [187] = 292, 338, 409, 610, and 676 (Q). In the other solvents [290], except for py, even at 10^{-6} M concentrations, broadening in the Q band region suggests dimerization occurs. Absorption and MCD spectra of CuTCRPc [214] illustrate clearly how dimerization results in a significant blue shift of the Q band, from 678 to 635 nm, yet little shift, or even a change in excited state character, in the B band region.

Na$_4$CuTSPc dissolves in DMSO, DMF, MeOH, and water, although in water, association is reported to take place above 10^{-6} M concentrations [76]. In DMF, the spectra give λ_{max} = 332, 604, and 666 nm, with Q log ϵ = 5.13. Similar spectral maxima were measured in DMSO and MeOH [76]. Achar et al. [181] report the spectrum of the tetraamine-substituted CuPc, with λ_{max} = 214, 300, 382, and 749 nm. Skorobogaty et al. [200] have reported the preparation, absorption, and EPR spectra for two isomers of Cu-(2-pyridyl-methylaminosulfonyl)$_4$Pc (named Cu-tpmaspc) and [Cu-(N-methyl-2-pyridiniomethylaminosulfonyl)$_4$Pc]$^{4+}$ (named [Cu-tmpmaspc]$^{4+}$). Each of these compounds exists as a monomer in DMF, but dimerizes as water is added, until aggregates begin to form with 80% H$_2$O. Q λ_{max} (ϵ) for monomeric 4,11,18,25-[Cu-tpmaspc] in DMF = 673 nm (15.7 × 10^4).

Cu-(CN)$_8$-Pc) in DMF exhibits an α-polymorph-like absorption spectrum in the Q band region, indicating that extensive aggregation has taken place [177]. While CuPc sublimes readily to form either crystals [49] or thin films on quartz slides, the spectral data from these matrices are very much broader than the solution data, as the electronic states are strongly coupled by Davydov effects [49]. Octa-substitution has also been reported by Wöhrle and Hundorf [291] with carboxylic acid, anhydride, imido, and amido groups being prepared from the Cu-(CN)$_8$-Pc. They have reported extensive spectral data for this range of complexes dissolved in sulfuric acid. For Cu-(CONH$_2$)$_8$-Pc (complex 7b in [291]) in conc. H$_2$SO$_4$, λ_{max} = 230, 295, 344, 390, 640, 670, 699, and 751 nm, with Q log ϵ = 5.13. Wöhrle's group has also reported polymerization reactions of MPc complexes, for example, Wöhrle and Schulte [289] describe spectral data in conc. H$_2$SO$_4$ for a series of polymeric CuPc complexes in which the Pc rings are linked through rigid oxy or phenoxy groups attached to the benzo moiety of the phthalocyanine ring.

xxviii. ZnPc

The vapor spectrum of ZnPc (Fig. 30) [56,57] resembles that of MgPc (Fig. 23), the typical phthalocyanine species. All five $\pi-\pi^*$ bands are resolved, and no additional bands are observed. Matrix isolated spectra for ZnPc have been reported by Bajema et al. [58] for the Q band and by Schatz' group [59]. The most extensive series of solution absorption and MCD spectra have been reported by Stillman and Thomson [26] and Nyokong et al. [65] (Fig. 46). The ZnPc complexes were chosen in these studies because the appearance of "additional" bands assigned usually as CT, in the spectra of many phthalocyanine complexes, for example [3,10], has complicated the assignment of the absorption spectrum of the phthalocyanines in the past. No CT bands are to be expected with ZnPc, yet the important chemistry involving axial ligands, and redox reactions of the ring [92], are still available with ZnPc. Homborg and Murray [229] have reported the preparation and spectra of [tetrabutylammonium][(F)ZnPc(−2)] in CH_2Cl_2. The absorption spectrum shows no indication that the complexation by the fluoride perturbs the $\pi-\pi^*$ spectrum of the ZnPc(−2) ring (Fig. 29).

The absorption and MCD data of (L)ZnPc complexes [3,65] show the development of twin bands in the B band region, as for $(L)_2Li_2Pc$ [26]. Many of the absorption and MCD spectra recorded for (L)ZnPc have been deconvoluted. Figure 18 shows the results of band deconvolution for the (CN^-)ZnPc complex [65].

Degenerate bands are found at 246 (assigned as L [21]), 282 (N), 336 (B_2), 368 (B_1), and 671 nm (Q). Figure 47 compares the energies of bands located by the fitting procedures for (imid)ZnPc and $(imid)_2MgPc$ measured from CH_2Cl_2 solutions. With the exception of a couple of weak bands and the Q_{vib} regions, the two sets of spectral data are arise from the same transitions.

These results have been discussed in more detail above. Lately, the absorption spectra of $(L)_2MgPc$ complexes have been used in our laboratory as the standard for a generic main group phthalocyanine absorption spectrum [9]. This is in part because MgPc dissolves in a wider range of solvents than many other phthalocyanine complexes.

In recent calculations reported by Khatib et al. [352] for ZnPc, the energies of transitions between many of the filled orbitals and the unoccupied orbitals $7e_g$, $8e_g$ and $13a_{1g}$, are tabulated. If an offset of ca. 9,700 cm^{-1} (1.2 eV) is added to each of the energies, the calculated energies can be compared with the energies of bands obtained by deconvolution. If we follow our current nomenclature of B_1 and B_2 for the two bands in the near UV region, then Khatib et al. [352] predict the following orbital assignments: Q $(2a_{1u} \rightarrow 7e_g)$ at 14,600 cm^{-1}, B_1 $(4a_{2u} \rightarrow 7e_g)$ at 26,150 cm^{-1}, B_2 $(1a_{1u} \rightarrow 7e_g)$ at 31,400 cm^{-1}, N $(2a_{1u} \rightarrow 8e_g)$ at 34,200 cm^{-1} and L $(3a_{2u} \rightarrow 7e_g)$ at 37,400 cm^{-1}. The next transition $(2a_{2u} \rightarrow 7e_g)$ is calculated to occur at 42,900 cm^{-1} and might be the C band. While the energy match is quite close, the absence of oscillator strengths and configuration interaction information, means that it is not possible to assess how accurately these energy calculations represent the bands observed in the spectrum of ZnPc.

xxix. ZnR_nPc: Peripheral Substitution

Spectral data have been reported from ZnPc octa-substituted by electron-withdrawing groups, e.g., CN^- [177], or electron-donating groups, e.g., butyl [180]. The absorption spectrum of $Zn-(CN)_8-Pc$ in DMF [177] is clearly from aggregates, as the spectral bands are very broad. Surprisingly, neither oxidation nor reduction resulted in much change in the spectrum [177].

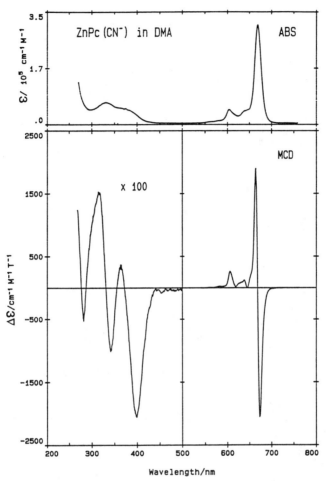

Figure 46 Absorption and MCD spectra of $[(CN)ZnPc]^-$ in DMA. Reproduced with permission from [6] (Fig. 2.5).

The spectral data for Zn-(butyl)$_8$-Pc [180] in DMF (solubility is reported as good in toluene, pyridine, DMF, and CCL$_4$, but poor in alcohols, ethers, and DMSO) are reported to be like those of unsubstituted ZnPc, with λ_{max} at 352, 622, and 678 nm. In concentrated H$_2$SO$_4$, protonation of the aza nitrogens results in a red shifts for the Q band, λ_{max} = 280, 310, 450, 729, and 827 nm [180]. The spectrum of tetraamine-substituted ZnPc in concentrated H$_2$SO$_4$ also exhibits a red-shifted Q band, λ_{max} = 214, 302, 384 and 742 nm [181]. Spectra of monomeric ZnTSPc can be obtained in water if a dissociating agent is used [202]. While low fractions of pyridine and DMF will dissociate dimeric ZnTSPc, the resultant monomeric Na$_4$ZnTSPc is photoinactive even in the presence of MV^{2+}. However, addition of the cationic micelle,

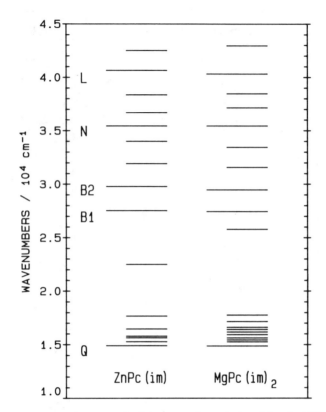

Figure 47 Comparison between the band energies calculated for (imid)ZnPc and (imid)₂MgPc. The longer bars indicate A terms were used in the fit, which identifies degenerate states. Reproduced with permission from [9] (Fig. 9).

hexadecyltrimethylammonium chloride (CTAC), will dissociate ZnTSPc aggregates, to give a normal, monomeric Q band spectrum, and retain photoactivity. Absorption and MCD spectral data for ZnTCRPc [187,214] in CHCl₃ exhibit band maxima at 347 and 680 nm for the monomer, with Q λ_{max} for the dimer (which forms when CH₃COOK is added) at 635 nm.

F. ABSORPTION SPECTRA OF SECOND AND THIRD ROW TRANSITION METAL PHTHALOCYANINES

i. RuPc

Because of its potential utility as a catalyst, similarity to FePc, and the special photochemical properties imparted by the low-spin Ru(II), Ru(II)Pc has been extensively studied [27,90,260,292–298].

Ru(II)Pc is frequently prepared from $Ru_3(CO)_{12}$, in which case axial coordination by CO occurs [298]. Figure 48 shows the absorption and MCD spectra of (CO)(DMF)RuPc in CH_2Cl_2. Notice the remarkably sharp band in the MCD spectrum near 350 nm, and the extremely well-resolved A term under the Q band; these two features are characteristic of coordination by CO. The $(pip)_2Ru(II)Pc$ spectra are quite different (Fig. 49). Because of the dependence of the absorption spectrum on axial ligands, and the propensity of Ru(II)Pc to abstract CO from the reaction mixture, even when $RuCl_3$ is used [296]; there has, in the past, been some difficulty confirming the exact product of the synthesis [296,298]. RuPc does not sublime readily, unlike many other phthalocyanine complexes. The CO is most easily removed photochemically [90,298] (Fig. 50). In chloroform, with THF added, the visible region spectrum shows maxima for (CO)(THF)RuPc at 581 and 642 nm; these values are also similar to those measured for (CO)(py)RuPc [296].

RuPc binds the usual range of axial ligands. Q band maxima reported by Nyokong et al. [90] include the complexes $(py)_2RuPc$, 622 nm, and (CO)(DMF)RuPc, 638 nm. Doeff and Sweigart [260] describe in detail formation of complexes with $P(OBu)_3$, PBu_3, pyridine, Me(imid), and Cl^-. From their data [260], they suggest that coordination with chlorine forms a complex best written as (Cl)Ru(II)Pc, rather than (Cl)Ru(III)Pc. This is unlike the chemistry of FePc where (Cl)Fe(III)Pc forms readily [283]. While chlorination of the π ring is quite possible, Doeff and Sweigart found no evidence from the ^{1}H NMR spectra to suggest that this had occurred [260]. No clear explanation was suggested as to how the redox states of the Ru, Cl, and the phthalocyanine ring could be reconciled. Dolphin et al. [298] have discussed possible oxidation routes for (L)RuPc. It was also suggested that oxidation at the metal is unlikely in these phthalocyanines [298].

$(L)_2RuPc$ typically [6,27,260,294] exhibits a Q band at 640 nm, with log $\epsilon = 4.8$ liters mol^{-1} cm^{-1}. Other bands are found near 270, 314, 376, 442, and 587 nm [27,90,294]. Figures 48 and 49 illustrate quite well how sensitive the B band region is to axial ligation. Note the poorly resolved Q band region due to the presence of additional bands that lie under the $\pi \rightarrow \pi^*$ singlet. Oxidation of Ru(II) to Ru(III) does not readily occur. With both chemical and electrochemical oxidations, the π ring oxidizes first to form the π cation radical species, Ru(II)Pc(-1) [27,90,298]. The assignment of the $\pi-\pi^*$ bands in the $(L)_2Ru(II)Pc(-2)$ spectrum is more difficult than for many phthalocyanines, because the spectrum is so dependent on the nature of the axial ligand (see Figs. 48 and 49). In particularly, CO results in apparent decoupling of the π ring system from the Ru(II) so that very intense Q band absorption and MCD signals are observed. At the same time, the sharp line-like band near 355 nm dominates the B band region of the MCD spectrum (Fig. 48). This band appears to be completely Ru–CO dependent, as we have not found a similar band, either as sharp or as intense, in any other phthalocyanine system. It is tempting to assign it as CT in character [6,27].

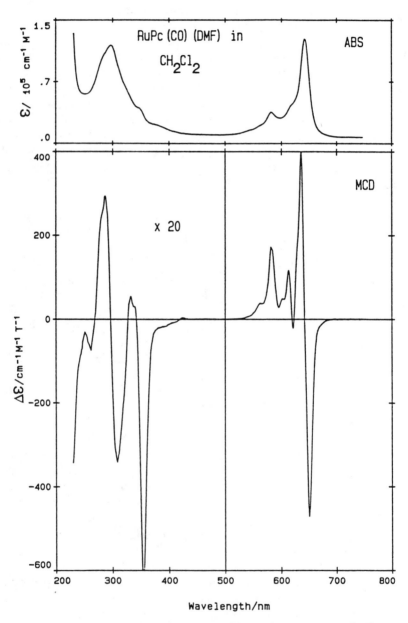

Figure 48 Absorption and MCD spectra of (CO)(DMF)RuPc in CH$_2$Cl$_2$. Reproduced with permission from [6] (Fig. 2.14).

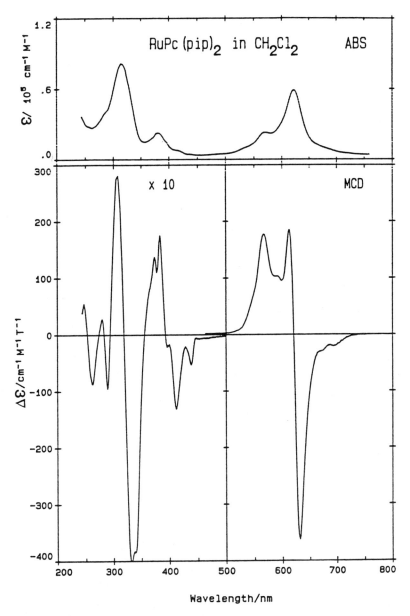

Figure 49 Absorption and MCD spectra of (pip)$_2$RuPc in CH$_2$Cl$_2$. Reproduced with permission from [6] (Fig. 2.12).

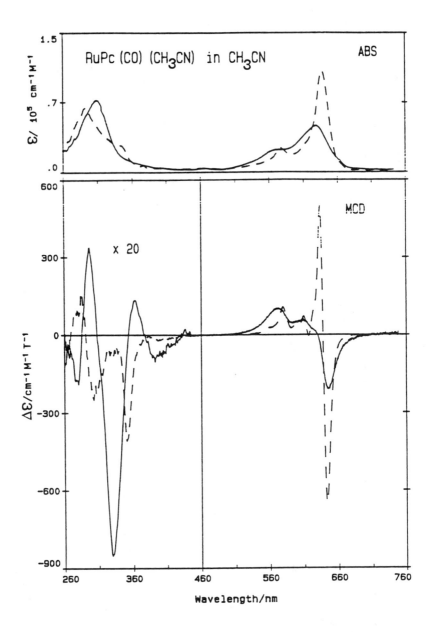

Figure 50 Absorption and MCD measured before (dashed line) and after (solid line) photolysis of (CO)(CH$_3$CN)RuPc in CH$_3$CN. The photolysis reaction, using visible region light, dissociates the CO from the (CO)(CH$_3$CN)RuPc to form the (CH$_3$CN)RuPc species. Note the loss of the A term in the Q band region as a result of the presence of new charge transfer bands in the near 650 nm. Reproduced with permission from [6] (Fig. 6.16).

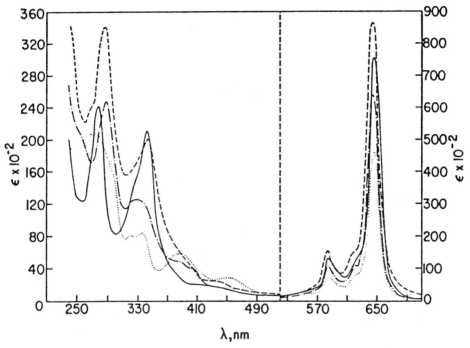

Figure 51 Absorption of acido-(Rh(III)Pc) complexes in CH_3CN: $(L)_2Rh(III)Pc$, L = H_2O (——), L = CH_3OH,Cl^- (----), L = CH_3OH,Br^- (–·–·) and L = CH_3OH,I^- (····). Reproduced with permission from [300] (Fig. 1).

ii. RhPc

Munz and Hanack [299] have reported the preparation and spectral properties of $(L)_2Rh(II)Pc$, where L = *n*-butylamine, pyridine, 4,4′-bipyridine, and pyrazine. In DMF solutions each gave a similar spectrum, for example, $(py)_2Rh(II)Pc$, λ_{max} = 326, 372, 598, 632, and 661 nm.

Ferraudi and co-workers have reported the photochemical properties of (acido)Rh(III)Pc complexes in acetonitrile [300,301] using UV-energy light. Spectra of $(CH_3OH)(L)Rh(III)Pc$, where L = Cl^-, Br^-, and I^-, are quite different when compared with the spectra observed for complexes such as (L)ZnPc. As an example, see the spectrum of $(CH_3OH)(Cl)Rh(III)Pc$, where Q λ_{max} = 645 nm (Fig. 51). While the extra bands are assigned as CT, there is also considerable interaction with the $\pi–\pi^*$ bands in the 300-nm region, much like the hyper effects reported for porphyrins [21]. Similar spectra have been reported by Munz and Hanack [302], for a series of $(L)(Cl^-)Rh(III)Pc$ complexes in $CHCl_3$ [L = pyridine, 4,4′-bipyridine, 2-methylpyrazine, and 1,4-diazabicyclo[2.2.2]octane (dabco)]. In $CHCl_3$ solutions each gave a similar spectrum, for example, $(dabco)(Cl)Rh(III)Pc$, λ_{max} = 283, 345, 589, 625, and 652 nm [302].

Hanack and Munz have reported synthesis of the polymeric [(CN)Rh(III)Pc]n, which involves chains of Rh(III)Pc rings linked through CN groups, from K[(CN)$_2$Rh(III)Pc] [303]. The monomeric (L)(CN)RhPc is formed in coordinating solvents, such as pyridine [303]. In DMF, two band maxima are reported, for K[(CN)$_2$Rh(III)Pc] λ_{max} = 343 and 651 nm, for (*n*-butylamine)(CN)RhPc, λ_{max} = 345 and 650 nm, and for [(Cl)Rh(III)Pc], λ_{max} = 344 and 644 nm.

iii. OsPc

Omiya et al. [296] have reported the preparation and X-ray study of (CO)(THF or pyridine)OsPc. The N–Os bond length averages 201 pm, with the Os directly coordinated to the CO and pyridine; the Os is slightly out of the plane of the ring. For (CO)(py)OsPc, λ_{max} in chloroform = 575 and 632 nm. Hanack et al. (unpublished data) have prepared (py)$_2$Os(II)Pc, λ_{max} = 308, 369, 563, and 616 nm.

iv. PtPc and PdPc

Like ZnTSPc, the absorption spectrum of monomeric PdTSPc can be obtained in water if a dissociating agent is used [202]. Na$_4$PdTSPc is photoinactive. However, as with ZnTSPc, addition of a cationic micelle, hexadecyltrimethylammonium chloride (CTAC), results in a normal, monomeric Q band spectrum being observed (λ_{max} at 330 and 659 nm), together with retention of photoactivity [202].

G. ABSORPTION SPECTRA OF LANTHANIDE AND ACTINIDE PHTHALOCYANINES

Lanthanide phthalocyanines frequently form as polymeric species, with the dimer, Ln(Pc)$_2$, for example, Sm(Pc)$_2$ [304] and Nd(Pc)$_2$ [305], and the trimer Ln$_2$(Pc)$_3$, dominating much of this chemistry. The most studied of the lanthanide phthalocyanines has been lutetium. The spectral properties of the Ln phthalocyanines are complicated, and it is clear that these species can involve both oxidation and reduction of the Pc ring to give π cationic or π anionic species, as well as aggregation into dimers and trimers; for example, see [104]. While neutral Ln(Pc)$_2$ complexes are characterized by a band at 665 nm, the reduced, π anion exhibits two bands near 620 and 700 nm. For example, for Dy(Pc)$_2$ in CH$_2$Cl$_2$ [104], in addition to the usual phthalocyanine features (Q λ_{max} = 665 nm), the spectrum exhibits a more intense B band, at 330 nm, and a band near 455 nm. Reaction with Et$_3$N resulted in dramatic changes in the spectrum, especially in the Q band region, with new bands at 320, 624, and 700 nm, with a loss of intensity at 450 nm [104]. The reaction

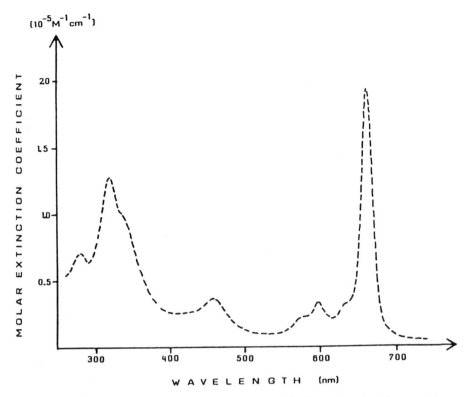

Figure 52 Absorption spectrum of $Tm(Pc)_2$ in CH_2Cl_2. Reproduced with permission from [315] (Fig. 2).

was characterized as

$$(H)Lu(Pc)_2 + Et_3N \rightarrow [Lu(Pc)_2]^- + Et_3NH^+$$

where the π anion radical species is formed [104].

Metallation with UO_2^+ results in formation of a superphthalocyanine, which involves five, rather than the usual four, pyrrole rings [29].

i. Dy, Er, Eu, Gd, La, Nd, Tb, Yb, Tm

Many spectral data have been published for phthalocyanine complexes which include these rare earths elements; a selection of papers includes [28,104,306–309]. M'Sadak et al. [104,309] and Kasuga et al. [308] have reported the preparation of several $(L)Ln(Pc)_2$ complexes.

Figure 52 shows the absorption spectrum for $Tm(Pc)_2$ in CH_2Cl_2. The B band region is particularly well resolved. The absorption spectrum of many $Ln(Pc)_2$ species extends into the near infrared region (Fig. 53). The spectral

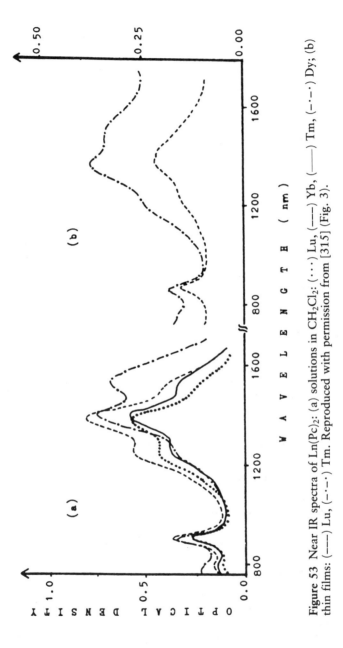

Figure 53 Near IR spectra of Ln(Pc)$_2$: (a) solutions in CH$_2$Cl$_2$: (···) Lu, (——) Yb, (——) Tm, (–·–·) Dy; (b) thin films: (——) Lu, (–·–·) Tm. Reproduced with permission from [315] (Fig. 3).

data for Gd, Eu, Nd, and La suggested that a "tri-decker" molecule was formed, $Ln_2(Pc)_3$, which did not involve an additional proton. Similar "tri-decker" structures were suggested for $Lu(Pc)_2$ following high-temperature sublimation [104], and $Y(Pc)_n$ following heating for several hours [308]. The spectrum of $Lu_2(Pc)_3$ exhibits a simpler spectrum than the other complexes, with a single band near 640 nm [104], while $Y_2(Pc)_3$ in chlorobenzene exhibits a Q band maximum at 633 nm, with a new band near 705 nm, assigned as the "X" band [308].

It would seem that MCD spectra could be useful in these assignments because of the significant changes in the molecular geometry and ring oxidation states between these different species [80]. Similarly complicated species that, in this case, involve a rare earth, a single phthalocyanine ring, and dibenzoylmethane (DBM), have been reported by Sugimoto et al. [28,307], for a range of rare earths. The absorption spectrum for (DMB)LnPc (where Ln = Tb and Er) exhibits bands at 336 and 679 nm, with no absorption in the 450-nm region. The spectrum closely resembles that of ZnPc in DCM [92]. A quite different spectrum is observed when protonation of the complex occurs. It is suggested that the anion radical forms [307].

ii. Lu-Phthalocyanines

$Lu(Pc)_2$ has a rich redox chemistry [37,38,95,96,104,310–313]. Recently, semiconductor properties have been reported for thin films made by cosubliming $LuPc_2$ and an electron acceptor [311]. Representative absorption spectra between 400 and 1000 nm have been reported for the neutral, π cationic, and π anionic species of $Lu(Pc)_2$ in CH_3Cl, DCM, CH_3CN, and DMF [95,96,311,314,315], as well as thin films [315,316,317]. Spectral data have also been reported for the octamethylsubstituted $Lu[(CH_3)_8$-$Pc]_2$ as thin films [37,317] and in solution [313]. Except for bands at 900 and near 1400 nm, for both complexes as thin films [316], and in CH_2Cl_2 solutions [315], the Q band region of $LuPc_2$ is similar [95,96,311,314,315] to that of MgPc [9]. The B band region is resolved in the data of Markovitsi et al. [315] for $Tm(Pc)_2$, the spectrum in green solutions of which is reported to be the same as $Lu(Pc)_2$, $Yb(Pc)_2$, and $Dy(Pc)_2$ [315]. It has been suggested, for example [315], that the bands near 450 and 900 nm in the absorption spectrum of the neutral $Lu(Pc)_2$ in DMF, DCM, and CH_3CN [95,315] are π–π, that arise from the Pc(-1) ring that is part of the Pc(-1)Lu(III)Pc(-2) "double-decker" complex. These bands would, therefore, resemble the bands observed for simpler π cation radical species, for example the $[ZnPc(-1)]^+$ complex [27]. The B band region exhibits considerable resolution between 270 and 380 nm [315], with $\lambda_{max} = 360$ nm). These workers [311,315] have also reported detailed spectral data for the region 800–1700 nm for both DCM solutions and thin films. The bands observed are redox state dependent, and it has been suggested that the

1382-nm band is charge transfer in nature [315]. Clearly, an MCD study would provide considerable detail that could be used in these comparisons.

Addition of octaalkyl groups greatly changes the physical properties of the phthalocyanine. Besbes et al. [37] report that these substituted $Lu(R_8Pc)_2$ complexes exhibit liquid crystal properties.

Using Langmuir–Blodgett techniques, a film of $Lu(R_8-Pc)_2$ was laid onto a Pt electrode, where the R substituents are the long chain alkyl, octyl (C_8), or dodecyl (C_{12}) chains [37]. Absorption spectra of the neutral, one-electron oxidized and one-electron reduced species have been measured [37]. The spectra of all three redox species resemble the spectral data of phthalocyanines in solution [311,313] rather than the broad spectrum expected for a film. This effect is observed presumably because the bulky chains reduced any possible intramolecular interactions [37,313].

iii. UPc and ThPc

Complexes similar to those reported by Sugimoto for the lanthanides [307] have been reported for U and Th by Guilard et al. [319]. In these 1:1 complexes between the actinide and a normal phthalocyanine ring (see below for "superphthalocyanines"), two acac chelates occupy the second half of the coordination sphere of the out-of-plane actinide. The absorption spectrum, for, typically, $(acac)_2ThPc$ in PhCN [319], has band maxima at 353, 617, and 684 nm. As above, this spectrum is close to that of ZnPc [92] or MgPc [9].

iv. $SPcUO_2$

The superphthalocyanine, $SPcUO_2$, is formed directly from o-cyanobenzene and UO_2Cl_2 [30]. Subsequent reactions of UO_2SPc with metallic salts (MX_2) yield the phthalocyanine species [30]. Three main bands are observed in the absorption spectrum of UO_2SPc, at about 420, 810, and 914 nm, for solutions in toluene, α-chloronaphthalene [29], or α-chloronaphthalene/DMF mixtures [30]. The absorption spectrum of UO_2SPc (Fig. 54) is considerably broader and red-shifted when compared with the spectra of ZnPc and MgPc.

H. SPECTRA FROM THE SOLID STATE

The solid state chemistry of phthalocyanines involves both crystals and polymorphic thin film material. Much of the spectroscopic data were first reported during the 1960s and 1970s. Some transmission spectra and many more reflectance spectra have been reported for crystals (which exist as the β polymorph) [50], however, because of intense absorption, single crystals are

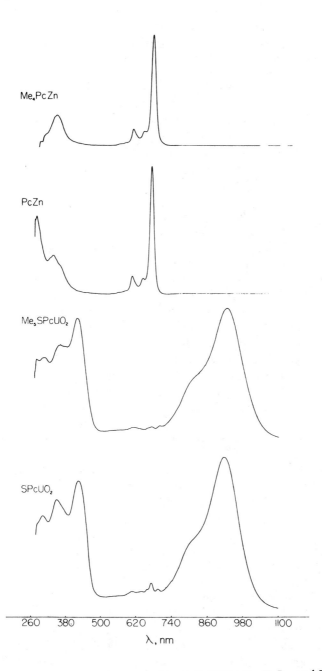

Figure 54 Absorption spectra in benzene of (top) Zn(CH$_3$)$_4$Pc, ZnPc, and UO$_2$-(CH$_3$)$_5$SPc and (bottom) UO$_2$SPc (SPc = superphthalocyanine). Reproduced with permission from [30] (Fig. 6).

usually too optically dense, and too small, for extensive spectroscopic study [50]. Few transmission spectra and no MCD data spectra are available directly from crystals. On the other hand, thin films are extremely easy to study because even a single monolayer provides enough absorption for the spectrum to be measured readily. The presence of the band structure is readily apparent from the purple reflection and the blue transmission of these films.

The preparation of a thin film may be achieved by simply subliming the phthalocyanine on to a quartz disk (see Figs. 8 and 24). Many phthalocyanines sublime readily near 300°C in a high vacuum [320]. The optical properties of these films on quartz disks have been extensively studied [51,52,53,54,213,239]. The thin film phthalocyanine complexes are chemically reactive. α-FePc and α-ZnPc react with pyridine to form complexes [276], and can be oxidized by Br_2.

Because spectroscopic measurements between liquid helium temperatures and 400 K have been carried out routinely on thin films, it might be expected that the analysis and assignment of the phthalocyanine absorption spectrum would proceed via the thin film species rather than via complexes in solution. That this is not the case stems from the extensive band broadening, arising from the Davydov effect, which is observed from the thin film species [52,213]. The Davydov coupling broadens and splits each of the transitions observed in solution, in much the same manner as described above for the solution dimers. Thus, while spectra of high quality have been reported for thin films, it has proven difficult to relate spectral effects observed in solution with transitions observed in the solid state [52]. It is important to note that while several authors have suggested that the solid state spectra actually resemble solution spectra, this is not the case. In general, solid state spectra for phthalocyanine complexes are very much broader than their respective solution spectrum.

The band structure of MPc films laid on to conducting electrodes have been exploited for use electrochemically (see, for example, Fan and Faulkner's report of the electrochemical properties of thin films of H_2Pc, ZnPc, and NiPc [321], the electrochromism of films of $Lu(Pc)_2$ [309,310,317], and the study of the reduction of O_2 by FePc films on gold electrodes [322]). Highly conducting, iodine-doped phthalocyanines have been prepared in the solid state that involve stacked phthalocyanine units [323]. Very complicated solid state spectra have been reported for these species, which suggest that partial oxidation of the π ring occurs when the iodine is doped into the phthalocyanine crystal; examples include $CoPcI_x$ [324] and $NiPcI_x$ [325]. Such compounds are expected to exhibit not only the characteristics of a molecular metal, but also new electrochemical and photochemical properties. The $Lu(Pc)_2$ complexes may also behave as molecular semiconductors [311]. It is not unexpected that such an exciting spectroelectrochemistry should be found for phthalocyanine complexes in the solid state, considering the properties already reported from solution chemistry. But these special properties of thin films remind us of the remarkably high molar absorbances, and how a sink of 18 π electrons can readily be oxidized or reduced with minimal changes to the metal binding.

A common theme in solid state spectral studies is to obtain spectral data that extend into the far-UV region. In this respect, spectra from thin films can be compared with those obtained from vapor phase [56,57,326] and matrix-isolated [58,59,141] phthalocyanines. Techniques which have been studied as a means of obtaining solvent-free absorption and magnetic circular dichroism spectra from the Q band to the far UV.

Examples of reflection spectral data from crystals include the reports for Co, Ni, Cu, Zn, and H_2Pc by Day and Williams [50] and Fielding and MacKay [326] for H_2, Ni, Cu, Zn, Cr, Co, and FePc (both crystal and vapor spectra are described).

i. Spectra of Phthalocyanines Deposited as Thin Films

There are three, well-characterized polymorphs in the solid state, each with a unique X-ray powder diffraction pattern, IR spectrum, [213,327–330] and absorption spectrum [212,331]. The α-polymorph can be formed by precipitation [330] or sublimation at low temperatures [52–54,213], β is formed from sublimation at higher temperatures [53,54] or following thermal treatment of α films [332], and for single crystals grown following sublimation under vacuum [50,218]. The χ-polymorph is formed following prolonged grinding [212,331].

All three species have been prepared as thin films and the molecular packing for each has been described. The lattices involve stacks of phthalocyanine or H_2Pc molecules that are tilted such that the interactions between the rings are stronger in the β phase than in the α phase, even though the ring-to-ring separation remains about constant at 340 pm, whereas the metal-to-metal distance is longer in β (at 480 pm), than in α (at 380 pm); see, for example, for CrPc [327] and for FePc and CoPc [328]. The packing in the β phase is also found in crystals and the optical spectra of crystals closely resemble those of β phase thin films [212]. This is quite different to the situation in the porphyrins where solution-like spectra are obtained from crystals. The α phase can be converted into the β phase by heating [333]. The spectral data observed for all three polymorphs is usually interpreted in terms of Davydov splitting of the main π–π* states according to the number of translationally inequivalent molecules in the unit cell. For β this is known to be two, so two-banded spectra are expected for metallophthalocyanines and four-banded spectra for H_2Pc. This is what is essentially observed experimentally [49,50]; analysis of MCD spectra suggests that the D_{4h} symmetry of the ring is maintained for phthalocyanines, and that in α-H_2Pc the effect of the two protons is lost such that again a D_{4h} spectrum is observed.

Transmission data for an extensive series of phthalocyanines have been reported by Lucia and Verderame [213] for H_2Pc, VOPc, CuPc, NiPc, CoPc, and ZnPc, for both α- and β-polymorphs, by Schechtman [54] for H_2Pc, CuPc, NiPc, ZnPc, and (Cl)CuPc; by Schechtman and Spicer [53] for α-polymorphic H_2Pc, CuPc, and ZnPc, and by Wagner et al. [320] for the β-polymorphs of H_2Pc, MgPc, CuPc, VOPc, ZnPc, and PbPc. Absorption and MCD data for a similar range of complexes were analyzed by Hollebone and Stillman [52] (spectra were reported for α-H_2, Co, Ni, and CuPc, see Figs. 8 and 24).

Lyons et al. [218] reported polarized absorption spectra for a single crystal in the β-polymorphic form, comparing these spectral data with those of H_2Pc in 1-ClN. Meier et al. [178] have reported the spectrum of $Zn(OCH_3)_8Pc$ in KBr, and the photoconductive spectra of $Zn(CN)_8Pc$ and $Zn(OCH_3)_8Pc$.

Grenoble and Drickamer [264] have studied the pressure dependence of the absorption spectra of H_2Pc and $(L)_2FePc$, L = py and 4-pic, in the solid state. The visible region spectrum for $(py)_2FePc$ is very broad at 20 kbar pressure, which is typical for the α- and β-polymorphic forms. Increasing the pressure shifts and Q band maximum from 675 nm (at 20 kbar) to 758 nm (at 140 kbar). In the UV region, increasing the pressure results in a red shift of the 416-nm band and also a red shift of a band out of the B band envelope, so that at 121 kbar, the absorption spectrum begins to resemble that of $[(CN^-)_2-Fe(II)Pc]^{2-}$. Grenoble and Drickamer characterize the pressure dependence of the optical and Mossbauer data in terms of an increase in the fraction of intermediate spin for the Fe(II) to 50% at 140 kbar.

The published spectra of thin films extend far into the UV region and cover all the major transitions responsible for the color of phthalocyanines. Phthalocyanine crystals and thin films are electro- and photoconductive, so many of the reports of absorption spectra from solid state matrices include information on conductivity [50,326]. These optical data relate directly to the phthalocyanine complexes cast or sublimed on to surfaces for use as electrodes [232], so today there may be new uses for the older literature. For example, the photocurrent action spectrum recorded in the Q band region for SnO_2/H_2Pc electrodes in 5 mM H_2O/p-benzoquinone [334] quite closely resemble the absorption spectrum of α-H_2Pc.

ii. Spectra of the α, β, and χ Polymorphic Phases of Phthalocyanines

The low-energy bands in the spectra of phthalocyanines as thin films are remarkably independent of the central metal. The major features are dependent on the polymorphic character [53,54,213,331]. In general α phase phthalocyanine complexes sublimed onto quartz exhibit an absorption spectrum [52,53,54] extending from 800 nm to below 200 nm. The Q band is characterized by a shoulder on the red edge. The β-polymorph exhibits a similar spectrum, except that in the Q region, the shoulder lies on the blue edge of the main band [53,54]. As described below, the χ-polymorph exhibits a much more resolved set of bands in the Q region, extending beyond 800 nm. We describe below selected examples of the spectra of α and β phthalocyanines.

iii. H_2Pc

Absorption spectra of all three polymorphic forms have been reported, α-H_2Pc (Fig. 24) [331], β-H_2Pc [331], and χ-H_2Pc [331]. The Xerox group [212] discovered that α-H_2Pc could be converted into χ-H_2Pc by milling for a week. The absorption spectrum of the χ-polymorph is quite different when compared with the spectra of the α and β-polymorphs, in that the main Q band envelope resolves completely into two bands [212,331]. The data were analyzed [212] in terms of exciton splitting of dimeric H_2Pc, such that four components derived from the original split Q_x and Q_y are observed (in cm^{-1}) at 12,500 (M$-$), 16,380 (M$+$), 15,490 (L$-$), and 17,150 (L$+$).

iv. VOPc

Huang and Sharp [239] reported Q band absorption and luminescence spectra from particulates embedded in a polymer film, and from an evaporated film. λ_{max} in the Q band for the film were 635, 660, 725 nm, spectral characteristic that are similar to those of the β-polymorph [213].

v. FePc

Ercolani et al. [335] have reported thin film spectra for β-FePc and two μ-oxo dimer isomers, Fe(III)Pc—O—Fe(III)Pc. The spectral data for thin films of the μ-oxo dimers are quite similar to the spectra of the β-polymorph of FePc. Savy et al. [322] report the spectral properties of a thin film of Fe(III)Pc on a gold electrode. In the Q band region, band maxima are 565 and 624 nm, with an envelope that is characteristic of α-FePc, and a band at 712 nm, with a band envelope characteristic of the β-polymorph.

vi. CoPc

Schechtman and Spicer [53] and Hollebone and Stillman [52] have reported similar absorption data. Figure 8 shows that the first 4 π–π^* bands of α-CoPc are at 121, 259, 332, and 623 nm; Schechtman and Spicer's spectra include two further bands, at 175 and 159 nm. These spectra are quite similar to those observed for H_2Pc, CuPc, and NiPc, although the band maxima and absorption coefficients of the 259- and 332-nm bands do appear to change from metal to metal, much like the effects observed for phthalocyanines in solution.

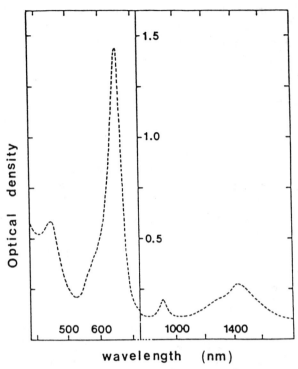

Figure 55 Absorption spectrum of a thin film (1 = 1000 Å) of Lu(Pc)$_2$. Reproduced with permission from [316] (Fig. 1).

vii. CuPc

The polarized reflectance spectrum for the Q region from crystalline CuPc shows the broadness observed for transmission spectra from α-CuPc as a thin film [49]. These authors assign the spectral data as arising from exciton splitting of the x/y pairs of the E_u π^* state. They observed both parallel (from A_u symmetry) and perpendicular (from B_u symmetry) polarization in Q region bands. Optical conductivity spectra for the Q band were very much better resolved, with a strong band, assigned to one of the Davydov components, that intensified at 30 K [49].

viii. Summary of the Solid-State Spectral Data

Generally, the spectra of phthalocyanines as thin films are readily accessible. However, these data do not offer the solvent-free envelope that had been hoped for. Films of sterically hindered species, like Lu(R$_8$-Pc)$_2$, where the R group is a long alkyl chain (Fig. 55), do exhibit solution-like absorption spectra from the solid state [316]. Redox chemistry of sublimed phthalocyanines has been demonstrated with the formation of cation radical species on the quartz surface. The analysis used in assigning the spectra from thin films is currently of use in the assignment of the spectra observed for dimers formed in solution either as a result of aggregation reactions or because the rings are tied together by connecting or bridging groups.

ACKNOWLEDGMENTS

Our research on the properties of phthalocyanines has been funded by NSERC of Canada, Imperial Oil Ltd., and the Academic Development Fund at the UWO. T.N. thanks CIDA for a scholarship during her graduate training at the UWO. M.J.S. would like to acknowledge the special contributions made to our work on phthalocyanines by Kathy Martin Creber, Edward Ough, Bob Kitchenham, Scott Kirby, Zbyszek Gasyna, and William Browett. We would also like to thank the many undergraduates, who have contributed to our general knowledge on this subject over the last decade. We would like to express our gratitude to Zbyszek Gasyna for his very considerable help in preparing the text and the figures, and to Professor Barry Lever for his helpful comments on the manuscript.

REFERENCES

1. R. P. Linstead, *J. Chem. Soc.*, (1934) 1016.
2. F. H. Moser and A. L. Thomas, *The Phthalocyanines*, Vol. I, *Properties*, CRC Press, Boca Raton, FL, 1983.
3. M. J. Stillman and A. J. Thomson, *J. Chem. Soc., Faraday Trans.*, II, 70 (1974) 790.
4. A. B. P. Lever, *Adv. Inorg. Radiochem.*, 7 (1965) 27.
5. J. R. Darwent, P. Douglas, A. Harriman, G. Porter and M. C. Richoux C., *Chem. Rev.*, 44 (1982) 83.
6. T. Nyokong, "Spectroscopic and Electrochemical Studies of Phthalocyanines." Ph.D. awarded 1986 by UWO.
7. M. J. Stillman, "Spectroscopic Studies of Metal Complexes." Ph.D. awarded 1973 by University of East Anglia, UK.
8. N. S. Hush and I. S. Woolsey, *Mol. Phys.*, 21 (1971) 465.
9. E. Ough, T. Nyokong, K. A. M. Creber and M. J. Stillman, *Inorg. Chem.*, 27 (1988) 2725.
10. A. B. P. Lever, S. R. Pickens, P. C. Minor, S. Licoccia, B. S. Ramaswamy and K. Magnell, *J. Am. Chem. Soc.*, 103 (1981) 6800.
11. B. W. Dale, *J. Chem. Soc., Faraday Trans.* (1969) 331.
12. R. Taube, *Pure Appl. Chem.*, 38 (1974) 427.
13. A. B. P. Lever, S. Licoccia, K. Magnell, P. C. Minor and B. S. Ramaswamy, *ACS Symp. Ser.*, 201 (1982) 237.

14. F. Cariati, D. Galizzioli, F. Morazzoni and C. Busetto, *J. Chem. Soc., Dalton Trans.*, (1975) 556.

15. C. Weiss, *J. Mol. Spectr.*, 44 (1972) 37.

16. L. I. Shipman, T. M. Cotton, J. R. Norris and J. J. Katz, *J. Am. Chem. Soc.*, 98 (1976) 8222.

17. D. Frackowiak, D. Bauman, H. Manikowski, W. R. Browett and M. J. Stillman, *Biophys. Chem.*, 28 (1987) 101.

18. M. Zerner, M. Gouterman and H. Kobayashi, *Theoret. Chim. Acta*, 6 (1966) 363.

19. J. W. Owens and C. J. O'Connor, *Coord. Chem. Rev.*, 84 (1988) 1.

20. A. Ceulemans, W. Oldenhof, C. Gorller-Walrand and L. G. Vanquickenborne, *J. Am. Chem. Soc.*, 108 (1986) 1155.

21. M. Gouterman, in *The Porphyrins*. Vol. III, Part A. *Physical Chemistry*, D. Dolphin, Ed., Academic Press, New York, 1978, pp. 1–165.

22. G. T. Byrne, R. P. Linstead and A. R. Lowe, *J. Chem. Soc.*, (1934) 1017.

23. R. P. Linstead and A. R. Lowe, *J. Chem. Soc.*, (1934) 1022.

24. C. E. Dent and R. P. Linstead, *J. Chem. Soc.*, (1934) 1027.

25. R. F. Ziolo, W. H. H. Gunther and J. M. Troup, *J. Am. Chem. Soc.*, 103 (1981) 4629.

26. M. J. Stillman and A. J. Thomson, *J. Chem. Soc., Faraday Trans., II*, 70 (1974) 805.

27. T. Nyokong, Z. Gasyna and M. J. Stillman, *ACS Symp. Ser.*, 321 (1986) 309.

28. H. Sugimoto, T. Higashi and M. Mori, *Chem. Lett.*, (1982) 801.

29. E. A. Cuellar and T. J. Marks, *Inorg. Chem.*, 20 (1981) 3766.

30. T. J. Marks and D. R. Stojakovic, *J. Am. Chem. Soc.*, 100 (1978) 1695.

31. P. Sayer, M. Gouterman and C. R. Connell, *Acc. Chem. Res.*, 15 (1982) 73.

32. A. T. Gradyushko, A. N. Sevchenko, K. N. Solovyov and M. P. Tsvirko, *Photochem. Photobiol.*, 11 (1970) 387.

33. P. C. Minor, M. Gouterman and A. B. P. Lever, *Inorg. Chem.*, 24 (1985) 1894.

34. J. W. Dodd and N. S. Hush, *J. Chem. Soc.*, (1964) 4607.

35. D. W. Clack and J. R. Yandle, *Inorg. Chem.*, 11 (1972) 1738.

36. G. Engelsma, A. Yamamoto, E. Markham and M. Calvin, *J. Am. Chem. Soc.*, 66 (1962) 2517.

37. S. Besbes, V. Plichon, J. Simon and J. Vaxiviere, *J. Electroanal. Chem.*, 237 (1987) 61.

38. P. Turek, P. Petit, J. J. Andre, J. Simon, R. Even, B. Boudjema, G. Guillaud and M. Maitrot, *J. Am. Chem. Soc.*, 109 (1987) 5119.

39. P. A. Forshey and T. Kuwana, *Inorg. Chem.*, 22 (1983) 699.

40. T. Inabe, J. G. Gaudiello, M. K. Moguel, J. W. Lyding, R. L. Burton, W. J. McCarthy, C. R. Kannewurf and T. J. Marks, *J. Am. Chem. Soc.*, 108 (1986) 7595.

41. D. W. DeWulf, J. K. Leland, B. L. Wheeler, A. J. Bard, D. A. Batzel, D. R. Dininny and M. E. Kenney, *Inorg. Chem.*, 26 (1987) 266.

42. D. Dolphin, B. R. James and H. C. Welborn, *ACS Symp. Ser.*, 23 (1981) 563.

43. S. E. Creager, S. S. Raybuck and R. W. Murray, *J. Am. Chem., Soc.*, 108 (1986) 4225.

44. L. A. Bottomley, L. Olson and K. M. Kadish, *ACS Symp. Ser.*, 201 (1982) 279.

45. I. Fujita, L. K. Hanson, F. A. Walker and J. Fajer, *J. Am. Chem., Soc.*, 105 (1983) 3296.

46. R. Langlois, H. Ali, N. Brasseur, J. R. Wagner and J. E. Lier, *Photochem. Photobiol.*, 44 (1986) 117.

47. J. D. Spikes, *Photochem. Photobiol.*, 43 (1986) 691.

48. E. Ben-Hur and I. Rosenthal, *Photochem. Photobiol.*, 43 (1986) 615.

49. Y. Iyechika, K. Yakushi and H. Kuroda, *Chem. Phys.*, 87 (1984) 101.

50. P. Day and R. J. P. Williams, *J. Chem. Phys.*, 37 (1962) 567.

51. B. R. Hollebone and M. J. Stillman, *Chem. Phys. Lett.*, 29 (1974) 284.

52. B. R. Hollebone and M. J. Stillman, *J. Chem. Soc., Faraday Trans.*, II, 74 (1978) 2107.

53. B. H. Schechtman and W. E. Spicer, *J. Mol. Spectr.*, 33 (1970) 28.

54. B. H. Schechtman, SU-SEL-68-087, Stanford Univ. Stanford Electronics Laboratories. Technical Report 5207-2. *Photoemission and Optical Studies of Organic Solids: Phthalocyanines and Porphrins*, Stanford University, Stanford, CA, 1968.

55. T-H. Huang, K. E. Rieckhoff and E. M. Voigt, *J. Chem. Phys.*, 77 (1982) 3424.

56. D. Eastwood, L. Edwards, M. Gouterman and J. Steinfeld, *J. Mol. Spectr.*, 20 (1966) 381.

57. L. Edwards and M. Gouterman, *J. Mol. Spectr.*, 33 (1970) 292.

58. L. Bajema, M. Gouterman and B. Meyer, *J. Mol. Spectr.*, 27 (1968) 225.

59. T. C. Van Cott, J. L. Rose, G. C. Misener, B. E. Williamson, A. E. Schrimp, M. E. Boyle and P. N. Schatz, *J. Phys. Chem*, 93 (1989) 2999.

60. L. K. Lee, N. H. Sabelli and P. R. LeBreton, *J. Phys. Chem.*, 86 (1982) 3926.

61. A. Henriksson, B. Roos and M. Sundbom, *Theoret. Chim. Acta*, 27 (1972) 303.

62. A. Henriksson and M. Sundbom, *Theoret. Chim. Acta*, 27 (1972) 213.

63. A. J. McHugh, M. Gouterman and C. Weiss, *Theoret. Chim. Acta*, 24 (1972) 346.

64. P. D. Hale, W. J. Pietro, M. A. Ratner, D. E. Ellis and T. J. Marks, *J. Am. Chem. Soc.*, 109 (1987) 5943.

65. T. Nyokong, Z. Gasyna and M. J. Stillman, *Inorg. Chem.*, 26 (1987) 1087.

66. D. Dolphin, Ed., *The Porphyrins*, Vol. III, Part A, *Physical Chemistry*, Academic Press, New York, 1978.

67. J. A. Shelnutt and M. R. Ondrias, *Inorg. Chem.*, 23 (1984) 1175.

68. F. H. Moser and A. L. Thomas, *Phthalocyanine Compounds*, Reinhold, New York, 1963.

69. R. S. McDonald and P. A. Wilks, *Anal. Spectr.*, 42 (1988) 151.

70. W. R. Browett and M. J. Stillman, *Comp. Chem.*, 11 (1987) 73.

71. M. S. Fischer, D. H. Templeton, A. Zalkin and M. Calvin, *J. Am. Chem. Soc.*, 93 (1971) 2622.

72. J. F. Kirner, W. Dow and W. R. Scheidt, *Inorg. Chem.*, 15 (1976) 1685.

73. J. P. Collman, J. L. Hoard, N. Kim, G. Lang and C. A. Reed, *J. Am. Chem. Soc.*, 97 (1975) 2676.

74. N. H. Sabell and C. A. Melendres, *J. Phys. Chem.*, 86 (1982) 4342.

75. A. M. Schaffer, M. Gouterman and E. R. Davidson, *Theoret. Chim. Acta*, 30 (1973) 9.

76. M. Whalley, *J. Chem. Soc.*, (1961) 866.

77. K. J. Reimer, M. M. Reimer and M. J. Stillman, *Can. J. Chem.*, 59 (1981) 1388.

78. L. Edwards, M. Gouterman and C. B. Rose, *J. Am. Chem. Soc.*, 98 (1976) 7638.

79. H. C. Longuet-Higgins, C. W. Rector and J. R. Platt, *J. Chem. Phys.*, 18 (1950) 1174.

80. K. A. Martin and M. J. Stillman, *Can. J. Chem.*, 57 (1979) 1111.

81. W. R. Browett and M. J. Stillman, *Biochim. Biophys. Acta*, 623 (1980) 21.

82. W. R. Browett and M. J. Stillman, *Biochim. Biophys. Acta*, 660 (1981) 1.

83. Z. Gasyna, W. R. Browett and M. J. Stillman, *Biochemistry*, 27 (1988) 2503.

84. W. R. Browett, Z. Gasyna and M. J. Stillman, *J. Am. Chem. Soc.*, 110 (1988) 3633.

85. H. B. Dunford and J. S. Stillman, *Coord. Chem. Rev.*, 19 (1976) 187.

86. J. H. Dawson and M. Sono, *Chem. Rev.*, 87 (1987) 1255.

87. N. Kobayashi, M. Koshiyama, K. Funayama, T. Osa, H. Shirai and K. Hanabusa, *J. Chem. Soc., Chem. Commun.* (1983) 913.

88. P. Day, H. A. O. Hill and M. G. Price, *J. Chem. Soc.*, (A) (1968) 90.

89. D. Stynes and B. R. James, *J. Am. Chem. Soc.*, 96 (1974) 2733.

90. T. Nyokong, Z. Gasyna and M. J. Stillman, *Inorg. Chim. Acta*, 112 (1986) 11.

91. W. R. Browett and M. J. Stillman, *Inorg. Chim. Acta*, 49 (1981) 69.

92. T. Nyokong, Z. Gasyna and M. J. Stillman, *Inorg. Chem.*, 26 (1987) 548.

93. P. S. Vincett, E. M. Voigt and K. E. Rieckhoff, *J. Chem. Phys.*, 55 (1971) 4131.

94. J. A. Menapace and E. R. Bernstein, *J. Chem. Phys.*, 87 (1987) 6877.

95. G. A. Corker, B. Grant and N. J. Clecak, *J. Electroanal. Chem.*, 126 (1979) 1339.

96. A. T. Chang and J. C. Marchon, *Inorg. Chim. Acta*, 53 (1981) L241.

97. A. Corsini and O. Herrmann, *Talanta*, 33 (1986) 335.

98. C. C. Leznoff, S. M. Marcuccio, S. Greenberg, A. B. P. Lever and K. B. Tomer, *Can. J. Chem.*, 63 (1985) 623.

99. W. A. Nevin, W. Liu, S. Greenberg, M. R. Hempstead, S. M. Marcuccio, M. M. Melnik, C. C. Leznoff and A. B. P. Lever, *Inorg. Chem.*, 26 (1987) 891.

100. C. C. Leznoff, H. Lam, W. A. Nevin, N. Kobayashi, P. Janda and A. B. P. Lever, *Angew. Chem.*, 26 (1987) 1021.

101. M. R. Hempstead, A. B. P. Lever and C. C. Leznoff, *Can. J. Chem.*, 65 (1987) 2677.

102. A. B. P. Lever, M. R. Hempstead, C. C. Leznoff, W. Liu, M. Melnik, W. A. Nevin and P. Seymour, *Pure Appl. Chem.*, 58 (1986) 1467.

103. K. Kasuga and M. Tsutsui, *Coord. Chem. Rev.*, 32 (1980) 67.

104. M. M'Sadak, J. Roncali and F. Garnier, *J. Chim. Phys.*, 83 (1986) 211.

105. O. T. E. Sielcken, M. M. van Tilborg, M. F. M. Roks, R. Hendriks, W. Drenth and R. J. M. Nolte, *J. Am. Chem. Soc.*, 109 (1987) 4261.

106. J. H. Dawson and D. M. Dooley, in *Iron Porphyrins*, Part 3, A. B. P. Lever and H. Gray, Eds., VCH, New York, 1988.

107. W. R. Browett and M. J. Stillman, *Biochim. Biophys. Acta*, 577 (1979) 291.

108. G. Sievers, P. M. A. Gadsby, J. Peterson and A. J. Thomson, *Biochim. Biophys. Acta*, 742 (1983) 637.

109. D. G. Eglinton, P. M. A. Gadsby, G. Sievers, J. Peterson and A. J. Thomson, *Biochim. Biophys. Acta*, 742 (1983) 648.

110. G. Sievers, P. M. A. Gadsby, J. Peterson and A. J. Thomson, *Biochim. Biophys. Acta*, 742 (1983) 659.

111. W. H. Woodruff, R. J. Kessler, N. S. Ferris, R. F. Dallinger, K. R. Carter, T. M. Antalis and G. Palmer, *ACS Symp. Ser.*, 201 (1982) 625.

112. D. A. Schooley, E. Bunnenberg and C. Djerassi, *Proc. Natl. Acad. Sci. U.S.A.*, 53 (1965) 579.

113. S. Suzuki, T. Yoshimura, A. Nakahara, H. Iwasaki, S. Shidara and T. Matsubara, *Inorg. Chem.*, 26 (1987) 1006.

114. J. H. Dawson, L. A. Andersson and M. Sono, *J. Biol. Chem.*, 257 (1982) 3606.

115. N. Foote, P. M. A. Gadsby, M. J. Berry, C. Greenwood and A. J. Thomson, *Biochem. J.*, 246 (1987) 659.

116. P. R. Griffiths, J. A. Pierce and G. Hongjin, *Transform Techniques in Chemistry*, P. R. Griffiths, Ed., Plenum, New York, 1978, Ch. 2, pp. 29–54.

117. D. Frackowiak, D. Bauman and M. J. Stillman, *Biochim. Biophys. Acta*, 681 (1982) 273.

118. A. F. Schreiner, J. D. Gunter, D. J. Hamm, I. D. Jones and R. C. White, *Inorg. Chem.*, 26 (1978) 151.

119. B. Briat, D. A. Schooley, R. Records, E. Bunnenberg and C. Djerassi, *J. Am. Chem. Soc.*, 89 (1967) 6170.

120. M. V. Belkov and A. P. Losev, *Spectr. Lett.*, 11 (1978) 653.

121. D. Frackowiak and M. J. Stillman, *Acta Phys. Polonica*, A67 (1985) 829.

122. J. D. Keegan, A. M. Stolzenberg, Y. C. Lu, R. E. Linder, G. Barth, A. Moscowitz, E. Bunnenberg and C. Djerassi, *J. Am. Chem. Soc.*, 104 (1982) 4305.

123. J. D. Keegan, A. M. Stolzenberg, Y. C. Lu, R. E. Linder, G. Barth, A. Moscowitz, E. Bunnenberg and C. Djerassi, *J. Am. Chem. Soc.*, 104 (1982) 4317.

124. R. M. Berger, D. R. McMillin and R. F. Dallinger, *Inorg. Chem.*, 26 (1987) 3802.

125. W. R. Browett and M. J. Stillman, *Biophys. Chem.*, 19 (1984) 311.

126. R. Gale, A. J. McCaffery and M. D. Rowe, *J. Chem. Soc., Dalton Trans.*, (1972) 596.

127. P. J. Stephens, W. Suetaka and P. N. Schatz, *J. Chem. Phys.*, 44 (1966) 4592.

128. R. A. Goldbeck, B. R. Tolf, E. Bunnenberg and C. Djerassi, *J. Am. Chem. Soc.*, 109 (1987) 28.

129. A. Kaito, T. Nozawa, T. Yamamoto, M. Hatano and Y. Orii, *Chem. Phys. Lett.*, 52 (1977) 154.

130. W. R. Browett, A. F. Fucaloro, T. V. Morgan and P. J. Stephens, *J. Am. Chem. Soc.*, 105 (1983) 1868.

131. Z. Gasyna, W. R. Browett and M. J. Stillman, *ACS Symp. Ser.*, 321 (1986) 298.

132. Z. Gasyna, W. R. Browett and M. J. Stillman, *Inorg. Chem.*, 24 (1985) 2440.

133. Z. Gasyna, W. R. Browett and M. J. Stillman, *Inorg. Chem.*, 23 (1984) 382.

134. N. Kobayashi, M. Koshiyama, Y. Ishikawa, T. Aso, H. Shirai and N. Hojo, *Chem. Lett.*, (1984) 1633.

135. N. Kobayashi, H. Shirai and N. Hojo, *J. Chem. Soc., Dalton Trans.*, (1984) 2107.

136. C. Gall and D. Simkin, *Can. J. Spectr.*, 18 (1973) 130.

137. U. Simonis, F. A. Walker, P. L. Lee, B. J. Hanquet, D. J. Meyerhoff and W. R. Scheidt, *J. Am. Chem. Soc.*, 109 (1987) 2659.

138. N. Kobayashi and Y. Nishiyama, *J. Phys. Chem.*, 89 (1985) 1167.

139. C. Gall and D. Simkin, *Can. J. Spectr.*, 18 (1973) 124.

140. P. N. Schatz and A. J. McCaffery, *Q. Rev. Chem. Soc.*, 23 (1969) 552.

141. I. N. Douglas, R. Grinter and A. J. Thomson, *Mol. Phys.*, 28 (1974) 1377.

142. W. Chen, K. E. Rieckhoff and E. Voigt, *Mol. Phys.*, 59 (1986) 355.

143. A. D. Buckingham and P. J. Stephens, *Annu. Rev. Phys. Chem.*, 17 (1967) 399.

144. P. J. Stephens, *J. Chem. Phys.*, 52 (1970) 3489.

145. B. Briat, *Nato Adv. Study Ser.*, (1975), 375.

146. J. Michl, *Pure Appl. Chem.*, 52 (1980) 1549.

147. J. Michl, *J. Am. Chem. Soc.*, 100 (1978) 6801.

148. D. M. Dooley and J. H. Dawson, *Coord. Chem. Rev.*, 60 (1984) 1.

149. M. Hatano and T. Nozawa, *Adv. Bio.*, 11 (1978) 95.

150. L. E. Vickery, *Methods Enzymol.*, LIV (1978) 284.

151. S. B. Piepho and P. N. Schatz, *Group Theory in Spectroscopy — With Applications to MCD*, Wiley, New York, 1983.

152. A. J. Thomson and M. K. Johnson, *Biochem. J.*, 191 (1980) 411.

153. J. I. Treu and J. J. Hopfield, *J. Chem. Phys.*, 63 (1975) 613.

154. Z. Gasyna, W. R. Browett and M. J. Stillman, *Inorg. Chem.*, 27 (1988) 4619.

155. E. Ough, Z. Gasyna and M. J. Stillman, *Inorg. Chem.*, to be published.

156. W. R. Browett and M. J. Stillman, *Comp. Chem.*, 11 (1987) 241.

157. C. H. Henry, S. E. Schnatterly and C. P. Slichter, *Phys. Rev.*, A583 (1965) 137.

158. P. J. Stephens, R. L. Mowery and P. N. Schatz, *J. Chem. Phys.*, 55 (1971) 224.

159. G. A. Osborne and P. J. Stephens, *J. Chem. Phys.*, 56 (1972) 609.

160. M. Vala, R. Pyzalski, J. Shakhsemampour, M. Eyring, J. Pyka, T. Tipton and J. C. Rivoal, *J. Chem. Phys.*, 86 (1987) 5951.

161. P. J. Stephens, *Chem. Phys. Lett.*, 2 (1968) 241.

162. M. J. Stillman, P. W. M. Jacobs, K. O. Gannon and D. J. Simkin, *Phys. Sol. Stat.*, 124 (1984) 261.

163. K. Schmitt, P. W. M. Jacobs and M. J. Stillman, *J. Phys. C*, 16 (1983) 603.

164. Z. Gasyna, W. R. Browett and M. J. Stillman, *ACS Symp. Ser.*, 321 (1986) 298.

165. V. E. Shashoua, *J. Am. Chem. Soc.*, 86 (1964) 2109.

166. P. N. Schatz, A. J. McCaffery, W. Suetaka, G. N. Henning and A. B. Ritchie, *J. Chem. Phys.*, 45 (1970) 722.

167. J. A. Shelnutt and V. Ortiz, *J. Phys. Chem.*, 89 (1985) 4733.

168. C. Misener, P. N. Schatz and M. J. Stillman, Unpublished data.

169. W. T. Simpson, *J. Chem. Phys.*, 17 (1949) 1218.

170. A. Dedieu, M. M. Rohmer and A. Veillard, in *Advances in Quantum Chemistry*, Vol. 16, P-O Lowdin, Ed., Academic Press, New York, 1982.

171. S. Basu, *Ind. J. Phys.*, 28 (1954) 511.

172. A. M. Schaffer and M. Gouterman, *Theoret. Chim. Acta*, 25 (1972) 62.

173. T. Nozawa, N. Kobayashi, M. Hatano, M. Ueda and M. Sogami, *Biochim. Biophys. Acta*, 626 (1980) 282.

174. M. H. Perrin, M. Gouterman and C. L. Perrin, *J. Chem. Phys.*, 50 (1969) 4137.

175. E. Ciliberto, K. A. Doris, W. J. Pietro, G. M. Reisner, D. E. Ellis, I. Fragala, F. H. Herbstein and M. A. Ratner, *J. Am. Chem. Soc.*, 106 (1984) 7748.

176. C. K. Jorgensen, *Prog. Inorg. Chem.*, 12 (1970) 101.

177. A. Louati, M. E. I. Meray, J. J. Andre, J. Simon, K. M. Kadish, M. Gross and A. Giraudeau, *Inorg. Chem.*, 24 (1985) 1175.

178. H. Meier, W. Albrecht, D. Wöhrle and A. Jahn, *J. Phys. Chem.*, 90 (1986) 6349.

179. H. Yanagi, Y. Ueda and M. Ashida, *Bull. Chem. Soc. Jpn.*, 61 (1988) 2313.

180. D. Wöhrle and V. Schmidt, *J. Chem. Soc., Dalton Trans.*, (1988) 549.

181. B. N. Achar, G. M. Fohlen, J. A. Parker and J. Keshavayya, *Polyhedron*, 6 (1987) 1463.

182. S. Lukac and J. R. Harbour, *J. Chem. Soc.*, (1982) 154.

183. A. R. Monahan, J. A. Brado and A. F. DeLuca, *J. Phys. Chem.*, 76 (1972) 1994.

184. W. Freyer, S. Dahne, L. Q. Minh and K. Teuchner, *Z. Chem.*, 26 (1986) 334.

185. K. Bernauer and S. Fallab, *Fasciculus VII*, XLV (1962) 2487.

186. N. Kobayashi, Y. Nishiyama, T. Ohya and M. Sato, *J. Chem. Soc., Chem. Commun.*, (1987) 390.

187. N. Kobayashi and A. B. P. Lever, *J. Am. Chem. Soc.*, 109 (1987) 7433.

188. J. M. Birchall, R. N. Haszeldine and J. O. Morley, *J. Chem. Soc.*, (C) (1970) 2667.

189. D. Wöhrle and G. Krawczyk, *Makromol. Chem.*, 187 (1986) 2535.

190. S. Gaspard, C. Giannotti, P. Maillard, C. Schaeffer and T. H. Tran-Thi, *J. Chem. Soc., Chem. Commun.*, (1986) 1239.

191. R. P. Seiders and J. R. Ward, *Anal. Lett.*, 17 (1984) 1763.

192. K. Bernauer and S. Fallab, *Fasciculus*, XLIV (1961) 1287.

193. L. C. Gruen and R. J. Blagrove, *Aust. J. Chem.*, 26 (1973) 319.

194. A. Harriman and M. C. Richoux, *J. Photochem.*, 14 (1980) 253.

195. D. J. Cookson, T. D. Smith, J. F. Boas and J. R. Pilbrow, *J. Chem. Soc., Dalton Trans.*, (1976) 1791.

196. D. J. Cookson, T. D. Smith, J. F. Boas, P. R. Hicks and J. R. Pilbrow, *J. Chem. Soc., Dalton Trans.*, (1977) 211.

197. W. A. Nevin, W. Liu, M. Melnik and A. B. P. Lever, *J. Electroanal. Chem.*, 213 (1986) 217.

198. G. Ferraudi, *Inorg. Chem.*, 18 (1979) 1005.

199. L. Trynda, H. Przywarska-Boniecka and T. Kosciukiewicz, *Inorg. Chim. Acta*, 135 (1987) 55.

200. A. Skorobogaty, T. D. Smith, G. Dougherty and J. R. Pilbrow, *J. Chem. Soc.*, (1985) 651.

201. G. Ferraudi and E. V. Srisankar, *Inorg. Chem.*, 17 (1978) 3164.

202. J. R. Darwent, I. McCubbin and G. Porter, *J. Chem. Soc.*, 78 (1982) 903.

203. S. Zecevic, B. Simic-Glavaski, E. Yeager, A. B. P. Lever and P. C. Minor, *J. Electroanal. Chem.*, 196 (1985) 339.

204. M. J. Camenzind and C. L. Hill, *Inorg. Chim. Acta*, 99 (1985) 63.

205. C. C. Leznoff, H. Lam, S. M. Marcuccio, W. A. Nevin, P. Janda, N. Kobayashi and A. B. P. Lever, *J. Chem. Soc., Chem. Commun.*, (1987) 699.

206. W. A. Nevin, W. Liu and A. B. P. Lever, *Can. J. Chem.*, 65 (1986) 855.

207. W. A. Nevin, M. R. Hempstead, W. Liu, C. C. Leznoff and A. B. P. Lever, *Inorg. Chem.*, 26 (1987) 570.

208. E. S. Dodsworth, A. B. P. Lever, P. Seymour and C. C. Leznoff, *J. Phys. Chem.*, 89 (1985) 5698.

209. M. Gouterman, D. Holten and E. Lieberman, *Chem. Phys.*, 25 (1977) 139.

210. M. Z. Zgierski, *Chem. Phys. Lett.*, 111 (1984) 553.

211. M. Z. Zgierski, *Chem. Phys. Lett.*, 124 (1986) 53.

212. J. H. Sharp and M. Lardon, *J. Phys. Chem.*, 72 (1968) 3230.

213. E. A. Lucia and F. D. Verderame, *J. Chem. Phys.*, 48 (1968) 2674.

214. Z. Gasyna, N. Kobayashi and M. J. Stillman, *J. Chem. Soc., Dalton Trans.*, in press, (1989).

215. A. R. Kane, J. F. Sullivan, D. H. Kenny and M. E. Kenney, *Inorg. Chem.*, 9 (1970) 1445.

216. A. R. Monahan, J. A. Brado and A. F. DeLuca, *J. Phys. Chem.*, 76 (1972) 446.

217. K. Y. Law, *J. Phys. Chem.*, 92 (1988) 4226.

218. L. E. Lyons, J. R. Walsh and J. W. White, *J. Chem. Soc.*, (1960) 167.

219. V. E. Bondybey and J. H. English, *J. Am. Chem. Soc.*, 101 (1979) 3446.

220. D. L. Ledson and M. V. Twigg, *Inorg. Chim. Acta*, 13 (1975) 43.

221. L. D. Rollmann and R. T. Iwamoto, *J. Am. Chem. Soc.*, 90 (1968) 1455.

222. S. M. Marcuccio, P. I. Svirskaya, S. Greenberg, A. B. P. Lever, C. C. Leznoff and K. B. Tomer, *Can. J. Chem.*, 63 (1985) 3057.

223. P. A. Barrett, D. A. Frye and R. P. Linstead, *J. Chem. Soc.*, (1938) 1157.

224. K. A. Martin and M. J. Stillman, *Inorg. Chem.*, 19 (1980) 2473.

225. H. Homborg and W. Kalz, *Z. Naturforsch.*, 33b (1978) 968.

226. H. Homborg and W. Kalz, *Z. Anorg. Allg. Chem.*, 514 (1984) 115.

227. H. Homborg and W. Kalz, *Z. Naturforsch.*, 33b (1978) 1063.

228. R. F. Ziolo and M. Extine, *Inorg. Chem.*, 20 (1981) 2709.

229. H. Homborg and K. S. Murray, *Z. Anorg. Allg. Chem.*, 517 (1984) 149.

230. H. Homborg, *Z. Anorg. Allg. Chem.*, 507 (1983) 35.

231. R. E. Linder, J. R. Rowlands and N. S. Hush, *Mol. Phys.*, 21 (1971) 417.

232. D. Belanger, J. P. Dodelet, L. H. Dao and B. A. Lombos, *J. Phys. Chem.*, 88 (1984) 4288.

233. P. Petelenz and M. Z. Zgierski, *Mol. Phys.*, 25 (1973) 237.

234. T. M. Mezza, N. R. Armstrong, G. W. Ritter, J. P. Iafalice and M. E. Kenney, *J. Electroanal. Chem.*, 137 (1982) 227.

235. I. W. Shim and W. M. Risen Jr., *J. Organomet. Chem.*, 260 (1984) 171.

236. B. L. Wheeler, G. Nagasubramanian, A. J. Bard, L. A. Schechtman, D. R. Dininny and M. E. Kenney, *J. Am. Chem. Soc.*, 106 (1984) 7404.

237. W. J. Pietro, T. J. Marks and M. A. Ratner, *J. Am. Chem. Soc.*, 107 (1985) 5387.

238. M. Gouterman, P. Sayer, E. Shankland and J. P. Smith, *Inorg. Chem.*, 20 (1981) 87.

239. T. H. Huang and J. H. Sharp, *Chem. Phys.*, 65 (1982) 205.

240. J. A. Elvidge and A. B. P. Lever, *J. Chem. Soc.*, (1961) 1257.

241. J. A. Elvidge and A. B. P. Lever, *Proc. Chem. Soc.*, (1959) 195.

242. A. Yamamoto, L. K. Phillips and M. Calvin, *Inorg. Chem.*, 7 (1968) 847.

243. A. B. P. Lever, P. C. Minor and J. P. Wilshire, *Inorg. Chem.*, 20 (1981) 2550.

244. A. B. P. Lever, J. P. Wilshire and S. K. Quan, *J. Am. Chem. Soc.*, 107 (1979) 3668.

245. N. T. Moxon, P. E. Fielding and A. K. Gregson, *J. Chem. Soc., Chem. Commun.*, (1981) 98.

246. A. B. P. Lever, J. P. Wilshire and S. K. Quan, *Inorg. Chem.*, 20 (1981) 761.

247. B. N. Figgis, E. S. Kucharski and G. A. Williams, *J. Chem. Soc., Dalton Trans.*, (1980) 1515.

248. B. Gonzales, J. Kouba, S. Yee, C. A. Reed, J. F. Kirner and W. R. Scheidt, *J. Am. Chem. Soc.*, 97 (1975) 3247.

249. L. H. Vogt, A. Zalkin and D. H. Templeton, *Inorg. Chem.*, 6 (1967) 1725.

250. J. G. Jones and M. V. Twigg, *Inorg. Chem.*, 8 (1969) 2120.

251. J. J. Watkins and A. L. Balch, *Inorg. Chem.*, 14 (1975) 2720.

252. D. W. Sweigart, *J. Chem. Soc., Dalton Trans.*, (1976) 1476.

253. F. Calderazzo, G. Pampaloni, D. Vitali, I. Collamati, G. Dessy and V. Fares, *J. Chem. Soc., Dalton Trans.*, (1980) 1965.

254. B. J. Kennedy, K. S. Murray, P. R. Zwack, H. Homborg and W. Kalz, *Inorg. Chem.*, 25 (1986) 2539.

255. L. A. Bottomley, C. Ercolani, J. N. Gorce, G. Pennesi and G. Rossi, *Inorg. Chem.*, 25 (1986) 2338.

256. W. D. Edwards, B. Weiner and M. C. Zerner, *J. Am. Chem. Soc.*, 108 (1986) 2196.

257. G. V. Ouedraogo, C. More, Y. Richard and D. Benlian, *Inorg. Chem.*, 20 (1981) 4387.

258. I. Collamati, *Inorg. Nucl. Chem. Lett.*, 17 (1981) 69.

259. L. A. Bottomley, J. N. Gorce, V. L. Goedken and C. Ercolani, *Inorg. Chem.*, 24 (1985) 3733.

260. M. M. Doeff and D. A. Sweigart, *Inorg. Chem.*, 20 (1981) 1683.

261. W. Kalz, H. Homborg, H. Kuppers, B. J. Kennedy and K. S. Murray, *Z. Naturforsch.*, 39B (1984) 1478.

262. J. Metz, O. Schneider and M. Hanack, *Inorg. Chem.*, 23 (1984) 1065.

263. G. Pennesi, C. Ercolani, P. Ascenzi, M. Brunori and F. Monacelli, *J. Chem. Soc.*, (1985) 1107.

264. D. C. Grenoble and H. G. Drickamer, *J. Chem. Phys.*, 55 (1971) 1624.

265. A. B. P. Lever and J. P. Wilshire, *Inorg. Chem.*, 17 (1978) 1145.

266. D. V. Stynes, *Inorg. Chem.*, 16 (1977) 1170.

267. J. G. Jones and M. V. Twigg, *Inorg. Nucl. Chem. Lett.*, 6 (1970) 245.

268. D. V. Stynes and B. R. James, *J. Am. Chem. Soc.*, 96 (1974) 2733.

269. J. G. Jones and M. V. Twigg, *Inorg. Chim. Acta*, 10 (1974) 103.

270. R. Adzic, B. Simic-Glavaski and E. Yeager, *J. Electroanal. Chem.*, 194 (1985) 155.

271. N. Kobayashi, K. Funayama, M. Koshiyama, T. Osa, H. Shirai and K. Hanabusa, *J. Chem. Soc., Chem. Commun.*, (1983) 915.

272. A. B. P. Lever, S. Licoccia and B. S. Ramaswamy, *Inorg. Chim. Acta*, 64 (1982) L87.

273. G. McLendon and A. E. Martell, *Inorg. Chem.*, 16 (1977) 1812.

274. S. M. Palmer, J. L. Stanton, N. K. Jaggi, B. M. Hoffman, J. A. Ibers and L. H. Schwartz, *Inorg. Chem.*, 24 (1985) 2040.

275. A. Van der Putten, A. Elzing, W. Visscher and E. Barendrecht, *J. Electroanal. Chem.*, 221 (1987) 95.

276. B. Stymne, F. X. Sauvage and G. Wettermark, *Spectra Chim. Acta*, 36A (1980) 397.

277. M. Tahiri, P. Doppelt, J. Fischer and R. Weiss, *Inorg. Chem.*, 27 (1988) 2897.

278. C. Weiss, H. Kobayashi and M. Gouterman, *J. Mol. Spectr.*, 16 (1965) 415.

279. G. Pennesi, C. Ercolani, G. Rossi, P. Ascenzi, M. Brunori and F. Monacelli, *J. Chem. Soc., Chem. Commun.*, (1985) 1113.

280. K. M. Kadish, L. A. Bottomley and J. S. Cheng, *J. Am. Chem. Soc.*, 100 (1978) 2731.

281. J. G. Jones and M. V. Twigg, *Inorg. Chim. Acta*, (1975) L15.

282. H. Homborg and W. Kalz, *Z. Naturforsch.*, 39B (1984) 1490.

283. B. J. Kennedy, G. Brain and K. S. Murray, *Inorg. Chim. Acta*, 81 (1984) L29.

284. E. N. Bakshi, C. D. Delfs, K. S. Murray, B. Peters and H. Homborg, *Inorg. Chem.*, 27 (1988) 4318.

285. C. Ercolani, M. Gardini, K. S. Murray, G. Pennesi and G. Rossi, *Inorg. Chem.*, 25 (1986) 3972.

286. T. D. Smith, C. H. Tan, D. J. Cookson and J. R. Pilbrow, *J. Chem. Soc., Dalton Trans.*, (1980) 1297.

287. J. Metz and M. Hanack, *Chem. Ber.*, 121 (1988) 231.

288. M. Hanack and R. Fay, *Recl. Trav. Chim. Pays-Bas*, 105 (1986) 427.

289. D. Wöhrle and B. Schulte, *Makromol. Chem.*, 189 (1988) 1167.

290. A. R. Koray, V. Ahsen and O. Bekaroglu, *J. Chem. Soc., Chem. Commun.*, (1986) 932.

291. D. Wöhrle and U. Hundorf, *Makromol. Chem.*, 186 (1985) 2177.

292. D. R. Prasad and G. Ferraudi, *Inorg. Chem.*, 22 (1983) 1672.

293. G. J. Ferraudi and D. R. Prasad, *J. Chem. Soc., Dalton Trans.*, (1984) 2137.

294. W. Kobel and M. Hanack, *Inorg. Chem.*, 25 (1986) 103.

295. F. Pomposo, D. Carruthers and D. V. Stynes, *Inorg. Chem.*, 21 (1982) 4245.

296. S. Omiya, M. Tsutsui, E. F. Meyer, I. Bernal and D. L. Cullen, *Inorg. Chem.*, 19 (1980) 134.

297. L. J. Boucher and P. Rivera, *Inorg. Chem.*, 19 (1980) 1816.

298. D. Dolphin, B. R. James, A. L. Murray and J. R. Thornback, *Can. J. Chem.*, 58 (1980) 1125.

299. X. Munz and M. Hanack, *Chem. Ber.*, 121 (1988) 235.

300. S. Muralidharan, G. Ferraudi and K. Schmatz, *Inorg. Chem.*, 21 (1982) 2961.

301. G. Ferraudi, S. Oishi and S. Muraldiharan, *J. Phys. Chem.*, 88 (1984) 5261.

302. X. Munz and M. Hanack, *Chem. Ber.*, 121 (1988) 239.

303. M. Hanack and X. Munz, *Syn. Metals*, 10 (1985) 357.

304. H. Sugimoto, T. Higashi, A. Maeda, M. Mori, H. Masuda and T. Taga, *J. Chem. Soc., Chem. Commun.*, (1983) 1234.

305. K. Kasuga, M. Tsutsui, R. C. Petterson, K. Tatsumi, N. Van Opdenbosch, G. Pepe and J. E. F. Meyer, *J. Am. Chem. Soc.*, 102 (1980) 4835.

306. K. Kasuga, M. Ando and H. Morimoto, *Inorg. Chim. Acta*, 112 (1986) 99.

307. H. Sugimoto, T. Higashi and M. Mori, *Chem. Lett.*, (1983) 1167.

308. K. Kasuga, M. Ando, H. Morimoto and M. Isa, *Chem. Lett.*, (1986) 1095.

309. M. M'sadak, J. Roncali and F. Garnier, *J. Electroanal. Chem.*, 189 (1985) 99.

310. G. C. S. Collins, and D. J. Schiffrin, *J. Electroanal. Chem.*, 139 (1982) 335.

311. M. Maitrot, G. Guillaud, B. Boudjema, J. J. Andre, H. Strzelecka, J. Simon and R. Even, *Chem. Phys. Lett.*, 133 (1987) 59.

312. D. Walton, B. Ely, G. Elliott, and J. C. Marchon, *J. Electrochem. Soc.*, 128 (1981) 2479.

313. F. Castaneda, C. Piechocki, V. Plichon, J. Simon and J. Vaxiviere, *Electro. Acta.*, 31 (1986) 131.

314. M. L'her, Y. Cozien and J. Courtot-Coupez, *J. Electroanal. Chem.*, 157 (1983) 183.

315. Markovitsi, T. Tran-Thi, R. T. Even and J. Simon, *Chem. Phys. Lett.*, 137 (1987) 107.

316. T. Tran-Thi, D. Markovitsi, R. Even and J. Simon, *Chem. Phys. Lett.*, 139 (1987) 207.

317. F. Castaneda, V. Plichon, C. Clarisse and M. T. Riou, *J. Electroanal. Chem.*, 233 (1987) 77.

318. J. Simon and J.-J. André in *Molecular Semiconductors: Photoelectrical Properties and Photo Cells*, Eds. J. M. Lehn and Ch. W. Rees, Springer-Verlag, NY (1985), ch. 3.

319. R. Guilard, A. Dormond, M. Belkalem, J. E. Anderson, Y. H. Liu and K. M. Kadish, *Inorg. Chem.*, 26 (1987) 1410.

320. H. J. Wagner, R. O. Loutfy and C. K. Hsiao, *J. Mat. Sci.*, 17 (1982) 2781.

321. F. R. Fan and L. R. Faulkner, *J. Am. Chem. Soc.*, 101 (1979) 4779.

322. M. Savy, C. Bernard and G. Magner, *Electrochim. Acta*, 20 (1975) 383.

323. R. S. Nohr, P. M. Kuznesof, K. J. Wynne, M. E. Kenney and P. G. Siebenman, *J. Am. Chem. Soc.*, 103 (1981) 4371.

324. J. Martinsen, J. L. Stanton, R. L. Greene, J. Tanaka, B. M. Hoffman and J. A. Ibers, *J. Am. Chem. Soc.*, 107 (1985) 6915.

325. C. J. Schramm, R. P. Scaringe, D. R. Stojakovic, B. M. Hoffman, J. A. Ibers and T. J. Marks, *J. Am. Chem. Soc.*, 102 (1980) 6702.

326. P. E. Fielding and A. G. MacKay, *Aust. J. Chem.*, 17 (1964) 750.

327. C. Ercolani, C. Neri and P. Porta, *Inorg. Chim. Acta*, 1 (1967) 415.

328. T. S. Srivastava, J. L. Przybylinski and A. Nath, *Inorg. Chem.*, 13 (1974) 1562.

329. A. N. Sidorov and I. P. Kotlyar, *Opt. Spectr.*, 11 (1961) 92.

330. M. T. Robinson and G. E. Klein, *J. Am. Chem. Soc.*, 74 (1952) 6294.

331. R. O. Loutfy, *Can. J. Chem.*, 59 (1981) 549.

332. J. H. Sharp and R. L. Miller, *J. Phys. Chem.*, 72 (1968) 3335.

333. R. A. Collins and K. A. Mohammed, *Thermochim. Acta*, 109 (1987) 397.

334. P. Leempoel, F. R. F. Fan and A. J. Bard, *J. Phys. Chem.*, 87 (1983) 2948.

335. C. Ercolani, M. Gardini, F. Monacelli, G. Pennesi and G. Rossi, *Inorg. Chem.*, 22 (1983) 2584.

336. M. Lachkar, A. De Cian, J. Fischer and R. Weiss, *New J. Chem.*, 12 (1988) 729.

337. P. J. Bernstein and A. B. P. Lever, in progress.

338. A. E. Cahill and H. Taube, *J. Am. Chem. Soc.*, 73 (1951) 2847.

339. D. Wöhrle and E. Preussner, *Makromol. Chem.*, 186 (1985) 2189.

340. D. Wöhrle, U. Marose and R. Knoop, *Makromol. Chem.*, 186 (1985) 2189.

341. J. F. Myers, G. W. R. Canham and A. B. P. Lever, *Inorg. Chem.*, 14 (1975) 461.

342. A. G. MacKay, *Aust. J. Chem.*, 26 (1973) 2425.

343. S. J. Edmondson and P. C. H. Mitchell, *Polyhedron*, 5 (1986) 315.

344. P. N. Moskalev, G. N. Shapkin and N. I. Alimova, *Russ. J. Inorg. Chem.*, 27 (1982) 794.

345. A. B. P. Lever, S. Licoccia, B. S. Ramaswamy, S. A. Kandil and D. V. Stynes, *Inorg. Chim. Acta*, 51 (1981) 169.

346. S. Fukuzumi and J. K. Kochi, *J. Am. Chem. Soc.*, 102 (1980) 2141.

347. D. R. Prasad and G. Ferraudi, *Inorg. Chem.*, 21 (1982) 4241.

348. D. Wöhrle, G. Meyer and B. Wahl, *Makromol. Chem.*, 181 (1980) 2127.

349. D. Wöhrle, V. Schmidt, B. Schumann, A. Yamada and K. Shigehara, *Ber. Bunsenges. Phys. Chem.*, 91 (1987) 975.

350. D. Wöohrle, U. Hundorf, G. Schulz-Ekloff and E. Ignatzek, *Z. Naturforsch*, 41b (1986) 179.

351. D. Wöhrle, G. Krawczyk and M. Paliuras, *Makromol. Chem.* 189 (1988) 1001.

352. N. el Khatib, B. Boudiema, M. Maitrot, H. Chermette and L. Porte, *Can. J. Chem.*, 66 (1988) 2313.

---

[1] Publication number 416 of the Photochemistry Unit, Department of Chemistry.

[2] Abbreviations: MPc, metallated phthalocyanes; peripheral substitution at the benzo groups is indicated by R, with tetra- and octasubstitution being most common, for tetrasulfonated, MTSPc; tetracarboxy, MTCPc; but tetraneopentoxy, MN_3Pc, and the bridged (B), trineopentoxy, MN_3BPc. Solvents include dimethyl sulfoxide (DMSO), dimethyl acetamide (DMA), dimethyl formamide (DMF), methylene chloride (DCM), pyridine (py), and acetonitrile (ACN), metallated tetraphenyl porphyrin (MTPP), and metallated octaethyl porphyrin (MOEP).

[3] We adopt the new ACS convention for labeling groups.

[4] Address enquiries to Radiation Chemistry Data Center, Radiation Laboratory, University of Notre Dame, Notre Dame, Indiana 46556-0768.

Table 1 Abbreviations Used Throughout Table 3

	Sorted by Name
4M2py	3,4-Dimethylpyridine
5Clpy	3,5-Dichloropyridine
5M2py	3,5-Dimethylpyridine
3Clpy	3-Chloropyridine
3CHOpy	3-CHOpy
3CNpy	3-Cyanopyridine
3-Opy	3-Hydroxypyridine
3Mpy	3-Methylpyridine
4Clpy	4-Chloropyridine
4CHOpy	4-CHOpyridine
4CNpy	4-Cyanopyridine
4-Opy	4-Hydroxypyridine
Mepy	4-Methylpyridine (4Mepy)
Ac^-	Acetate, CH_3COO^-
Acet	Acetone

(continued)

Table 1 *(continued)*

ACN	Acetonitrile
acac	Acetylacetonate
NH	Ammonia
Bnz	Benzene
BzCN	Benzonitrile
BCarb	Bicarbonate solution
bpy	Bipyridine
Br^-	Bromide ion
CO	Carbon monoxide
CCl	Carbon tetrachloride
TFAA	CF_3COOH
Cl^-	Chloride ion
ClBz	Chlorobenzene
$CHCl_3$	Chloroform
ClNa	Chloronaphthalene
xtl	Crystal; all other solid forms
CN^-	Cyanide ion
CyHx	Cyclohexane
DCB	Dichlorobenzene
DCM	Dichloromethane (methylene chloride)
DGE	Diethyl ether
DMOE	Dimethoxyethane
DMA	Dimethyl acetamide
DMF	Dimethyl formamide
DMSO	Dimethyl sulfoxide
Dpm	Dipivaloylmethane
EtOH	Ethanol (EtOH)
F^-	Fluoride ion
HDZ	Hydrazine (N_2H_4)
OH^-	Hydroxide ion
Imid	Imidazole
I_2	Iodine
MeOH	Methanol
MeIm	Methyl imidazole
MTN	Methyl nitrate (CH_3NO_3)
MTHF	Methyl THF
nBUT	N-Butylamine
NMD	N-Methylimidazole
Nujl	Nujol mull
H_3PO_4	Phosphoric acid
PIP	Piperidine
pyz	Pyrazine
py	Pyridine
film	Sublimed thin film
H_2SO_4	Sulfuric Acid
THF	Tetrahydrofuran
SCN^-	Thiocyanate
tol	Toluene
POBu	Tri-n-butyl phosphite
PBu_3	Tri-n-butylphosphine
TCM	Trichloromethane ($CHCl_3$)

Table 1 *(continued)*

ET$_3$N	Triethylamine
TFA	Trifluoroacetate (CF$_3$COO$^-$)
Vap	Vapor phase
H$_2$O	Aqueous solution

Sorted by Abbreviation

3-Opy	3-Hydroxypyridine
3CHOpy	3-CHOpy see REF 534
3Clpy	3-Chloropyridine
3CNpy	3-Cyanopyridine
3Mpy	3-Methylpyridine
4-Opy	4-Hydroxypyridine
4CHOpy	4-CHOpyridine
4Clpy	4-Chloropyridine
4CNpy	4-Cyanopyridine
4M2py	3,4-Dimethylpyridine
5Clpy	3,5-Dichloropyridine
5M2py	3,5-Dimethylpyridine
Ac$^-$	Acetate, CH$_3$COO$^-$
acac	Acetylacetonate
Acet	Acetone
OH$^-$	Hydroxide ion
PBu$_3$	Tri-n-butylphosphine
PIP	Piperidine
POBu	Tri-n-butyl phosphite
py	Pyridine
pyz	Pyrazine
SCN$^-$	Thiocyanate
ACN	Acetonitrile
BCarb	Bicarbonate solution
Bnz	Benzene
bpy	Bipyridine
Br$^-$	Bromide ion
BzCN	Benzonitrile
CCl	Carbon tetrachloride
TFAA	CF$_3$COOH
CHCl$_3$	Chloroform
Cl$^-$	Chloride ion
ClBz	Chlorobenzene
ClNa	Chloronaphthalene
CN$^-$	Cyanide ion
CO	Carbon monoxide
CyHx	Cyclohexane
DCB	Dichlorobenzene
DCM	Dichloromethane (methylene chloride)
DGE	Diethyl ether
DMA	Dimethyl acetamide
DMF	Dimethyl formamide
DMOE	Dimethoxyethane

(continued)

Table 1 *(continued)*

DMSO	Dimethyl sulfoxide
Dpm	Dipivaloylmethane
ET_3N	triethylamine
EtOH	Ethanol (EtOH)
F^-	Fluoride ion
film	Sublimed thin film
H_2O	Water/aqueous solution
HDZ	Hydrazine (N_2H_4)
H_3PO_4	Phosphoric acid
H_2SO_4	Sulfuric Acid
TCM	Trichloromethane ($CHCl_3$)
TFA	Trifluoroacetate (CF_3COO^-)
THF	Tetrahydrofuran
tol	Toluene
vap	Vapor phase
Xtl	Crystal phase; also all solid phases
I_2	Iodine
Imid	Imidazole
MeIm	Methyl imidazole
MeOH	Methanol
Mepy	4-Methylpyridine (4Mepy)
MTHF	Methyl THF
MTN	Methyl nitrate (CH_3NO_3)
nBUT	N-Butylamine
NH_3	Ammonia
NMD	N-Methylimidazole
Nujl	Nujol mull

Table 2 Coding Used in the Spectral Data Table (Table 3) for the Phthalocyanine Complexes[a,b]

Code No.	Solv	Axial ligand	Pc Ox	M Ox
1	Vap	None	-2	$+1$
2	Film	CN^-	-1	$+2$
3	DCM	Imid	0	$+3$
4	DMA	py	-3	$+4$
5	DMF	mepy	-4	
6	ClN	DMSO	-5	
7	DMSO	NH_3	-6	
8	py	CO		
9	CCl_4	SFA		
10	SFA	TFA		
11	THF	HDZ		
12	EtOH	Cl^-		
13	TCM	OH^-		
14	xtl	F^-		
15	H_2O	Br^-		

Table 2 *(continued)*

Code No.	Solv	Axial ligand	Pc Ox	M Ox
16	HPO$_4$	H$^+$		
17	BNz	O$_2^-$		
18	MePy	PiP		
19	DEE	bPy		
20	PiP	ACN		
21	ACN	DMF		
22	DCB	DMA		
23	MTN	THF		
24	BCB	I$_2$		
25	KBr	DPM		
26	CHCl	3Mpy		
27	CFC	4M$_2$py		
28	HDz	5M$_2$py		
29	DMOE	3Opy		
30	ClBz	4Opy		
31	PhCN	3Clpy		
32	CyHx	4Clpy		
33	Nujl	3COpy		
34	MeOH	4COpy		
35	tol	3CNpy		
36	MTHF	4CNpy		
37	Acet	5Clpy		
38		Et$_3$N		
39		OAc		
40		MeIm		
41		CFC		
42		But		
43		H$_2$O		
44		acac		
45		MeOH		
46		ClBz		
47		Ac$^-$		
48		SCN$^-$		
49		pyz		
50		PBu$_3$		
51		POBu		

[a] Elements are coded with their atomic number.

[b] Abbreviations for the table headings (note: the first six headings are numerical to aid automatic sorting routines retrieve specific data at a later date): Coding information: Solv, solvent; Ele, element; Ax Lg, axial ligand; Pc Ox, oxidation state of the Pc; M Ox, oxidation state of the central metal.

Table 3 Band Centers from the Absorption Spectra of the Phthalocyanine Complexes^{a,b,c}

Central Element	Z	Solv	Ax Lg	Pc Ox	M Ox	M^{n+}	Solv	Ax Lig	ML$_n$ (n=)	Ox Pc	Band positions	References and comments
H₂Pc	1	2	1	1	—	α-H₂	Film	—	—	-2	690sh 620 570 375 328 207	[52] MCD deconv 8K
											Note: Observed band centers	
	1	2	1	1	—	α-H₂	Film	—	—	-2	723 694 615 569 373 326 303 291 277 261 242 230 220 209	[52] absn+MCD Figs; 8K
											Notes: Davydov splitting: Q(00)=2070 cm⁻¹ spectral deconvolution results	
	1	2	1	1	—	α-H₂	Film	—	—	-2	680 600	[212] Fig
	1	2	1	1	—	β-H₂	Film	—	—	-2	700 650	[212] Fig
	1	2	1	1	—	x-H₂	Film	—	—	-2	800 645 610 583	[212]; see also [331]; Fig
											Note: Exciton splitting gives pairs of bands: for M⁻/M⁺, 800/645 and L⁻/L⁺, 610/583	[212]
	1	2	1	1	—	α-H₂	Film	—	—	-2	1130 681 614 326 288 207 177 163	[53] Fig and Tab
	1	2	1	1	—	β-H₂	Film	—	—	-2	716 645	[213] Fig and Tab
											Note: Spectra for β-ZnPc, CoPc, NiPc, and CuPc are shown in [213]	
	1	14	1	1	—	H₂	xtl	—	—	-2	750 620	[320] Fig
	1	1	1	1	—	H₂	vap	—	—	-2	620 340	[56] T=470°C Fig and Tab
	1	1	1	1	—	H₂	vap	—	—	—	686Q 622 340B 280 270N 240 220 210C	[57] T=503°C Fig and Tab
	1	1	1	1	—	H₂	vap	—	—	—	686Q 622 340B 280 270N 240 220 210C	[57] T=503°C Fig and Tab
	1	26	1	1	—	H₂	CHCl	—	4	-2	700 Qₓ 660 Qᵧ 345	[214] MCD Fig and Tab
											Notes: Monomeric species; Subs: tetra (15-crown-5): H₂TCRPc See the paper for extensive deconvolution data for absorption and MCD spectra	
	1	26	1	1	—	H₂	CHCl	—	4	-2	640 (A term in MCD) 335	[214] MCD Fig and Tab
											Notes: Dimeric species; Subs: tetra (15-crown-5): H₂TCRPc See the paper for extensive deconvolution data for absorption and MCD spectra	
	1	6	1	1	—	H₂	CLN	—	—	-2	698Q 663	[57] Tab
	1	6	1	1	—	H₂	CiN	—	—	-2	670Q	[56] Tab
	1	6	1	1	—	H₂	CiN	—	—	-2	Results of deconvolution calculations on MCD and absorption spectra 699(17,160) Qₓ 664(15,560) Qᵧ 648 636 603	[26] MCD Fig. [7] MCD deconv
	1	6	1	1	—	H₂	CiN	—	—	-2	690 654 637 625 595	[212] Fig
	1	6	1	1	—	H₂	CiN	—	—	-2	698 665 638 602 554 350	[76] Tab
											Note: See also data for Cu, Ni, Co, Fe, Zn, Mg, Pd, and Sn(Pc)₂	
	1	4	1	1	—	H₂	DMA	—	—	-2	693Qₓ 660Qᵧ 638 633 600 360 333 305 285	[80] MCD Fig 2 banded Q from Li₂Pc with H⁺

						spectral data						conditions	charge	λ	reference
1	5	1	1	H₂	DMF						−2	696 648 628 426			[348] Fig and Tab

Note: Subs: H₂(CN)₈Pc

| 1 | 37 | 1 | 1 | H₂ | Acetone | | | | | | −2 | 712 662 428 406 | | | [348] Tab |

Notes: Subs: H₂(CN)₈Pc; also for Fe, Co, Ni, Cn, Zn

| 1 | 4 | 7 | 1 | H₂ | DMSO NH₃ 6? | | | | | | −2 | 669Q 638sh 603Q(0-1) 380 367 | | | [80] MCD Fig |

Note: The 380 and 367 nm transitions are degenerate

D_{4h} Pc(−2)

H₂Pc + NH₃
[98] Subs: R = CH(CH₃)
[98] Subs: R = CH₂CH(CH₃)₂
[98] Subs: R = CH₂C(CH₃)₃
[98]

1	3	1	1	H₂	DCM	—	—	—	—	—	−2	706 670 640 608 388 342			
1	3	1	1	H₂	DCM	—	—	—	—	—	−2	706 670 642 610 390 343			
1	3	1	1	H₂	DCM	—	—	—	—	—	−2	705 670 642 608 390 341			
1	3	1	1	H₂	DCM	—	—	—	—	—	−2	700 668 636 612 384 336			

Note: Binuclear; Subs: R₁ = CH₃

| 1 | 3 | 1 | 1 | H₂ | DCM | — | | | | | −2 | 700 670 638 610 384 332 | | | [98] |

Notes: Binuclear; Subs: R = CH₂CH₃
All data from [98] in a Table
See similar spectra from clam-shell binuclear H₂Pc's

| 1 | 3 | 1 | 1 | H₂ | DCM | — | | | | | −2 | 708 676 642 620 388 336 | | | [208] Figs |

Note: Binuclear; Subs: R = (tBu)₄

| 1 | 3 | 1 | 1 | H₂ | DCM | — | | | | | −2 | 698 666 638 618 384 334 | | | [222] |

Note: Binuclear; Subs: R = H

| 1 | 3 | 1 | 1 | H₂ | DCM | — | | | | | −2 | 698 664 644 604 342 292 | | | [222] |

Subs: R = CH₂CH₂; Si(CH₃)₃

| 1 | 3 | 1 | 1 | H₂ | DCM | — | | | | | −2 | 698 664 644 604 342 290 | | | [222] |

Subs: R = CH₂CH₂, C(CH₃)₃

| 1 | 3 | 1 | 1 | H₂ | DCM | — | | | | | −2 | 702 668 642 620 340 292 | | | [222] |

Note: Binuclear; Subs: R = CH₂CH₂Si(CH₃)₃

| 1 | 3 | 1 | 1 | H₂ | DCM | — | | | | | −2 | 702 666 644 618 340 292 | | | [222] |

Note: Binuclear; Subs: R = CH₂CH₂C(CH₃)₃

| 1 | 3 | 1 | 1 | H₂ | DCM | — | | | | | −2 | 706 672 646 620 344 296 | | | [222] |

Note: Binuclear; Subs: R = CH₂C(CH₃)₃

| 1 | 3 | 1 | 1 | H₂ | DCM | — | | | | | −2 | 708 672 644 620 342 296 | | | [222] |

Notes: Binuclear; Subs: R = CH₂C(CH₃)₃; bridge = C₄H₈
All data from [222] in a Table

(continued)

Table 3 Band Centers from the Absorption Spectra of the Phthalocyanine Complexes[a,b,c] *(continued)*

Central Element	Z	Solv	Ax Lg	Pc Ox	M Ox	M^{n+}	Solv	Ax Lig	ML_n (n=)	Ox Pc	Band positions	References and comments
	1	3	1	1	—	H$_2$	DCM	—	4	−2	698 662 640 600	[182] Fig and Tab
											Notes: Subs: (H$_2$)$_2$-(t-Bu)$_4$Pc; also dissolved in vesicles	
	1	26	1	1	—	H$_2$	CHCl	—	—	−2	700 (ε=33,400) Q$_x$ 662 Q$_y$ 645 601 421 347	[187] Fig and Tab
											Note: Peripherally substituted with four 15-crown-5-ether groups = H$_2$TCRPc; monomer	
	1	26	1	1	—	H$_2$	CHCl	—	—	−2	639 (ε=17,500) 402	[187] Fig and Tab
											Note: Peripherally substituted with four 15-crown-5-ether groups = H$_2$TCRPc; dimer when K$^+$ added; see also [214]	
	1	7	1	1	—	H$_2$	DMSO	—	—	−2	698 (log ε=5.1) 668 (5.1) 608 364 342	[221] Tab; H$_2$TSPc
Al	13	37	1	1	3	ClAl	FLM	—	5	−2	690 350 (tabulated by Minor et al.)	[33] Tab
	13	2	14	1	3	Al^{3+}	FLM	—	6	−2	726 697 667 620 455 425 365 283 277	[229] Fig; a film at 10K
	13	3	14	1	3	Al^{3+}	DCM	—	6	−2	662 633 596 365 340 280	[229] Fig
	13	3	14	1	3	Al-O-Al	DCM	—	6	−2	700 630 575 385 327 282	[229] μ-oxo dimer Fig
Ag	47	35		1	1	Ag$^+$	tol			−2	679Q 606w 400sh 350B	
											Note: Ag(neopentoxy)$_4$Pc; G. Fu, Y. S. Fu, and A. B. P. Lever, unpublished	
As(III)	33	5	1	1	3	As^{3+}	DMF	—	5	−2	580 340	[31] Tab
											Note: Identified as "anomalous" in the paper	
Cd	48	7	1	1	2	Cd^{2+}	DMSO	—	4	−2	677Q 612 340B	[57]
Ce(IV)	58	3	1	1	4	Ce^{4+}	DCM	Pc TPP	6	−2 −2	[822 621 331, from MPc] [567 400, from MTPP]	[336] Fig and Tab
											Note: (TPPCe(IV)Pc) dimer	
	58	3	1	1	4	Ce^{4+}	DCM	Pc TPP	6	−1 −2	691vw 479 387s	[336] Fig and Tab
											Note: [TPPCe(IV)Pc]$^+$ dimeric radical cation; g = 2.0023	
Co Co(I)	27	5	1	1	1	Co^{1+}	DMF	—	4	−2	694 633 467 428 312	[35] Fig; green soln
	27	4	1	1	1	Co^{1+}	DMA	—	4	−2	704 641 467 426 311 (DMA + 2.5% pyridine)	[88] Fig and Tab

27	15	1	1	Co1+	H2O	—	4	−2	690 448 312 (aqueous KOH at pH 13)	[88] Tab; Subs: Co(I)TSPc
27	22	13	1	Co1+	DCB	—	4	−2	708(ϵ=45,000) 675 643 600 560 520 471 435 350 313	[207] Fig and Tab; Subs: Co(I) TNPc Subs
27	24	1	1	Co1+	BCB	—	4		649 467 308	[138] Fig and Tab; MCD

Note: Subs = Co(I)tetracarboxyPc; in aqueous bicarbonate solution at pH 9

27	7	6	1	Co1+	DMSO	—		−2	Results of deconvolution calculations on MCD and absorption spectra 702Q 669 640 533 432 387 320 293	[3] MCD Fig [7] MCD deconv
27	22	1	1	Co1+	DCB	—	1	−2	713(ϵ=51,900) 649(18,100) 476(39,500) 352	[337] Tab; Subs: CoTNPc
27	22	1	1	Co1+	DCB	—	1	−2	684(ϵ=7,940) 413(13,600) 346	[337] Tab; Subs: CoTNPc
27	22	1	1	Co1+	DCB	—	1	−2	766(ϵ=43,900) 397(14,100) 329	[337] Subs: CoTNPc+2H+Tab
27	15	?	1	Co1+	H2O	—	6	−2	680 610 sh 540sh 448 310sh	[197] Tab; spectral deconvoluti [52] deconv 8K data

Notes: Co(I)TSPc(−2); at pH 7; see paper for pH dependence

27	22	—	1	Co1+	DCB	—	1	−2	1040 708 (ϵ=93,900) 643 590(34,900) 472 350 312	[99] Fig and Tab; Subs

Notes: A tetranuclear molecule: $[Co(I)TrNPc]^{4-}$; pentaerythritol bridge; weak aggregation

Co(II)

27	2	1	1	α-Co2+	Film	—		−2	688 615 600 323 283 208	[52] MCD deconv

Note: Observed band centres

27	2	1	1	α-Co2+	Film	—		−2	Deconvolution results 696 625 545 404 368 335 332 282 288 247 252 208	[52] deconv absn+MCD Fig

Davydov splitting: Q(00)=2175 cm^{-1}
α- and β-CoPc exhibits near-IR absorption

27	2	1	2	Co2+	Film	—	4	−2	657Q 600 313B 240L 210C	[14] 800–2700 nm Fig
27	1	1	2	Co2+	vap	—	4	−2	670Q 604	[57] T=559°C Fig and Tab
27	6	1	2	Co2+	CLN	—	4	−2	672Q 348B	[139] Fig
27	6	1	2	Co2+	CLN	—	4	−2	665Q 600	[13] Tab
27	3	4	2	Co2+	DCM	py	4	−2	674 608 386 328	[27] Photochem Fig
27	3	1	2	Co2+	DCM	—	4	−2	680 620 390 335	[98] Subs: R=CH(CH3)2
27	3	1	2	Co2+	DCM	—	4	−2	680 620 386 320	[98] Subs: R=CH2C(CH3)3
27	3	1	2	Co2+	DCM	—	4	−2	670 626 386 320	[98] Binuclear, Subs: R=CH3
27	3	1	2	Co2+	DCM	—	4	−2	668 626 380 320	[98]

Notes: Binuclear; Subs R1=CH2CH3 All data from [98] in a Table

27	26	1	2	Co2+	CHCl	—	4	−2	672 608 330 290 (compound 20)	[222] Tab

Notes: Mononuclear; Subs: [−CH2CH2C(CH3)3]4

27	3	1	2	Co2+	DCM	—	4	−2	680 626 336 296 (compound 28)	[222] Tab

Notes: Binuclear; Subs as above, with bridge=C2H4 All data from [222] in a Table

(continued)

Table 3　Band Centers from the Absorption Spectra of the Phthalocyanine Complexes[a,b,c] (continued)

Central Element	Z	Solv	Ax Lg	Pc Ox	M Ox	M^{n+}	Solv	Ax Lig	ML$_n$ (n=)	Ox Pc	Band positions	References and comments
Co^{2+}	27	26	1	1	2	Co^{2+}	CHCl	—	4	-2	672 628 326 290 (compound 7c)	[222] Tab
											Note: Binuclear; Subs: [—OCH$_2$C(CH$_3$)$_3$]$_4$; bridge: catechol	
Co^{2+}	27	22	1	1	2	Co^{2+}	DCB	—	4	-2	678 645 612 380 330	[207] Subs Fig
Co^{2+}	27	4	1	1	2	Co^{2+}	DMA	—	4	-2	667 599 326	[88] Fig and Tab
Co^{2+}	27	5	1	1	2	Co^{2+}	DMF	—	4	-2	668 (ϵ=109,600) 606 380 326	[207] Fig and Tab; Subs: Co(II)TNPc
Co^{2+}	27	10	?	1	2	Co^{2+}	H$_2$SO$_4$	—	—	-2	738 388 298 212 (30 N sulfuric acid)	[181] Tab
											Note: Subs: Co-tetraamine-Pc·2H$_2$O	
Co^{2+}	27	15	?	1	2	Co^{2+}	H$_2$O	—	6	-2	665sh 622(max) 570sh 320sh	[197] Fig and Tab; Subs
											Notes: At pH 7; CoTSPc is aggregated in aqueous solution; see paper for pH dependence	
Co^{2+}	27	22	1	1	2	Co^{2+}	DCB	—	4	-2	674 (ϵ=102,000) 638 306 (ϵ=141,000)	[100] Tab
											Notes: Co^{2+} dimeric species, linked by a naphthalene group, per. subs=tetra–CH$_2$C(CH$_3$)$_3$; See paper for Cu and Zn derivatives; spectra of Co$^+$·Co^{2+} and Co$^+$·Co$^+$ shown	
Co^{2+}	27	22	1	1	2	Co^{2+}	DCB	—	4	-2	678 (ϵ=135,000) 645 612 382 (24,900) 330 (65,400)	[206] Fig and Tab; Subs
											Notes: Monomeric, Co(II)tetra(neopentoxy)Pc; see paper for binuclear species; Dimeric [Co(II)TNPc], has λ_{max} = 625 nm (ϵ=82,000)	
Co^{2+}	27	26	1	1	2	Co^{2+}	CHCl	—	4	-2	670 Q 293	[214] MCD Fig and Tab
											Notes: Monomeric species: weak bands near 380 nm and rising absorption to 290 nm; Subs tetra (15-crown-5): CoTCRPc	
Co^{2+}	27	26	1	1	2	Co^{2+}	CHCl	—	4	-2	630 (A term in MCD) 295	[214] Absorption/MCD Fig and Tab
											Note: Dimeric species; rising absorption to 290 nm; Subs tetra (15-crown-5): CoTCRPc	
Co^{2+}	27	22	—	1	2	Co^{2+}	DCB	—	4	-2	676 (96,100) 625sh 380 326	[99] Fig and Tab; Subs
											Note: A tetranuclear molecule: [Co(II)TrNPc]$_4$; pentaerythritol bridge; not aggregated	
Co^{2+}	27	24	1	1	2	Co^{2+}	Bicarb	—	4	-2	653 625 292	[138] Fig and Tab; MCD
											Note: Subs=Co(II)tetracarboxyPc; in aqueous bicarbonate solution at pH 9	
Co^{2+}	27	7	6	1	2	Co^{2+}	DMSO	—	4	-2	663Q 597 329	[57] Tab
Co^{2+}	27	7	6	1	2	Co^{2+}	DMSO	—	6	-2	657(ϵ=125,000Q 596(31 800) 327(78 900) 298(63,000) 270(57,000)	[3] MCD Fig; [7] MCD Fig
Co^{2+}	27	8	4	1	2	Co^{2+}	py	—	6	-2	656(ϵ=117 500)Q 599(33 700) 330 sh	[3] MCD Fig; [7] MCD Fig

Column header over spectral data: **Results of deconvolution calculations on MCD and absorption spectra**

					Metal	Solvent	Axial	n	z	λ (nm) (ε) / Notes	Reference
27	7	6	1	2	Co²⁺	DMSO	—	6	−2	657Q 605(0-1) 329 291 263	[3] MCD Fig; [7] MCD deconv
27	8	4	1	2	Co²⁺	py	py	6	−2	658 597 332	[262] CoPc(py)₂; Tab
27	8	4	1	2	Co²⁺	py	py	6	−2	657 630 597 416 331	[262] (t-Bu)₄-CoPc; Tab
										Note: Similar spectra are found for Subs = (OCH₃)₈	
27	8	4	1	2	Co²⁺	py	py	6	−2	677 614 457 425 375 342	[262] (NO₂)₄-CoPc Tab
27	22	1	1	2	Co²⁺	DCB			−2	702(ε=60,800) 379(29,600) 347	[337] Subs: CoTNPc H⁺ Tab
27	22	1	1	2	Co²⁺	DCB			−2	733(ε=94,700) 700(66,900) 383(36,300)	[337] Tab; Subs: CoTNPc/2H⁺
27	22	1	1	2	Co²⁺	DCB			−2	772(ε=94,300) 729(488,800) 685(30,700) 403(28,500) 341	[337] Tab; Subs: CoTNPc/3H⁺
27	22	1	1	2	Co²⁺	DCB			−2	669(ε=85,000) 603(27,400) 347	[337] Tab; Subs: CoTNPc Cl⁻
27	26	1	1	2	Co²⁺	CHCl	—		−2	668 (ε=59,400) 608 400 329 297	[187] Fig and Tab
										Note: Peripherally substituted with four 15-crown-5 ether groups = Co(II)TCRPc; monomer	
27	26	1	1	2	Co²⁺	CHCl	—		−2	627 (ε=41,200) 571 390 300	[187] Fig and Tab
										Note: Peripherally substituted with four 15-crown-5-ether groups = Co(II)TCRPc; dimer with K⁺ added	
27	15	1	1	2	Co²⁺	H₂O			−2	670 626 (CoTSPc in 0.1 NaOH + O₂ = adduct)	[193] Fig; Subs
27	15	1	1	2	Co²⁺	H₂O			−2	663 626 (CoTSPc in 20% ethanol = monomer)	[193] Fig; Subs
27	15	1	1	2	Co²⁺	H₂O			−2	620 (CoTSPc in conc. KCl = dimer)	[193] Fig; Subs
27	7	1	1	2	Co²⁺	DMSO			−2	663 (log ε = 5.11) 600 332	[221] Tab; CoTSPc
Co(III)											
27	37	—	1	3	Co³⁺	—	—	6	−2	682 368 (tabulated by Minor et al.)	[33] Tab
27	5	1	1	3	Co³⁺	DMF	—	4	−2	676 (ε=147,900) 610 355 340	[207] Fig and Tab
										Note: Subs: Co(III)TNPc	
27	21	2	1	3	Co³⁺	ACN	CN⁻	6	−2	678 645 610 440 360 355	[207] Fig and Tab
										Note: Subs: Co(III)TNPc	
27	3	2	1	3	Co³⁺	DCM	CN⁻	6	−2	665 636 601 416 392 340 325 272	[261] Fig and Tab; [282]
27	3	13	1	3	Co³⁺	DCM	OH⁻	6	−2	663 630 596 550 350 335 295 265	[282]
										Note: (Tetrabutylammonium)(OH)₂Co(III)Pc	
27	3	14	1	3	Co³⁺	DCM	F⁻	6	−2	663 630 596 550 340 285 265	[282] TBACo(F)₂Pc
										Note: (Tetrabutylammonium)(F)₂Co(III)Pc	
27	3	12	1	3	Co³⁺	DCM	Cl⁻	6	−2	663 630 596 550 445 395 335 285 275 270	[282] TBACo(Cl)₂Pc
										Note: (Tetrabutylammonium)(Cl)₂Co(III)Pc	
27	3	15	1	3	Co³⁺	DCM	Br⁻	6	−2	663 630 596 520 430 400 330 285	[282] TBACo(Br)₂Pc
										Notes: (Tetrabutylammonium)(Br)₂Co(III)Pc; All data from [282] includes figures and a table	

(continued)

Table 3 Band Centers from the Absorption Spectra of the Phthalocyanine Complexes[a,b,c] *(continued)*

Central Element	Z	Solv	Ax Lg	Pc Ox	M Ox	M^n+	Solv	Ax Lig	ML_n (n=)	Ox Pc	Band positions	References and comments
	27	7	2	1	3	Co3+	DMSO	—	6	−2	673(ε=187,200)Q 607(34,200) 427(23 600)	[3] MCD Fig
	27	7	2	1	3	Co3+	DMSO	—	6	−2	352(54 500) 284(73,000)	[7] MCD Fig
											Results of deconvolution calculations on MCD and absorption spectra	[3] MCD Fig
											674Q 644 630 609 432 393 347 320 282	[7] MCD deconv
	27	22	1	1	3	Co3+	DCB	—	1	−2	677(ε=13400) 607(31800) 369(30900)	[337] Subs: CoTNPc Tab
	27	22	1	1	3	Co3+	DCB	—	1	−2	726(ε=83100) 697(66100) 626sh 383(29300)	[337] Subs: CoTNPc 2H+ Tab
	27	15	?	1		Co3+	H2O	—	6	−2	666 (max) 632 600sh 345sh 332	[197] Fig and Tab; Subs

Note: At pH 7; CoTSPc is aggregated in aqueous solution; see paper for pH dependence

Central Element	Z	Solv	Ax Lg	Pc Ox	M Ox	M^n+	Solv	Ax Lig	ML_n (n=)	Ox Pc	Band positions	References and comments
	27	22	—	1	3	Co3+	DCB	—	4	−2	676(ε=149,000) 614 (60,200) 395 336	[99] Fig and Tab; Subs

Note: A tetranuclear molecule: [Co(III)TrNPc]4+; pentaerythritol bridge; aggregated

Central Element	Z	Solv	Ax Lg	Pc Ox	M Ox	M^n+	Solv	Ax Lig	ML_n (n=)	Ox Pc	Band positions	References and comments
	27	26	—	1	3	Co3+	pyz/Cl		6	−2	667 637 602 565 429 350 327 277	[287] Tab

Note: (Pyrazine)(Cl)CoPc   268 262 254

Central Element	Z	Solv	Ax Lg	Pc Ox	M Ox	M^n+	Solv	Ax Lig	ML_n (n=)	Ox Pc	Band positions	References and comments
	27	26	—	1	3	Co3+	Mepyz/Cl		6	−2	669 642 605 564 427 350 330 280	[287] Tab

Note: (Methylpyrazine)(Cl)CoPc

Central Element	Z	Solv	Ax Lg	Pc Ox	M Ox	M^n+	Solv	Ax Lig	ML_n (n=)	Ox Pc	Band positions	References and comments
	27	26	19	1	3	Co3+	bpy/Cl		6	−2	667 642 603 564 427 349 330	[287] Tab

Note: (Bipyridine)(Cl)CoPc

Central Element	Z	Solv	Ax Lg	Pc Ox	M Ox	M^n+	Solv	Ax Lig	ML_n (n=)	Ox Pc	Band positions	References and comments
Cu	29	1	1	1	2	Cu2+	vap	—	4	−2	656 600 340	[56] 510°C Fig and Tab
	29	1	1	1	2	Cu2+	VAP	—	4	−2	658Q 600 325B 276N 240L 218C	[57] 534°C Fig and Tab
	29	2	1	1	2	α-Cu	Film	—	—	−2	1127 708 620 335 264 214 175 159	[53] Fig and Tab
	29	2	1	1	2	α-Cu	Film	—	—	−2	693 623 581 332 259 212	[52] MCD deconv

Note: Observed band centers

Central Element	Z	Solv	Ax Lg	Pc Ox	M Ox	M^n+	Solv	Ax Lig	ML_n (n=)	Ox Pc	Band positions	References and comments
	29	2	1	3	2	α-Cu	Film	—	—	−2	Deconvolution results 730 691 633 591 394 368 333 291 272 259... ... 213 192	[52] deconv absn + MCD Fig

Note: Davydov splitting Q(00) =1675 cm^{-1}

Central Element	Z	Solv	Ax Lg	Pc Ox	M Ox	M^n+	Solv	Ax Lig	ML_n (n=)	Ox Pc	Band positions	References and comments
	29	15	1	1	2	Cu2+	H2O	—	—	−2	625	[338] Tetrasulfonated
	29	15	1	1	2	Cu2+	H2O / H3PO4	—	—	−2	720 525	[338] Tetrasulfonated

Note: Data in [338] from a figure

Central Element	Z	Solv	Ax Lg	Pc Ox	M Ox	M^n+	Solv	Ax Lig	ML_n (n=)	Ox Pc	Band positions	References and comments
	29	6	1	1	2	Cu2+	ClN	—	4	−2	678 648 611 588 567 526 510 350	[4] Tab
	29	6	1	1	2	Cu2+	ClN	—	4	−2	678 610	[57] Tab and Fig
	29	17	1	1	2	Cu2+	BNZ	—	4	−2	750 685	[4]

				Metal	Solvent		Data	Charge	References
29	6	1	1	Cu²⁺	ClN	4	678Q 350B	−2	[13] Tab
29	3	1	1	Cu²⁺	DCM	4	677 638 614 384 338	−2	[98] R=CH(CH₃)₂
29	3	1	1	Cu²⁺	DCM	4	682 620 384 337	−2	[98] Subs: R=CH₂CH(CH₃)₂
29	3	1	1	Cu²⁺	DCM	4	680 614 380 338	−2	[98] Subs: R=CH₂C(CH₃)₃
29	3	1	1	Cu²⁺	DCM	4	682 618 384	−2	[98]

Note: Binuclear; Subs: R=CH(CH₃)₂

29	17	1	1	Cu²⁺	BNZ	4	745 695	−2	All data from [98]: Table; [4] Cu-R₄-Pc; dimer
29	14	1	1	Cu²⁺	xtl	4	750 700	−2	[4] Cu-R₄-Pc
29	3	1	1	Cu²⁺	DCM	4	676 628 384 336	−2	[98] Tab

Note: Binuclear; Subs R'=CH₃; R=CH₂C(CH₃)₃

29	6	1	1	Cu²⁺	ClN	4	690 645 610	−2	[230]
29	7	1	1	Cu²⁺	DMSO	4	671Q 607 327B	−2	[49] Fig
29	10	1	—	Cu²⁺	H₂SO₄	—	821 779 446	−2	[57] Tab; [188] Cu(F₁₆)Pc Tab
29	26	1	1	Cu²⁺	CHCl	—	676 (ε=129,400) 610 409 338 292	−2	[187] Fig and Tab

Note: Peripherally substituted with 4 15-crown-5 ether groups=Cu(II)TCRPc; monomer

| 29 | 26 | 1 | | Cu²⁺ | CHCl | — | 635 (ε=74,200) 389 328 | −2 | [187] Fig and Tab |

Note: Peripherally substituted with four 15-crown-5 ether groups=Cu(II)TCRPc; dimer with K⁺ added

| 29 | 5 | 1 | | Cu²⁺ | DMF | — | 673 | −2 | [200] Fig |

Notes: For Cu-4,11,18,25-tetra(2-pyridylmethylaminosulfonyl)Pc=Cu-tpmasPc. See paper for data from pyridine and chloroform

| 29 | 26 | 1 | | Cu²⁺ | CHCl | — | 680 (ε=47,200) 647 (23,000) 615 382 340 250 | −2 | [290] Fig and Tab |

Note: Peripherally substituted with four 15-crown-5 ether groups=Cu(II)TCRPc; monomeric

| 29 | 7 | 1 | | Cu²⁺ | DMSO | — | 677 (log ε=5.44) 609 350 | −2 | [221] Tab; CuTSPc |
| 29 | 10 | ? | | Cu²⁺ | H₂SO₄ | — | 749 382 300 214 (30 N sulfuric acid) | −2 | [181] Tab |

Note: Subs: Cu-tetraamine-Pc·2H₂O

| 29 | 26 | 1 | | Cu²⁺ | CHCl | 4 | 700 Q 340 290 | −2 | [214] MCD Fig and Tab |

Notes: Monomeric species; weak bands between 350 and 450 nm. Subs tetra (15-crown-5): CuTCRPc

| 29 | 26 | 1 | | Cu²⁺ | CHCl | 4 | 635 (A term in MCD) 330 | −2 | [214] Absorption/MCD Fig and Tab |

Notes: Dimeric species; weak bands between 350 and 450 nm. Subs tetra (15-crown-5): CuTCRPc

| 29 | 6 | 1 | 1 | Cu²⁺ | ClN | 4 | 678 648 611 588 567 526 510 350 | −2 | [289] |
| 29 | 5 | 1 | 1 | Cu²⁺ | DMF | 4 | 688 645 348 | −2 | [289] Tab |

(continued)

Table 3 Band Centers from the Absorption Spectra of the Phthalocyanine Complexes[a,b,c] *(continued)*

Central Element	Z	Ax Lg	Solv	Pc Ox	M Ox	M^{n+}	Solv	Ax Lig	ML_n (n=)	Ox Pc	Band positions	References and comments
						Note: Cu-octacarbonitrile-Pc = Cu-(CN)$_8$Pc						
	29	1	5	1	2	Cu^{2+}	DMF		4	-2	674 606 344	[289] Tab
						Note: Compound 5a = Cu-tetraphenoxy-Pc						
	29	1	10	1	2	Cu^{2+}	H_2SO_4		4	-2	790 746 698 634 438 303 254 222 210	[289] Tab
	29	1	10	1	2	Cu^{2+}	H_2SO_4		4	-2	728 696 660 630 385 345 290 241 220	[289] Tab
						Note: Cu-octacarbonitrile-Pc; see also [291,339,340]						
	29	1	10	1	2	Cu^{2+}	H_2SO_4		4	-2	838 740 530 432 304 216	[289] Tab
						Note: Compound 5a = Cu-tetraphenoxy-Pc; see also refs. [291,339,340]						
Cr(II)	24	1	1	1	2	Cr^{2+}	VAP	—	.4	-2	664Q 600 315B 263N 240L	[57] T=549°C Fig and Tab
	24	1	5	1	2	Cr^{2+}	DMF	—	4	-2	843CT 613Q 588 568 543 481CT 386CT 342	[10] Fig and Tab
						Note: Subs tetrasulfonated; see also [13]						
	24	1	7	1	2	Cr^{2+}	DMSO	—	4	-2	685Q 622 344B	[57] Tab
	24	4	8	1	2	Cr^{2+}	py	py	6	-2	687(ε=70,800) 632 560 525 500(5,000) 345(38 900)	[240] Tab
	24	1	5	1	2	Cr^{2+}	DMF	—	4	-2	675Q 348B	[13] Electrochem Tab; Subs
Cr(III)	24	12	2	2	3	Cr^{3+}	Film	Cl$^-$	6	-1	797 753 680 576 445 362 280	[341] Tab TSPcCr(III)Na$_3$
	24	—	37	1	3	Cr^{3+}	—	—			676 344 268 (tabulated by Minor et al.)	[33] Tab
	24	1	5	1	3	Cr^{3+}	DMF	—	4	-2	1266CT 1170CT 1039CT 912(trip-mult) 696Q 664 626 498CT 356	[10] Fig and Tab
						Note: Subs: tetrasulfonated; see also [13]						
	24	1	30	1	3	Cr^{3+}	ClBz	OH$^-$	4	-2	689(ε=83,200) 621 502 347(41,700)	[240] Tab
	24	15	34	1	3	Cr^{3+}	MeOH	H$_2$O	6	-2	676(ε=147,910) 610 502 477(9,800) 344(39 800)	[240] Tab
	24	2, 13	34	1	3	Cr^{3+}	MeOH	CN$^-$ OH$^-$	6	-2	677 611 515 345 321 308 271	[240] Tab
	24	2	34	1	3	Cr^{3+}	MeOH	CN$^-$	6	-2	675 615 518 490 380 345 322 309 258	[240] Tab
	24	48	34	1	3	Cr^{3+}	MeOH	SCN$^-$	6	-2	682 617 511 349 298 270	[240] Tab
	24	12	34	1	3	Cr^{3+}	MeOH	Cl$^-$ +HCl	6	-2	682(ε=141,250) 615 507 495(8,500) 355 347 282	[240] Tab
	24	1	5	1	3	Cr^{3+}	DMF	—	4	-2	696Q 356B	[13] Electrochem Tab

Element						Ion	Solvent				Data	Ref.
Dy												
						Note: The electronic configurations suggested for the different redox states of Lu(Pc)₂ complexes apply for [313,315]:						
						[DyPc(−1)Pc(−2)]; [DyPc(−2)Pc(−2)]⁻ ; [DyPc(−1)Pc(−1)]⁺						
	66	3	—	1	3	Dy³⁺	DCM	dimer	6	−2	Q as for Tm; 790 900 (ε=7,000) 1260 1404 1536	[315] Fig and Tab
	66	3	—	1	3	Dy³⁺	DCM	dimer	6		660(180,000) 450	[309] Fig
						Note: DyPc(−2)Pc(−1)						
						Note: [Pc(−2)DyPc(−2)]H; see also for PcNdPcH; see paper for redox chemistry						
	66	3	—	1	3	Dy³⁺	DCM	dimer	6	−2	665(180,000) 455 "green form"	[104] Fig
						Note: [Pc(−2)DyPc(−2)]H; see paper for pH titrations						
										−2		
										−2	700 624	
						Note: [Pc(−2)DyPc(−2)]⁻ ; a blue solution obtained by addition of base, or reduction						
Er												
	68	26	—	1	3	Er³⁺	Cl₃C	—	—	−2	679Q 336 Er³⁺·Pc(dbm)(dbmH) 8 coordinate	[28] Fig
										−2	725 679 Er³⁺·Pc(dbm) 6 coordinate	[28] Fig
						Note: dbm = 1,3-diphenyl-1,3-propanedionato						
Eu												
						Note: The electronic configurations suggested for the different redox states of Ln(Pc)₂ complexes [13,315]:						
						[EuPc(−1)Pc(−2)]; [EuPc(−2)Pc(−2)]⁻ ; [EuPc(−1)Pc(−1)]⁺						
	63	3	—	1	3	Eu³⁺	DCM	dimer	6	−2	697 654	[104] Fig
										−2	PcEuPcH; see paper for pH titration	
										−2	640; typical of dimeric species Pc-M-Pc	
										−2		
						Note: [Pc(−2)EuPc(−2)]⁻ ; the "blue form," forms with addition of base or reduction						
Fe(I)												
	26	8	4	1	1	Fe¹⁺	Py	5		−2	800 (ε=11,880) 650 565 500 360 320	[265] Fig and Tab
	26	15	?	1	1	Fe¹⁺	H₂O	6		−2	696 610sh 494 320	[197] Fig and Tab; Subs
						Notes: At pH 7; there is little indication that Fe(I)TSPc aggregates in aqueous solution; see paper for pH dependence						
Fe(II)												
	26	2	1	1	2	Fe²⁺	Film			−2	α-polymorphic film: 712	[322] Fig
	26	2	1	1	2	Fe²⁺	Film			−2	β-polymorphic film: 624 565 470	[322] Fig

(continued)

Table 3 Band Centers from the Absorption Spectra of the Phthalocyanine Complexes[a,b,c] (continued)

Central Element	Z	Solv	Ax Lg	Pc Ox	M Ox	M^{n+}	Solv	Ax Lig	ML_n (n=)	Ox Pc	Band positions	References and comments
Fe^{2+}	26	22	1	1	2	Fe^{2+}	DCB	—	6	-2	820(A) 758 654Q 448sh 329(A)	[3] MCD Fig
											(A): degenerate transitions	
Fe^{2+}	26	3	3	1	2	Fe^{2+}	DMC	Im	6	-2	659Q 428C 345B 312L	[6] MCD Fig; [27] Photochem Fig
Fe^{2+}	26	7	6	1	2	Fe^{2+}	DMSO	DMSO	6	-2	654Q 320B	[6]
Fe^{2+}	26	4	1	1	2	Fe^{2+}	DMA	—	4	-2	666Q 322B	[6]
Fe^{2+}	26	4	7	1	2	Fe^{2+}	DMA	HDZ	6	-2	666Q 438CT 350B	[6] Fig
Fe^{2+}	26	4	2	1	2	Fe^{2+}	DMA	CN⁻	6	-2	666Q 424CT 394B 304N	[6] Fig
Fe^{2+}	26	4	3	1	2	Fe^{2+}	DMA	Im	6	-2	659Q 404CT 340B	[6] Fig
Fe^{2+}	26	4	8	1	2	Fe^{2+}	DMA	CO	6	-2	664Q 428CT 320N	[6] Fig
Fe^{2+}	26	8	4	1	2	Fe^{2+}	py	py	6	-2	654Q 410CT 324N	[6] Fig
												All data from [6]: Table
Fe^{2+}	26	9	4	1	2	Fe^{2+}	CCl4	—	4	-2	648 592 413	[276] Fig
Fe^{2+}	26	9	4	1	2	Fe^{2+}	CCl4	py	6	-2	680 651 615 593 567 522 416	[276] Fig
Fe^{2+}	26	9	4	1	2	Fe^{2+}	CCl4	py	5	-2	652 628 592 413	[276] Fig
Fe^{2+}	26	7	3	1	2	Fe^{2+}	DMSO	Im	6	-2	659 590 420	[258] Fig
Fe^{2+}	26	7	3	1	2	Fe^{2+}	DMSO	Im	—	-2	618 590	[258] Oxo-dimer Fig
Fe^{2+}	26	3	3	1	2	Fe^{2+}	DCM	CN⁻	6	-2	667 640 605 509 485 455 425 391 310	[261] Fig
Fe^{2+}	26	3	11	1	2	Fe^{2+}	DCM	HDZ	6	-2	660 633 597 417 327	[261] ($N_2H_4)_2$Pc(–2) Fig
Fe^{2+}	26	6	1	1	2	Fe^{2+}	CLN	—	4	-2	686	[139] MCD Fig
Fe^{2+}	26	6	1	1	2	Fe^{2+}	CLN	—	4	-2	675Q 358B	[13] Subs: 4dodecy Tab
Fe^{2+}	26	4	1	1	2	Fe^{2+}	DMA	—	4	-2	650 600 430	[265] Fig
Fe^{2+}	26	24	1	1	2	Fe^{2+}	BCB	—	4	-2	678 615 440 325	[138] Fig and Tab; MCD

Note: Subs = Fe(II)tetracarboxy-Pc; in aqueous bicarbonate solution at pH 9

Central Element	Z	Solv	Ax Lg	Pc Ox	M Ox	M^{n+}	Solv	Ax Lig	ML_n (n=)	Ox Pc	Band positions	References and comments
Fe^{2+}	26	7	1	1	2	Fe^{2+}	DMSO	—	4	-2	655Q 595	[257] Fig
Fe^{2+}	26	7	18	1	2	Fe^{2+}	DMSO	pip	6	-2	663Q 642 602 434CT	[257] Fig
Fe^{2+}	26	26	3	1	2	Fe^{2+}	CLF	Im	6	-2	663Q 640 602 430CT	[257] Fig
Fe^{2+}	26	26	26	1	2	Fe^{2+}	CLF	3mpy	6	-2	655Q 630 595 415CT	[257]
Fe^{2+}	26	26	5	1	2	Fe^{2+}	CLF	4mpy	6	-2	655Q 630 595 413CT	[257]
Fe^{2+}	26	26	27	1	2	Fe^{2+}	CLF	4M₂P	6	-2	656Q 630 595 413CT	[257]
Fe^{2+}	26	26	28	1	2	Fe^{2+}	CLF	5M₂P	6	-2	656Q 630 595 415CT	[257]
Fe^{2+}	26	7	29	1	2	Fe^{2+}	DMSO	3OPy	6	-2	656Q 630 595 412CT	[257] Fig
Fe^{2+}	26	7	30	1	2	Fe^{2+}	DMSO	4OPy	6	-2	662Q 635 602 432CT	[257] Fig
Fe^{2+}	26	26	31	1	2	Fe^{2+}	CLF	3Clp	6	-2	653Q 627 592 446CT 407CT	[257] Fig
Fe^{2+}	26	7	32	1	2	Fe^{2+}	DMSO	4Clp	6	-2	680Q 642 613 555CT	[257] Fig
Fe^{2+}	26	26	27	1	2	Fe^{2+}	CLF	5Clp	6	-2	650Q 625 591 465CT 397CT	[257] Fig
Fe^{2+}	26	7	33	1	2	Fe^{2+}	DMSO	4Cpy	6	-2	655Q 630 595 410CT	[257] Fig
Fe^{2+}	26	7	34	1	2	Fe^{2+}	DMSO	4Cpy	6	-2	654Q 630 595 520CT 410CT	[257] Fig
Fe^{2+}	26	26	35	1	2	Fe^{2+}	CLF	3CNp	6	-2	651Q 625 591 482CT 402CT	[257] Fig
Fe^{2+}	26	26	36	1	2	Fe^{2+}	CLF	4CNP	6	-2	652Q 625 591 527CT 405CT	[257] Fig
												All data from [257]: Table
Fe^{2+}	26	3	3	1	2	Fe^{2+}	DCM	Im	6	-2	650 600 430 330	[87] Fig
Fe^{2+}	26	7	6	1	2	Fe^{2+}	DMSO	DMSO	6	-2	640 580	[250] Fig

26	7	3	1	2	Fe²⁺	DMSO	Im	6	-2	650 640 590 430	[250] Fig
26	8	4	1	2	Fe²⁺	py	—	6	-2	654(ε = 136,000)Q 592(35,200) 468(22,200) 332(82,000)	[3] MCD Fig / [7] MCF Fig
26	3	2	1	2	Fe²⁺	DCM	CN⁻	6	-2	775 685 645 620 600 540 500 420 400 317 279	[261] Fig and Tab
26	3	2	1	2	Fe²⁺	DCM	CN⁻	6	-2	783 690 651 595 555 510 412 317	[261] Fig and Tab
26	8	4	1	2	Fe²⁺	Py	Py	4	-2		
26	1	1	1	2	Fe²⁺	VAP	—	4	-2	676Q 340B 242L 212C	[57] T = 560°C Fig and Tab
26	7	1	1	2	Fe²⁺	DMSO	—	4	-2	656Q 600 320B	[57] Tab
26	7	1	1	2	Fe²⁺	DMSO	—	4	-2	656 598 328	[11] Tab and Fig / [3] MCD Fig
26	7	4	1	2	Fe²⁺	DMSO	Py	6	-2	656 633 595 415 357 330	[11] Tab / [3] MCD Fig
26	7	3	1	2	Fe²⁺	DMSO	Im	6	-2	662 637 602 429 346 323	[11] Tab
26	7	7	1	2	Fe²⁺	DMSO	NH₃	6	-2	671 610 440 352 336 321	[11] Tab and Fig / [3] MCD Fig
26	7	42	1	2	Fe²⁺	DMSO	BUT	6	-2	669 645 610 440 353 333 321	[135]
26	7	18	1	2	Fe²⁺	DMSO	PiP	6	-2	667 606 435 351 338 319	[11] Tab
26	7	2	1	2	Fe²⁺	DMSO	CN⁻	6	-2	669 645 610 452 424 393 379 311	[11] Fig and Tab / [3] MCD Fig
26	8	4	1	2	Fe²⁺	py	py	6	-2	655 595 415 333	[262] FePc(py)₂ Tab
26	8	4	1	2	Fe²⁺	py	py	6	-2	657 630 597 416 331	[262] (t-Bu)₄-FePc Tab
26	8	4	1	2	Fe²⁺	py	py	6	-2	Similar spectra are found for (CH₃)₈ and (OCH₃)₈ 679 650 618 437 357	[262] Tab / [262] Tab

Note: Subs Fe(II)-Cl₁₆-Pc(py)₂

26	30	—	1	2	Fe²⁺	CH₃	Li⁺	6	-2	707 (log ε: 4.96) 675 639 540(4.75) 479(4.94)	[277] Tab

Notes: A σ-alkyl-Fe(II)Pc
See paper for similar spectral data for R = C₂H₅ and [CH(CH₃)₂]

26	8	4	1	2	Fe²⁺	py	py	6	-2	655 (log ε = 5.07) 593 513 (4.32) 332	[269] Tab
26	3	11	1	2	Fe²⁺	DCM CN⁻	HDZ	6	-2	663 635 600 425 367 310	[261] Fig

Note: Fe(II)Pc(N₂H₄)(CN⁻)

26	15	?	1	2	Fe²⁺	H₂O		6	-2	668 635sh 610sh 435 325	[197] Fig and Tab; Subs

Note: At pH 7; Fe(II)TSPc may be aggregated in aqueous solution; see paper for pH dependence

26	26	?	1	2	Fe²⁺	CHCl	—	6	-2	663(ε=115,300) 599 320	[251] Tab

Note: for (C₆H₅NO)FePc(n-C₄H₉NH₂)

26	26	?	1	2	Fe²⁺	CHCl		6	-2	661(ε=128,900) 598 424 342	[251] Tab

Note: for (CH₃C₅H₃N₂)₂FePc
Similar spectra are reported for a range of aromatic N, and P, binding ligands

[251]

(continued)

Table 3 Band Centers from the Absorption Spectra of the Phthalocyanine Complexes^{a,b,c} *(continued)*

Central Element	Z	Solv	Ax Lg	Pc Ox	M Ox	M^{n+}	Solv	Ax Lig	ML_n (n=)	Ox Pc	Band positions	References and comments
Fe(III)	26	37	—	1	3	Fe^{3+}	—	—	—	−2	668 326 (Tabulated by Minor et al.)	[33] Tab
	26	15	1	1	3	Fe^{3+}	H_2O	—	4	−2	820 693 620 349	[135] Subs: R = CO_2H; Fig
	26	15	1	1	3	Fe^{3+}	H_2O	—	4	−2	642 590 329	[135] Subs: R = CO_2H; Fig
	26	15	1	1	3	Fe^{3+}	H_2O	—	4	−2	646 329 270	[135] Fig

Subs: R_1 = H R = $OCC_6H_3(CO_2H)_2$

Central Element	Z	Solv	Ax Lg	Pc Ox	M Ox	M^{n+}	Solv	Ax Lig	ML_n (n=)	Ox Pc	Band positions	References and comments
	26	3	1	1	3	Fe^{3+}	DCM	—	4	−2	690 620 370 320	[135] Subs: R = CO_2H; Fig
	26	15	1	1	3	Fe^{3+}	H_2O	—	4	−2	680 615 420 320	[135] Subs: R = CO_2H, R' = H; Fig
		9					H_2SO_4						Fig
	26	3	3	1	3	Fe^{3+}	DCM	Im	6	−2	689Q 554 355B 282L	[6] [27] Photochem; Fig
	26	3	2	1	3	Fe^{3+}	DCM	CN^-	6	−2	775 685 645 620 600 540 500 420 400 317 279	[261] $[CN)_2Pc(-2)]^-$; Fig
	26	10	2	1	3	Fe^{3+}	H_2SO_4	CN^-	6	−2	765 685 550 480 423 395 303 220	[261] Fig; $[(CH)(HSO_4)Fe(III)Pc(-2)]^-$
		9	9					H_2SO_4					
	26	3	2	1	3	Fe^{3+}	DCM	CN^-	6	−2	783 690 651 595 555 510 412 317	[261] $[(CN)(py)Pc(-2)]$ Fig.
	26	8	4	1	3	Fe^{3+}	py	py	4	−2	840 640 320	[87] High-spin MCD; Fig
	26	3	1	1	3	Fe^{3+}	DCM	—	4	−2	670 660 610 550 400	[265] Fig
	26	3	1	1	3	Fe^{3+}	DMA	—	4	−2	690 617 552 410 329	[134] Low-spin Fig and Tab substituted
	26	3	1	1	3	Fe^{3+}	DCM	—	4	−2	817 738 650 337	[134] High spin Fig and Tab monomer
	26	11	1	1	3	Fe^{3+}	THF	—	4	−2	636(ϵ=34,673) 333(43,651)	[272] Tab

Note: μ-oxobis(tetradodecylsulfonanamidiphthalocyanine); a μ-oxo-bridged Fe(III) dimer TdPcFe(III)-O-Fe(III)PcTd; 4-sulfonated

Central Element	Z	Solv	Ax Lg	Pc Ox	M Ox	M^{n+}	Solv	Ax Lig	ML_n (n=)	Ox Pc	Band positions	References and comments
	26	24	1	1	3	Fe^{3+}	BCB	—	4	−2	640 560 329	[138] Fig and Tab; MCD

Note: Subs = Fe(III)tetracarboxyPc; in aqueous bicarbonate solution at pH 9

Central Element	Z	Solv	Ax Lg	Pc Ox	M Ox	M^{n+}	Solv	Ax Lig	ML_n (n=)	Ox Pc	Band positions	References and comments
	26	3	3	1	3	Fe^{3+}	DCM	Im	5	−2	660 560 320	[87] Low-spin MCD; Fig
	26	10	—	1	3	Fe^{3+}	H_2SO_4	—		−2	806 500	[188] Tab Fe(F_{16})Pc
	26	8	8	1	3	Fe^{3+}	py	py		−2	658(ϵ = 18,200) 626 537 [Fe(IV)Pc–N–Fe(III)Pc]	[259] Electrochem Fig
				3	3	Fe^{4+}						
	26	8	8	1	3	Fe^{3+}	py	py		−2	655(ϵ = 42 600) 627 594 [Fe(III)Pc–N–Fe(III)Pc]$^-$	[259] Electrochem Fig
	26	15			3	Fe^{3+}		H_2O		6	−2	670sh 632 580sh 326	[197] Fig and Tab; Subs

Notes: At pH 7; Fe(III)TSPc mainly aggregated in aqueous solution; see paper for pH dependence μ-oxo dimers expected at high pH

Central Element	Z	Solv	Ax Lg	Pc Ox	M Ox	M^{n+}	Solv	Ax Lig	ML_n (n=)	Ox Pc	Band positions	References and comments
	26	6	12	1	3	Fe^{3+}	ClN	OH^-	6	−2	690 360	[254] Fig and Tab

Note: Use of bulky cation, B^+
Note: Similar spectra reported for X = F^-, Cl^-, Br^-, I^-, and μ-oxo dimer [254] Fig and Tab

26	3	2	1	3	Fe³⁺	DCM	CN⁻	6	−2	775 685 540 425 400 322	[254] Fig and Tab
											[254] Fig and Tab

Note: Similar spectra reported for X = NCO⁻ and NCS⁻; cation = bis(triphenylphosphine)N⁺

Fe(IV)

26	8	8	1	4	Fe⁴⁺	py	py		−2	648(ε=30,000) 634 574 [Fe(IV)Pc–N–Fe(IV)Pc]⁺	[259] Electrochem Fig
26	8	8	4	4	Fe⁴⁺	py	py		−2	658(ε=18,200) 626 537 [Fe(IV)Pc–N–Fe(III)Pc]	[259] Electrochem Fig
26	8	8	1	3	Fe³⁺	py	py				
26	15	?	3	4	Fe⁴⁺	H₂O	H₂O	6	−2	810vw 610sh 494 320sh	[197] Fig and Tab; Subs

Note: At pH 7; FeTSPc may be aggregated in aqueous solution; see paper for pH dependence

Ge

32	8	4		2	Ge²⁺	py	—		−2	655 445 330	[31] Tab
32	32	—	1	4	GeO²⁺	CyHx	—		−2	668(ε=480,000)Q 353(78,000) 343(65,000) 330	[215] Tab

Note: Monomer

32	32	—	1	4	GeO²⁺	CyHx	—	6	−2	631(ε=280,000) 331(89,000)	[215] Tab

Note: Dimer with Ge–O–Ge links

Li

3	25		1		2Li⁺	KBr	Ag	5(?)	−2	812 745 672 645 607 582 565 482 435	[227] Fig and Tab

Note: In KBr at 10K, AgLiPc(−2)

3	25		1		2Li⁺	KBr	Cu	5(?)	−2	678 643 610 585 567 522 455 425	[227] Fig and Tab

Note: In KBr at 10K, CuLiPc(−2)

All data from [227]; Fig and Tab

3	7	6	1		2Li⁺	DMA	DMA	6	−2	667(ε=63 300)Q 641 604 370 332(22 300)	[26] MCF Fig [224] absn ⁺H⁺ Fig [7] deconv
3	7	6	1		2Li⁺	DMSO	DMSO	6	−2	Deconvolution results: 669Q 642 631 609 391 344 290 268	absn+MCD Fig [26] MCD Fig
3	7	2	1		2Li⁺	DMSO	CN⁻	6	−2	665(76 100)Q 636 600 378 326(13 600)	[7] deconv
3	7	2	1		2Li⁺	DMSO	CN⁻	6	−2	Deconvolution results: 667Q 640 605 387 325 283 271 261	absn+MCD
3	3	1	1		Li⁺ NR₄⁺	DCM	—	1	−2	667(ε=230,000) 636 619 580 568 480 430 380 330 267 240	[225] Fig and Tab

Notes: (NR₄⁺)Li⁺Pc(−2); R=buryl, tri-n-dodecyl-buryl
See paper for spectra from 10K film and KBr pellet

3	8	1	1		Li⁺ H⁺	py	—	1	−2	667 645 605 442 385 327	[226] Fig and Tab

Notes: H⁺Li⁺Pc(−2)
See paper for spectra from KBr at 10K

(continued)

Table 3 Band Centers from the Absorption Spectra of the Phthalocyanine Complexes[a,b,c] *(continued)*

Central Element / Z	Solv	Ax Lg	Pc Ox	M Ox	M^{n+}	Solv	Ax Lig	ML$_n$ (n=)	Ox Pc	Band positions	References and comments
Lu(III)											
										Note: The electronic configurations suggested for the different redox states of Ln(Pc)$_2$ complexes	
										[LuPc(−1)/Pc(−2)]; [LuPc(−2)Pc(−2)]$^-$; [LuPc(−1)Pc(−1)]$^+$	
71	3	1	1	3	Lu^{3+}	DCM	—	—	−2	630 600 581 450	[96] Lu(Pc)$_2$ Fig
71	3	1	1	3	Lu^{3+}	DCM	—	—	−2	1500 1382 1248 906 880 660 600 460 316	[315] Fig
										Note: LuPc(−2)Pc(−1)	
71	3	1	1	3	Lu^{3+}	DCM	—	—	−2	700 620 350	[315] Fig
										Note: LuPc(−2)Pc(−2)	
71	3	1	1	3	Lu^{3+}	DCM	—	—	−2	855 700 450 350	[315] Fig
										Note: LuPc(−1)Pc(−1)	
71	3	1	2	3	Lu^{3+}	DCM	—	—	−1	690 610 450 430	[96] Lu(Pc)$_2$ Fig
71	3	1	5	3	Lu^{3+}	DCM	—	—	−4	680 510 410 350	[96] Lu(Pc)$_2$Fig
71	3	1	4	3	Lu^{3+}	DCM	H$^+$	—	−3	700 620 460	[96] Lu(Pc)$_2$ Fig
71	15	16	4	3	Lu^{3+}	H$_2$O	H$^+$	5	−2	720 660 460	[4] (Pc)$_2$LuH Fig
71	15	1	4	3	Lu^{3+}	H$_2$O	H$^+$	4	−2	720 625 570	[4]
71	2	16	1	3	Lu^{3+}	Film	H$^+$	5	−2	680 470	[4]
71	2	16	4	3	Lu^{3+}	Film	H$^+$	5	−2	660 600 460	[4]
71	2	1	4	3	Lu^{3+}	Film	H$^+$	4	−2	720 625	[4]
71	2	16	1	3	Lu^{3+}	Film	H$^+$	5	−2	720 500	[4]
										Note: All data from [4] in a Table	
71	2	1	4	3	Lu^{3+}	Film	—	4	−2	1423 1280 915 670 440; mixed valence band at 1423 nm	[316] Fig / [315] see Yb/Tm/Dy Fig and Tab / [313]
71	3	1	2	3	Lu^{3+}	DCM	—	—	−2/−1	600(ε=155,000Q; "green form"; radical cation [PcLuPc]$^{.+}$	see also: [314] [311] [85][104]
71	3	1	2	3	Lu^{3+}	DCM	—	—	−2/−2 −1/−1 −2/−1	620(ε=125,000); "blue form" 690(ε=47,000) 472(47,000) "orange form"; diradical cation 668(ε=120,000)Q; "green form"; radical cation [PcLuPc]$^-$	Electrochem [313] Lu[Pc(CH$_2$OC$_8$H$_{17}$)$_8$]$_2$ alkyl Subs Fig
71	3	1	2	3	Lu^{3+}	DCM	—	—	−2/−2 −1/−1 −2/−1	627(ε=106,000); "blue form" 701(ε=49,500) 488(49,500) "orange form"; diradical cation 671(ε=120,000)Q; "green form"; radical cation [PcLuPc]$^-$	[313] Lu[Pc(CH$_2$OC$_{12}$H$_{25}$)$_8$]$_2$ alkyl Subs
71	3	1	2	3	Lu^{3+}	DCM	—	—	−2/−2 −1/−1	704 631(ε=106,000); "blue form" 704(ε=44 500) 492(49 500) "orange form"; radical cation	

					Solvent	Ligand	CN	Charge	Spectral data	Reference
Mg(II)										
Mg²⁺	12	1	1	1	vap	—	4	−2	665Q 600 340	[56] T = 535°C Fig and Tab
Mg²⁺	12	1	1	1	Vap	—	4	−2	666Q 610 332B 280N	[57] T = 570°C Fig and Tab
Mg²⁺	12	14	1	1	xtl	—	4		820 700 630	[320] Fig
Mg²⁺	12	37	—	1	DCM	H₂O	6	−2	671 340 (tabulated by Minor et al.)	[33] Tab
Mg²⁺	12	3	43	1	DCM	H₂O	6	−2	670Q 642 606 345 282	[9] MCD Fig
Mg²⁺	12	3	43	1	DCM	H₂O	6	−2	Deconvolution data from absorption and MCD data 382 670Q 607 (refer to paper for Q$_{vib}$ bands) 361(A)B₁ 338(A)B₁ 316 301 281(A) 268 260 246(A) 320	[9] MCD deconv

Note: (A) indicates a degenerate transition

					Solvent	Ligand	CN	Charge	Spectral data	Reference
Mg²⁺	12	3	3	1	DCM	imid	6	−2	672Q 643 607 345 281	[9] MCD Figs
Mg²⁺	12	3	3	1	DCM	imid	6	−2	Deconvolution data from absorption and MCD data 388 672Q 609 (refer to paper for Q$_{vib}$ bands) 364(A) 339(A) 317 299 282(A) 269 260 248(A) 233	[9] MCD deconv

Note: (A) indicates a degenerate transition
These data also apply closely to ligands:
methyl-imidazole, pyridine, methyl-pyridine, CN⁻

					Solvent	Ligand	CN	Charge	Spectral data	Reference
Mg²⁺	12	3	—	1	DCM	L	6	−2	The fitting programs identified five degenerate transitions that can be associated with π–π* bands: 672Q 363 338 282 247	[9] MCD Figs
Mg²⁺	12	8	4	1	py	py	6	−2	674	[56] Tab
Mg²⁺	12	11	1	1	THF	—	—	−2	1000 800 755 685 490 370 335 290 270 A very broad band, 1200–500 nm	[230] MgPc·HCl Fig Suspension?
Mg²⁺	12	11	1	1	THF	—	4	−2	667Q 637 603 578 557 430 358 340 281 266	[230] MgPc·2H₂O Fig
Mg²⁺	12	7	1	1	DMSO	—	4	−2	650 620 350	[31] Fig
Mg²⁺	12	7	1	1	DMSO	—	4	−2	672Q 607 345B	[57] Tab
Mg²⁺	12	6	1	1	CLN	—	4	−2	678Q 612	[57] Tab
Mn(II)										
Mn²⁺	25	6	1	1	CLN	Cl⁻	4	−2	654 520	[139] Fig
Mn²⁺	25	2	12	1	Film	Cl⁻	5	−1	1374 764 680 525 365 275	[341] Tab
Mn²⁺	25	5	38	1	DMF	—	6	−1	917CT 862(trip-mult) 672Q 641 581 380CT 332	[10] Fig and Tab
Mn²⁺	25	4	17	1	DMA	—	5	−2	705(42 657) 678 634 495 417 355 295	[246] O₂ adduct
Mn²⁺	25	4	40	1	DMA	NMD	6	−2	816 710 666 620 570 475 320	[246] (L = NMeIm) Fig
Mn²⁺	25	37	—	1	—	—	—		680 (tabulated by Minor et al.)	[33] Tab
Mn²⁺	25	15	1	1	H₂O	—		−2	718 636 (MnTSPc; a mixture of monomer and dimer)	[195] Fig
Mn²⁺	25	15	1	1	H₂O	—		−2	718 (5% DMF: H₂O; monomer forms)	[195] Fig
Mn²⁺	25	8	4	1	py	py	5	−2	880 835 660 643 467	[36] Fig and Tab

Note: Photochemistry [36]; spectra of MnPc complexes in [36,241,242]

					Solvent	Ligand	CN	Charge	Spectral data	Reference
Mn(III)										
Mn³⁺	25	12	13	1	EtOH	OH⁻	6	−2	650 590 550 390 320	[272] Fig

(continued)

Table 3 Band Centers from the Absorption Spectra of the Phthalocyanine Complexes^{a,b,c} (continued)

Central Element	Z	Solv	Ax Lg	Pc Ox	M Ox	M^{n+}	Solv	Ax Lig	ML$_n$ (n=)	Ox Pc	Band positions	References and comments
	25	5	39	1	2	Mn³⁺	DMF	OAC	5	-2	1311CT 1074CT 716Q 681 645 497CT 420 368 1200CT 937(trip-mult)	[10] Subs: tetrabutyl; See also [13] Fig and Tab
	25	4	13	1	3	Mn³⁺	DMA	—	5	-2	717(ε=48 978) 646 504 359	[246] Mn(III)PcOH
	25	4	1	1	3	Mn³⁺	DMA	—	5	-2	710w 620s 570 320	[246] Fig
						Note: Dimeric; PcMn(III)-O-Mn(III)Pc						
Mo(IV)												
	42	1	—	1	4	MoO²⁺	vap	—	5	-2	695 625	[342] Tab
	42	8	41	1	4	MoO²⁺	py	py	6	-2	710 676 652	[343] Tab
	42	30	41	1	4	MoO²⁺	ClBz	—	5	-2	703 643	[343] Tab
Nd												
	60	3	45	1	3	Nd³⁺	DCM	MeOH		-2/-2	670sh 636strong 340strong	[306] Fig
						Note: Characterized as the "blue" Pc(-2)NdPc(-2)H dimer						
	60	3	45 47	1	3 1	Nd³⁺	DCM	MeOH Ac		-2/-2	670strong 330	[306] Fig
						Note: Characterized as monomeric PcNdCH₃COO						
	60	3	—	1	3	Nd³⁺	DCM	dimer	6	-2/-2	674 460	[104] Fig
						Note: PcNdPcH; see paper for pH titration						
										-2/-2	640; typical of the dimeric species Pc-M-Pc	
						Note: [Pc(-2)NdPc(-2)]⁻; the "blue form", form with addition of base or reduction						
	60	3	45	1	3	Nd³⁺	DCM	MeOH		-2/-1	676strong 470 326strong	[306] Fig
						Notes: Characterized as Pc(-2)NdPc(-1), the "green form"						
						Dimer with one ring oxidized by p-benzoquinone						
Ni												
	28	1	1	1	2	Ni²⁺	VAP	—	4	-2	651Q 594 328B 235L 210C	[57] T=517°C Fig and Tab
	28	2	1	1	2	Ni	film	—	—	-2	693sh 629s	[213] Fig and Tab
	28	2	1	1	2	Ni	film	—	—	-2	696sh 677sh 624 569 334 286 269 249 207	[52] MCD Fig
	28	2	1	1	2	Ni	film	—	—	-2	Deconvolution results using absorption and MCD 728 687 626 577 398 366 335 299 287 273 250 205	[52] MCD deconv Fig
						Davydov splitting: Q(00)=2125 cm⁻¹						
	28	6	1	1	2	Ni²⁺	CLN	—	4	-2	669Q 604	[57] Tab
	28	6	1	1	2	Ni²⁺	CLN	—	4	-2	670(147,000)Q 642 603 580 remainder obscured by solvent	[26] MCF Fig; [7] MCD

					Ion		Medium		Charge	Data	Ref
	28	2	1	1	Ni²⁺	2	ClN	—	4	Deconvolution results: 670Q 655 642 606	[7] deconv [57] absn Tab [33] Tab
	28	37	—	1	Ni²⁺	2	—	—	-2	671 351 (tabulated by Minor et al.)	[33] Tab
	28	7	1	1	Ni²⁺	2	DMSO	6	-2	673(ε=182,000)Q 644 606 336(24,000)	[221] Fig and Tab; NiTSPc

Note: See paper [221] for spectra in DMF, MeOH, H₂O; tetrasulfonated

| | 28 | 26 | 1 | 1 | Ni²⁺ | — | CHCl | 1 | -2 | 667 (ε=70,000) 638 603 401 | [187] Fig and Tab |

Note: Peripherally substituted with four 15-crown-5 ether groups = Ni(II)TCRPc; monomer

| | 28 | 26 | 1 | 1 | Ni²⁺ | — | CHCl | 1 | -2 | 630 (ε=47,000) 385 360 | [187] Fig and Tab |

Note: Peripherally substituted with four 15-crown-5 ether groups = Ni(II)TCRPc; dimer with K⁺ added

| | 28 | 10 | ? | 1 | Ni²⁺ | 2 | H₂SO₄ | — | -2 | 738 380 302 208 (30 N sulfuric acid) | [181] Tab |

Note: Subs: Ni-tetraamine-Pc·2H₂O

| | 28 | 26 | 1 | 1 | Ni²⁺ | 1 | CHCl | 4 | -2 | 669 Q 291 | [214] MCD Fig and Tab |

Notes: Monomeric species; weak bands between 350 and 450 nm
Subs tetra (15-crown-5): NiTCRPc
See the paper for deconvolution data for absorption and MCD spectra

| | 28 | 26 | 1 | 1 | Ni²⁺ CHCl | | | 4 | -2 | 632 (A term in MCD) 290 | [214] MCD Fig and Tab |

Notes: Dimeric species; weak bands between 350 and 450 nm; Subs: tetra (15-crown-5): NiTCRPc
See the paper for extensive deconvolution data for absorption and MCD spectra

Os	76	26	—	1	Os	2	CHCl	—	4	632 575 [(CO)(py)OsPc in chloroform]	[296] Tab
	76	11	—	1	Os	2	THF	—	4	636 576 [(CO)(THF)OsPc in THF]	[296] Tab
P	15	12	1	1	P³⁺	3	EtOH	—	-2	651 438 325	[31] Tab
	15	12	1	1	P³⁺	3	EtOH	—	-2	656 441 325	[31] Fig and Tab

Note: Subs ((tetra-*t*-butyl))Pc

	15	8	1	1	P³⁺	3	py	—	-2	655 626 597 442 435 413	[238] Fig and Tab
	15	8	1	1	P⁵⁺	5	py	—	-2	653 593 413	[238] Fig and Tab
Pb	82	13	1	1	Pb	2	xtl	—	4	790 710 650 460	[320] Fig
	46	7	1	1	Pb	2	DMSO	—	4	700 630 450 350	[31] Fig and Tab
	82	1	1	1	Pb	2	VAP	—	4	698Q 640 333B 280N 245L 207C	[57] Fig and Tab
	82	7	1	1	Pb	2	DMSO	?	4	702Q 633 336B	[57]
	82	6	1	1	Pb²⁺	2	ClN	—	4	714 430 342	[31] Tab
Pd	46	15	1	1	Pd	1	H₂O	—	4	660 640 580	[202] PdTSPc; Fig

(continued)

Table 3 Band Centers from the Absorption Spectra of the Phthalocyanine Complexes[a,b,c] (continued)

Central Element	Z	Solv	Ax Lg	Pc Ox	M Ox	M^{n+}	Solv	Ax Lig	ML_n (n=)	Ox Pc	Band positions	References and comments
Pd	46	6	1	1		Pd	ClN			−2	660 633 595 577 557 347	[76] Tab
Pr	59	2	1	?	3	Pr^{3+}	FLM	—	—	?	680 620 505	[344]
	59	2	1	?	3	Pr^{3+}	FLM	—	—	?	660 320	[344] Fig
						Note: Diphthalocyanine data in [344]						
Rh(II)	45	5	42	1	2	Rh^{2+}	DMF	but	6	−2	661 633 598 371 327 (butylamine)$_2$RhPc	[299] Tab
	45	5	4	1	2	Rh^{2+}	DMF	py	6	−2	661 632 598 372 326	[299] Tab
	45	5	19	1	2	Rh^{2+}	DMF	bpy	6	−2	661 630 598 370 325 (similarly for L=pyrazine)	[299] Tab
Rh(III)	45	11 / 15	12	1	3	Rh^{3+}	THF / H_2O	Cl^-	6	−2	650Q	[345] Tab
	45	5	—	1	3	Rh^{3+}	DMF	Cl^-	5	−2	644Q 344	[303] Tab
	45	5	2	1	3	Rh^{3+}	DMF	CN^-	6	−2	651Q 343 [(CN^-)Rh(III)Pc]K	[303] Tab
	45	5	42	1	3	Rh^{3+}	DMF	CN^-		−2	650Q 345 (CN^-)(n-butylamine)Rh(III)Pc	[303] Tab
	45	21	12	1	3	Rh^{3+}	ACN	Cl^-		−2	645Q 345	[301] Fig photochem
	45	21	12	1	3	Rh^{3+}	ACN	L	6	−2	650 Q for all (L)$_2$RhPc complexes	[300] Fig
						Notes: L=H_2O, H_2O; CH_3OH, Cl; CH_3OH, Br; CH_3OH, I						
						See paper for details of 225–500 nm region						
	45	26	12	1	3	Rh^{3+}	CHCl	Cl^-	5	−2	654 625 590 341 280	[302] Tab
						Note: Similar spectral data are reported for L=py, bpy, dabco, mepyz, dma						
Ru	44	18	5	1	2	Ru^{2+}	MePy	MePy	6	−2	623Q 565 378 310	[6] Tab
	44	3	5	1	2	Ru^{2+}	DCM	MePy	6	−2	621Q 566 377 314	[6] Fig and Tab
	44	3	5	1	2	Ru^{2+}	DCM	MePy	6	−2	622(ε=67,000) 560 375 310	[298,346] Fig and Tab
	44	3	4	1	2	Ru^{2+}	DCM	Py	6	−2	626Q 565 387B 316N 284L	[6] Fig and Tab
	44	19	4	1	2	Ru^{2+}	DEE	Py	6	−2	618Q 564 374 312	[6] Tab
	44	8	4	1	2	Ru^{2+}	Py	Py	6	−2	622Q 568 378 313	[6] Tab
	44	36	4	1	2	Ru^{2+}	MTHF	Py	6	−2	625Q	[90] Fig photochem
	44	5	4	1	2	Ru^{2+}	DMC	Py	6	−2	622Q 566 375	[6] Tab
	44	3	4	1	2	Ru^{2+}	DCM	Py	6	−2	620 375 300	[347]
	44	4	4	1	2	Ru^{2+}	DCM	Py	6	−2	622(ε=74,000)Q 567 375 315(120,000)	[298,346] Fig and Tab
	44	3	18	1	2	Ru^{2+}	DCM	PiP	6	−2	622Q 568 440 420 384 314	[6] Fig and Tab
	44	20	18	1	2	Ru^{2+}	PiP	PiP	6	−2	622Q 566	[6] Tab
	44	3	19	1	2	Ru^{2+}	DCM	bpy	6	−2	622Q 566 374 314	[6] Tab
	44	4	19	1	2	Ru^{2+}	DMA	bpy	6	−2	622Q 562 376 298	[6] Tab
	44	3	19	1	2	Ru^{2+}	DCM	bpy	6	−2	621 567 375	[346]

44	4	2		1	2	Ru²⁺	DMA	CN⁻	6	−2	620Q 564 432 412 352	[6] Tab
44	21	2		1	2	Ru²⁺	ACN	CN⁻	6	−2	619Q 562 432 414 294	[6] Fig and Tab
44	21	20		1	2	Ru²⁺	ACN	ACN	6	−2	630Q 570 312	[6] Fig and Tab
44	3	6		1	2	Ru²⁺	DCM	DMSO	6	−2	640Q 579 305	[6] Fig and Tab
44	7	6		1	2	Ru²⁺	DMSO	DMSO	6	−2	632Q 580 313	[90] Fig MCD
44	19	6		1	2	Ru²⁺	DEE	DMSO	6	−2	634Q 574 312	[6] Tab
44	3	3		1	2	Ru²⁺	DCM	DMA	6	−2	637(ϵ=88,000)Q 578 381 318	[298] Tab
44	3	22		1	2	Ru²⁺	DMA	DMA	6	−2	624Q 566 378 314	[6] Tab
44	4	22		1	2	Ru²⁺	DMA	DMA	6	−2	626Q 570 386 312	[6] Tab
44	3	3		1	2	Ru²⁺	DCM	Im-	6	−2	622Q 566 440 380 312	[6] Fig and Tab
44	3	3		1	2	Ru²⁺	DMC	Im-	6	−2	621(90,000)Q 566 312	[298] Tab
44	3	21		1	2	Ru²⁺	DCM	DMF	6	−2	624Q 566 377 314	[6] Tab
44	5	21		1	2	Ru²⁺	DMF	DMF	6	−2	622Q 566	[6] Tab
44	7	21		1	2	Ru²⁺	DMSO	DMF	6	−2	624Q 570 378 314	[6] Tab
44	3	21		1	2	Ru²⁺	DCM	DMF	6	−2	622(ϵ=68,000)Q 568 382 312	[298] Tab
44	3	5	8	1	2	Ru²⁺	DCM	MePy / CO	6	−2	642Q 578 348 294	[90] Fig photochem
44	18	5		1	2	Ru²⁺	MePy	MePy	6	−2	643Q 580	[6] Tab
44	3	5	8	1	2	Ru²⁺	DCM	MePy / CO	6	−2	640(ϵ=152,000) 618 580 342 295	[298,346] Fig and Tab
44	4	4		1	2	Ru²⁺	DMA	Py	6	−2	642Q 582 345 291	[6] Tab
44	8	4		1	2	Ru²⁺	Py	Py	6	−2	644Q 584	[6] Fig and Tab
44	3	4	8	1	2	Ru²⁺	DCM	Py / CO	6	−2	642Q 580 300	[6] MCD Fig and Tab; photochem, electrochem
44	17	8		1	2	Ru²⁺	BNZ	CO	6	−2	640 340 325	[347] Fig
44	3	4	8	1	2	Ru²⁺	DCM	Py / CO	6	−2	637 614 578 342	[346] Fig
44	3	4	8	1	2	Ru²⁺	DCM	bpy / CO	6	−2	643Q 580 345	[6] Tab
44	19	19	8	1	2	Ru²⁺	DEE	bpy / CO	6	−2	636Q 575 345 290	[6] MCD
44	4	19	8	1	2	Ru²⁺	DMA	bpy / CO	6	−2	640Q 578	[6] MCD
44	3	19	8	1	2	Ru²⁺	DCM	bpy / CO	6	−2	640 615 579 345	[346] Fig
44	7	6	8	1	2	Ru²⁺	DMSO	DMSO / CO	6	−2	646Q 584 346	[6] MCD
44	3	6	8	1	2	Ru²⁺	DCM	DMSO / CO	6	−2	645(ϵ=123,000) 583 344 293	[298] Tab
44	3	22		1	2	Ru²⁺	DCM	DMA / CO	6	−2	643Q 583 300	[6]
44	5	22	8	1	2	Ru²⁺	DMF	DMA / CO	6	−2	642Q 580 345 295	[6]

(continued)

Table 3 Band Centers from the Absorption Spectra of the Phthalocyanine Complexes[a,b,c] (continued)

Central Element	Z	Solv	Ax Lg	Pc Ox	M Ox	M^{n+}	Solv	Ax Lig	ML_n (n=)	Ox Pc	Band positions	References and comments
	44	19	22 8	1	2	Ru^{2+}	DEE	DMA	6	−2	637Q 576 345 295	[6]
	44	8	22 8	1	2	Ru^{2+}	Py	DMA CO	6	−2	644Q 582 345 310	[6]
	44	4	22 8	1	2	Ru^{2+}	DMA	DMA CO	6	−2	641Q 580 346 316	[6] Fig
	44	7	22 8	1	2	Ru^{2+}	DMSO	DMA	6	−2	645Q 583 357 295	[6] All data from [6]: Tab
	44	3	21 8	1	2	Ru^{2+}	DCM	DMF CO	6	−2	642Q 582 297 250	[90] Fig MCD
	44	7	21 8	1	2	Ru^{2+}	DMSO	DMF	6	−2	644Q 582 315	[6]
	44	5	21 8	1	2	Ru^{2+}	DMF	DMF CO	6	−2	642Q 579 344 293	[6] [27] Photochem
	44	21	21 8	1	2	Ru^{2+}	ACN	DMF CO	6	−2	637Q 577 345 294	[6]
	44	19	21 8	1	2	Ru^{2+}	DEE	DMF CO	6	−2	637Q 575 345	[6]
	44	4	21 8	1	2	Ru^{2+}	DMA	DMF CO	6	−2	640Q 580 345	[6] All data from [6]: Tab
	44	3	21 8	1	2	Ru^{2+}	DCM	DMF CO	6	−2	638(ϵ=120,000) 575 344 291	[6,298] Tab
	44	17	—	1	2	Ru^{2+}	BNz	—	4	−2	645 584	[4]
	44	17	—	1	2	Ru^{2+}	BNz	—	4	−2	632 580 378	[4] PcRu·6PHNH$_3$
	44	17	—	1	2	Ru^{2+}	BNz	—	4	−2	632 582 379	[4] PcRu·6CH$_3$C$_6$H$_4$NH$_2$ Data in [4]: Tab
	44	11	23	1	2	Ru^{2+}	THF	—	4	−2	627	[296] Fig
	44	26	—	1	2	Ru^{2+}	CHCl	—	4	−2	642 581 (CO)(THF)RnPc in chloroform	[296] Tab
	44	26	—	1	2	Ru^{2+}	CHCl	—	4	−2	624 581 (CO)(py)RnPc in chloroform	[296] Tab
	44	3	20	1	2	Ru^{2+}	DCM	ACN	6	−2	633(ϵ=90,000)Q 578 312	[298] Fig and Tab
	44	11	23 8	1	2	Ru^{2+}	THF	THF	6	−2	640 580	[296]
	44	5	21 8	2	2	Ru^{2+}	DMF	DMF$^+$ CO	6	−2	700 510 400	[27] photochem, electrochem Fig
	44	3	4 8	2	2	Ru^{2+}	DCM	py$^+$ CO	6	−2	637Q 342	[13] Electrochem Fig
	44	4	4	2	2	Ru^{2+}	DCM	py	6	−2	621Q 315	[13] Electrochem
	44	35	51	1	2	Ru^{2+}	tol	POBu	6	−2	638(ϵ=77,000) 566 402(10,000)	[260] Tab
	44	35	50	1	2	Ru^{2+}	tol	PBu3	6	−2	631(ϵ=46,000) 572 435 361(33,000)	[260] Tab
	44	35	4	1	2	Ru^{2+}	tol	py	6	−2	622(ϵ=84,000) 567 375(26,000)	[260] Tab
Note: Chlorinated ring suggested												
	44	35	40	1	2	Ru^{2+}	tol	MeIm	6	−2	622(ϵ=79,000) 567 423 384(27,000)	[260] Tab
Note: Chlorinated ring suggested												

44	26	49	1	2	Ru²⁺	ClBz	L2	6	−2	641 587 442 376 314 268	[294] Tab

Note: L₂ = pyrazine, similarly for Mepyz, Etpyz, and tButpyz, see paper

44	26	49	1	2	Ru²⁺	ClF	L2	6	−2	626 575 446 370 314 272	[294] Tab

Note: L₂ = 4,4′-bipyridine; similarly for pyridazine and pyrimidine, see paper

14	8	12	1	4	Si⁴⁺	Py	Cl⁻	6	−2	699Q 627 367 314	[175]
14	8	13	1	4	Si⁴⁺	Py	OH⁻	6	−2	671Q 641 604	[175]
14	11	13	1	4	Si⁴⁺	THF	OH⁻	6	−2	667Q 636 602 377 363 318	[175]
14	13	13	1	4	Si⁴⁺	TCM	OH⁻	6	−2	674Q 644 605 357	[175]
14	17	13	1	4	Si⁴⁺	Bnz	OH⁻	6	−2	672Q 642 605 355	[175]

Note: All data from [175] in a Table

14	17	–	1	4	Si⁴⁺	Bnz	(OR)	6	−2	668.5 638 604 353 330	[175] Fig

Note: SiPc(OR)₂, R = Si[(C(CH₃)₃(CH₃)₂]⁻, monomer

14	17	–	1	4	Si⁴⁺	Bnz	(OR)	6	−2	(690W) 633 606 574 554 332	[175] Fig

Note: ROSiPcOSiPcOR, R = Si[(C(CH₃)₃(CH₃)₂]⁻, dimer

14	3	–	1	4	SiO	DCM	poly	6	−2	700 (v weak) 620Q (v broad, no Q$_{vib}$) 330 275	[41] Fig

Note: Trimer; (n-C₆H₁₃)₃SiO(SiPcO)₃Si(n-C₆H₁₃)₃; same for tetramer

14	33	–	1	4	Si⁴⁺	Nujl	–	6	−2	742 670 388 276 229	[235] Tab

Note: Monomer, SiPcL₂, L = Cl

14	33	–	1	4	Si⁴⁺	Nujl	–	6	−2	742 648 388 368 289 207	[235] Tab

Note: Polymer, (LSiPcL)ₙ, L = HC≡CC₆H₄C≡CH⁻

14	33	–	1	4	Si⁴⁺	Nujl	–	6	−2	625Q-very broad 320	[40] Fig

Note: Polymer, (SiPcO)ₙ; see also SiPc(−1) data with I₂

14	32	–	1	4	SiO	CyHx	–	6	−2	665(ε=460,000)Qs 351(78,000) 342(76,000) 331	[215] Fig and Tab

Note: Monomer, very well resolved Q

14	32	–	1	4	SiO	CyHx	–	6	−2	dimer, Si-O-Si links: 630(ε=270,000) 329(120,000)	[215] Dimer, Fig and Tab

Note: Trimer, Si-O-Si-O-Si: 618(ε=160,000) 327(150,000)
[215] Tetramer: 615(ε=170,000)s + broad 326(180,000)

[215] Trimer Fig and Tab
Tetramer Tab

14	11	–	1	4	SiO	THF	–	6	−2	665(ε=253,300)Q 352(84,100) 336 287 265 257 221	[8] Fig

Note: Monomer, very well resolved Q

Si

(continued)

Table 3 Band Centers from the Absorption Spectra of the Phthalocyanine Complexes^{a,b,c} (continued)

Central Element	Z	Solv	Ax Lg	Pc Ox	M Ox	M^{n+}	Solv	Ax Lig	ML$_\pi$ (n=)	Ox Pc	Band positions	References and comments
	14	11	—	1	4	SiO	THF	—	6	−2	dimer, Si-O-Si links: 751w 683w 628(ε=222,800) 604 574 554 331(ε=101,200) 282 271 216	[8] Dimer, Fig
	14	3	—	1	4	SiO	DCM	—	6	−2	772Q 732 704 686 655 636	[236] Fig
	Note: Monomer (with low % dimer) Si-naphthalocyanine											
	14	3	—	1	4	SiO	DCM	—	6	−2	668Qs 638 612 601 575 557 350 270	[236] Fig
	Note: Monomer SiPc(OR)$_2$, R=OSi(nC$_6$H$_{13}$)$_3$											
	14	3	—	1	4	SiO	DCM	—	6	−2	620s	[236] Fig
	Note: Dimer, RO(SiPcO)$_2$R											
Sn	50	1	12	1	?	Sn	VAP	Cl	6	−2	693Q 640 323B 280N 250L	[57] T=506°C Fig and Tab
	50	7	12	1	?	Sn	CLN	Cl	6	−2	701Q 630 363B	[57] Tab
	50	1	1	1	?	Sn	VAP	—	4	−2	690Q 625 327B 270N	[57] T=482°C Fig and Tab
	50	7	1	1	?	Sn	DMSO	—	4	−2	677Q	[57] Tab
	50	6	1	1	?	Sn	CLN	—	4	−2	631Q 575	[57] Tab
	50	29	1	1	?	Sn	DMOE	—	4	−2	622Q 575 337B 290N	[57] Dimer Tab
	50	10	—	1	4	Sn^{4+}	H$_2$SO$_4$	—	4	−2	792 756 710 407	[188] Tab
	Note: (Cl)$_2$Sn—(F$_{16}$)Pc											
	50	30	1	1	4	Sn^{4+}	CBz	—		−2	774 626 575 338 (data for Sn(Pc)$_2$)	[76] Tab
	50	12	1	1	2	Sn^{2+}	EtOH	—	4	−2	682 359	[31] Tab
Tb	65	2	1	1	3	Tb^{3+}	FLM	—	—	—	660 600 460	[344] Diphthalocyanine Fig; stacked dimer
	65	2	24	1	3	Tb^{3+}	FLM	+I$_2$	—	—	670 500	[344] Fig; stacked dimer
	65	2	24	1	3	Tb^{3+}	FLM	+I$_2$	—	—	720 500 420	[344] Fig; monomer Pc(−1)
	65	17	25	1	3	Tb^{3+}	BNZ	DPM	6	−2	660 640 610 370	[307] Fig
	Note: Tb^{3+}Pc(dipivaloylmethane)$_2$ with Bu$_4$NBH$_4$, added to form the M^{3+}Pc(−2)Pc(−2) complex											
Th	90	30	43	1	6	Th^{6+}	PhCN	acac	8?	−2	684(ε=207,000) 617 353 for (acac)$_2$ThPc	[319] Fig and Tab
Ti	22	1	1	1	4	TiO^{2+}	VAP	—	4	−2	676Q 616 337B 266N 250N 250L 210C	[57] T=535°C; Fig and Tab
	22	6	1	1	4	TiO^{2+}	CLN	—	4	−2	698Q 630	[57] Tab
	22	5	1	1	4	TiO^{2+}	DMF	—	4	−2	693Q 353	[13] Tab
	Note: Electrochemical studies; subs											

Element						Ion	Phase	Dimer		Charge	Spectral data	Ref
Tm	69	3	—	1	3	Tm³⁺	DCM	dimer	6	-2	1500 1392 1260 906(ε=8,000) 880 660(ε=190,000)Q 460 335 320 275	[315] Fig and Tab
Note: All dimers = M(Pc)₂												
	69	2		1	3	Tm³⁺	Film	dimer		-2	no data shown for 250–700 nm; 908 1280 1415 1490	[315] thin film Fig and Tab
U SPc(−2) (superphthalocyanine)	92	17	—	1	6	UO₂²⁺	Bnz		6?	-2	914 (ε=66,700)Q-very broad 810 424 345 300	[29] SuperPc Tab [30] Fig
	92	35	1	1	6	UO₂²⁺	tol			-2	912 (ε=66,700)Q-very broad 810sh 420	[29] Fig and Tab
	92	35	1	1	6	UO₂²⁺	tol			-2	922 810sh 420; for Subs (4-Me)₅	[29] Tab
	92	35	1	1	6	UO₂²⁺	tol			-2	938(ε=69,800) 820sh 417(69,000); for (4,5-Bu₂)₅	[29] Fig and Tab; Subs
	92	30	43	1	6	U⁶⁺	PhCN	acac	8?	-2	689(ε=166,000) 620 351 for (acac)₂UPc	[319] Fig and Tab
V	23	2	1	1	2	VO²⁺	Film			-2	836 737 644 360	[213] Tab and Fig
	23	2	1	1	4	VO²⁺	Film	—	—	-2	840 735 635 (VOPc embedded in a polymer film)	[239] Fig
	23	2	1	1	4	VO²⁺	Film	—	—	-2	725 660 635	[239] Fig
	23	13	1	1	4	VO²⁺	xtl	—	—	-2	810 700 630	[320] Fig
	23	2	1	1	4	VO²⁺	Film	—	—	-2	800sh 694 650	[217] Fig
Note: VO-t-Bu₁,₄-Pc in polystyrene												
	23	2	1	1	4	VO²⁺	Film	—	—	-2	810 694 650	[217] Fig
Note: VO-t-Bu₁,₄-Pc in polystyrene, after ethyl acetate treatment												
	23	6	1	1	4	VO²⁺	CLN	—	—	-2	700 630	[239] + dimer Fig
	23	3	1	1	4	VO²⁺	DCM	—	—	-2	690 660 620	[239] + dimer Fig
	23	11	1	1	4	VO²⁺	THF	—	—	-2	680 650 618	[239] + dimer Fig
	23	1	1	1	4	VO²⁺	VAP	—	—	-2	671Q 620 333B 280N 244L 220C	[57] T=495°C Fig and Tab
	23	6	1	1	4	VO²⁺	CLN	—	—	-2	700Q 631	[57]
	23	7	1	1	4	VO²⁺	DMSO	—	—	-2	685Q 618 346B 290N	[57] Tab
	23	5	1	1	4	VO²⁺	DMF	—	—	-2	699Q 345	[13] Tab; Subs
Y	39	46	—	1	3	Y³⁺	ClBz	—	—	-2	(CH₃COO)YPc: 665Q; Y₂Pc₃: 705 633	[308] Fig
Yb	70	3	—	1	3	Yb³⁺	DCM	dimer	6	-2	Q as for Tm; 880 906 (ε=10,000) 1252 1386 1500	[315] Tab and Fig all dimers = Pc2M
Zn	30	1	1	1	2	Zn²⁺	VAP	—	—	-2	661Q 600 327B 276N 240L 220C	[57] T=556°C; Fig and Tab
	30	1	1	1	2	Zn²⁺	vap	—	—	-2	660 600 340	[56] T=530°C; Fig and Tab
	30	2	1	1	2	Zn²⁺	FLM	—	—	-2	700 600 460 340	[347] Fig.

(continued)

Table 3 Band Centers from the Absorption Spectra of the Phthalocyanine Complexes[a,b,c] *(continued)*

Central Element Z	Solv	Ax Lg	Pc Ox	M Ox	M^{n+}	Solv	Ax Lig	ML_n $(n=)$	Ox Pc	Band positions	References and comments
30	2	1	1	2	$\alpha\text{-}Zn^2$	Film	—	—	−2	720 620 335 288 210 182 163	[53] Fig and Tab
30	6	1	1	2	Zn^{2+}	ClN	—	4	−2	672	[56] Tab
30	37	—	1	2	Zn^{2+}	—	—	5	−2	681 347 (tabulated by Minor et al.)	[33] Tab
30	4	2	1	2	Zn^{2+}	DMA	CN^-	5	−2	669Q 386 331 274	[6] MCD Fig and Tab
30	4	2	1	2	Zn^{2+}	DMA	CN^-	5	−2	Results of deconvolution calculations on MCD and absorption spectra 669Q 651 639 629 606 569 386(A) 358 331(A) 300 274(A) 258(A)	[65] MCD deconv See the paper for A/B terms

Notes: The Q and bands marked (A) are degenerate
See the paper for more details

Central Element Z	Solv	Ax Lg	Pc Ox	M Ox	M^{n+}	Solv	Ax Lig	ML_n $(n=)$	Ox Pc	Band positions	References and comments
30	3	3	1	2	Zn^{2+}	DCM	Im	5	−2	671Q 363B$_1$ 336B$_2$ 282N 246L	[6] MCD Fig and Tab [92] Electrolysis Fig
30	3	3	1	2	Zn^{2+}	DCM	Im	5	−2	Results of deconvolution calculations on MCD and absorption spectra 671Q 655 640 633 608 566 368(A) 336(A) 313 294 282(A) 273 261 246(A) 235	[65] MCD deconv See the paper for A/D values

Notes: The Q and bands marked (A) are degenerate
See the paper for more details

Central Element Z	Solv	Ax Lg	Pc Ox	M Ox	M^{n+}	Solv	Ax Lig	ML_n $(n=)$	Ox Pc	Band positions	References and comments
30	3	4	1	2	Zn^{2+}	DCM	py	5	−2	670Q 340 280	[6] MCD Fig and Tab [27] Photochem Fig [92] Photolysis Fig
30	3	5	1	2	Zn^{2+}	DCM	mepy	5	−2	670Q 350 285	[6] MCD Tab
30	4	1	1	2	Zn^{2+}	DMA	—	4	−2	670Q 380 330 280	[6] MCD Fig and Tab
30	7	1	1	2	Zn^{2+}	DMSO	—	5	−2	672(ϵ=294,000)Q 607(46,100) 345(75,000)	[3] MCD Fig and Tab
30	7	2	1	2	Zn^{2+}	DMSO	—	6	−2	676(ϵ=307,000)Q 609(44,000) 332(63,200)	[7] MCD Fig [3] MCD Fig
30	7	6	1	2	Zn^{2+}	DMSO	—	6	−2	Results of deconvolution calculations on MCD and absorption spectra 671Q 648 609 392 350 317	[7] MCD Fig [3] MCD Fig [7] MCD deconv
30	7	2	1	2	Zn^{2+}	DMSO	CN^-	6	−2	Results of deconvolution calculations on MCD and absorption spectra 675Q 649 612 396 334 298 276 267	[3] MCD Fig [7] MCD deconv
30	3	14	1	2	Zn^{2+}	DCM	—	5	−2	667 641 335 275	[338] Fig and Tab
30	3	1	1	2	Zn^{2+}	DCM	—	4	−2	682 614 384 342	[229] Fig [98] Tab

Note: Subs: R = CH(CH$_3$)$_2$

Central Element Z	Solv	Ax Lg	Pc Ox	M Ox	M^{n+}	Solv	Ax Lig	ML_n $(n=)$	Ox Pc	Band positions	References and comments
30	3	1	1	2	Zn^{2+}	DCM	—	4	−2	680 614 384 344	[98] Tab

Note: Subs: R = CH$_2$C(CH$_3$)$_3$

Ele	No.	Ax Lg	Ox	M Ox	Solv	Ax Lg	λ_{max} (nm)	Pc Ox	Coding	Note
30	15	1	4	Zn^{2+}	H_2O	—	670 650 600	-2	[202] Fig, ZnTSPc	
30	15	1	4	Zn^{2+}	H_2O	—	750 730 680	-2	[202] Fig; Subs; naphthalocyanine	*Note:* HexadecafluoroPc: $Zn\text{-}F_{16}Pc$
30	10	—	1	Zn^{2+}	H_2SO_4	—	864 804 771 513 very broad Q band	-2	[188] Fig	*Note:* HexadecachloroPc: $ZnCl_{16}Pc$
30	35	1	1	Zn^{2+}	Tol/acetone	—	673 ($\epsilon=189{,}000$) 607 557 424(398,000) 349	-2	[190] Fig and Tab	*Note:* Mixed ZnTTP-Zn-tetrabutylPc dimer
30	26	1	—	Zn^{2+}	CHCl	—	677 ($\epsilon=96{,}700$) 610 420 352	-2	[187] Fig and Tab	*Note:* Peripherally substituted with four 15-crown-5 ether groups = Zn(II)TCRPc; monomer
30	26	1	—	Zn^{2+}	CHCl	—	635 ($\epsilon=51{,}000$) 580 342	-2	[187] Fig and Tab	*Note:* Peripherally substituted with four 15-crown-5 ether groups = Zn(II)TCRPc; dimer with K^+ added
30	26	1	4	Zn^{2+}	CHCl	—	680Q 347	-2	[214] MCD Fig and Tab	*Note:* Monomeric species; weak bands between 350 and 450; Subs: tetra (15-crown-5): ZnTCRPc
30	26	1	4	Zn^{2+}	CHCl	—	635 (A term in MCD) 347	-2	[214] MCD Fig and Tab	*Note:* Dimeric species; Subs; tetra (15-crown-5): ZnTCRPc
30	10	1	2	Zn^{2+}	H_2SO_4	—	742 384 302 214 (30 N sulfuric acid)	-2	[181] Tab	*Note:* Subs: Zn-tetraamine-Pc: $2H_2O$
30	5	?	2	Zn^{2+}	DMF	—	667 602 335	-2	[180] Tab	
30	10	?	2	Zn^{2+}	H_2SO_4	—	783 696 434 304 230 204	-2	[180] Tab	
30	5	?	2	Zn^{2+}	DMF	—	678 622 352	-2	[180] Tab	*Note:* Compound (1); Zn-2,3,9,10,16,17,23,24-octabutoxy-Pc; Compound (1)
30	10	?	2	Zn^{2+}	H_2SO_4	—	827 729 450 310 280	-2	[180] Tab	*Note:* Compound (1); Zn-2,3,9,10,16,17,23,24-octabutoxy-Pc; Compound (1)
30	5	?	2	Zn^{2+}	DMF	—	694 654 400 345	-2	[180]Tab	*Note:* Compound (1); Zn-2,3,9,10,16,17,23,24-octacyano-Pc; Compound (ZnL^2)
30	10	?	2	Zn^{2+}	H_2SO_4	—	756 735 697 663 390 227 207	-2	[180] Tab	*Note:* Compound (1); Zn-2,3,9,10,16,17,23,24-octacyano-Pc; Compound (ZnL^2)

[a] Elements are coded with their atomic number.

[b] Abbreviations for the table headings (note: the first six headings are numerical to aid automatic sorting routines retrieve specific data at a later date): Coding information: Solv, solvent; Ele, element; Ax Lg, axial ligand; Pc Ox, oxidation state of the Pc; M Ox, oxidation state of the central metal.

[c] The data reported are for the unsubstituted monomeric species, unless otherwise stated.

Table 4 Comparison of Spectra Data Recorded in H_2SO_4

Cation	Medium										Ref.
Co^{2+}	H_2SO_4	738	388	298	212						[181]
Cu^{2+}	H_2SO_4	821	779	446							[188]
Note: $Cu(F_{16})Pc$											
Cu^{2+}	H_2SO_4	749	382	300	214						[181]
Cu^{2+}	H_2SO_4	790	746	698	634	438	303	254	222	210	[289]
Cu^{2+}	H_2SO_4	728	696	660	630	385	345	290	241	240	[289]
Note: Cu-octacarbonitrile-Pc											
Cu^{2+}	H_2SO_4	838	740	530	432	304	216				[289]
Note: Cu-tetraamine-Pc·$2H_2O$											
Fe^{3+}	H_2SO_4/CN^-	765	685	550	480	423	395	303	220		[261]
Fe^{3+}	H_2SO_4	806	500								[188]
Note: $Fe(F_{16})Pc$											
Ni^{2+}	H_2SO_4	738	380	302	208						[181]
Sn^{4+}	H_2SO_4	792	756	710	407						[188]
Note: $(Cl)_2Sn\text{-}(F_{16})Pc$											
Zn^{2+}	H_2SO_4	820	780	700	458	very broad Q					[188]

									Note	Ref.
Zn^{2+}	H_2SO_4	864	804	771	513			very broad Q band	Note: hexadecafluoroPc: $Zn\text{-}F_{16}Pc$	[188]
Zn^{2+}	H_2SO_4	742	384	302	214			(30 N sulfuric acid)	Note: hexadecachloroPc: $ZnCl_{16}Pc$	[181]
Zn^{2+}	DMF	667	602	335						[180]
Zn^{2+}	H_2SO_4	783	696	434	304	230	204			[180]
Zn^{2+}	DMF	678	622	352					Note: Zn-tetraamine-$Pc \cdot 2H_2O$	[180]
Zn^{2+}	H_2SO_4	827	729	450	310	280			Note: compound (1); Zn-2,3,9,10,16,17,23,24-octabutoxy-Pc	[180]
Zn^{2+}	DMF	694	654	400	345				Note: compound (1); Zn-2,3,9,10,16,17,23,24-octabutoxy-Pc	[180]
Zn^{2+}	H_2SO_4	756	735	697	663	390	340	227 207	Note: compound (1); Zn-2,3,9,10,16,17,23,24-octacyano-Pc	[180]

Note: compound (1); Zn-2,3,9,10,16,17,23,24-octacyano-Pc

4

Photochemical Properties of Metallophthalo-cyanines in Homogeneous Solution

G. Ferraudi

A. INTRODUCTION

There has been a number of theoretical studies about the electronic structure of metallophthalocyanines which more or less account for some of these compounds electronic properties [1,2]. Although this subject will be reviewed in connection with the spectroscopy of phthalocyanines, it is convenient to give here a simple description of the orbitals' ordering. The diagrams in Figs. 1 and 2 show the typical energy levels reported for various phthalocyanines [1]. They give only a semiquantitative description of the electronic structure in the best of cases. For phthalocyanines with metal ions of the Ia or IIa families, the mixing between metal and ligand orbitals is negligible and the electronic transitions take place between ligand-centered orbitals. A number of intense features in the absorption spectrum appear at wavelengths in the UV and VIS regions and have been assigned to transitions between bonding and antibonding orbitals, e.g., $\pi\pi^*$ states [1–4]. Similar electronic transitions are observed in related macrocycles with a high degree of unsaturation in the ligand, e.g., porphyrins and porphyrazines [5,6]. One important difference between porphyrins and phthalocyanines is, however, the existence of four imino nitrogens bridging pyrrole units in the phthalocyanine macrocycle. The filled nonbonding orbitals, localized in the azomethine groups, give rise to $n\pi^*$ excited states which have been associated with some of the phthalocyanines' redox photochemistry and with electronic transitions that increase the width of the Q and Soret bands [1,7,8].

Another change, brought into these molecules when methine groups are replaced by azomethine groups, is a reduction in the size of the phthalocyanine's coordination sphere and a corresponding increase of the metal–ligand interaction. In the phthalocyanines, the shift of charge transfer bands to higher energies (with respect to those in the porphyrins) is a consequence of such intense metal–ligand interactions. Whenever charge transfer and ligand-centered transitions overlap, the spectrum, in particular at wavelengths between the Soret and Q bands, grows in complexity (Fig. 3) [9,10].

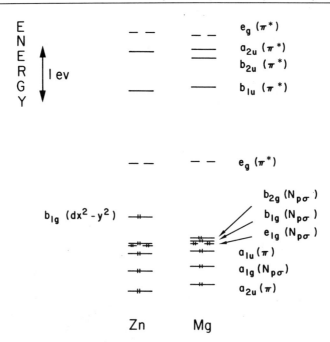

Figure 1 Energy levels in metallophthalocyanines of representative elements, Zn(II) and Mg(II). Adapted from Ref. [1].

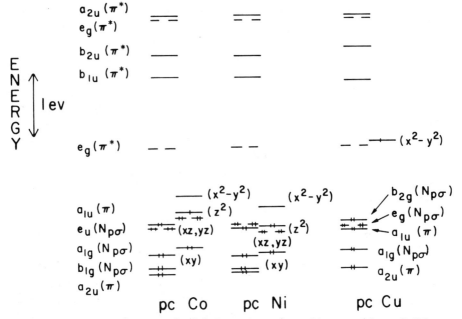

Figure 2 Energy levels in metallophthalocyanines of transition metal ions, Co(II), Ni(II), and Cu(II). Adapted from Ref. [1].

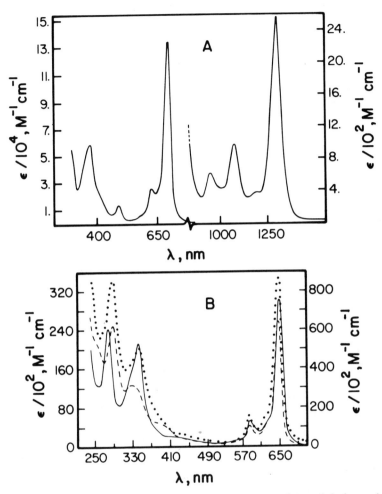

Figure 3 Absorption spectra of characteristic transition metal ion phthalocyanines: (A) spectrum of $((t\text{-Bu})_4Pc)Mn(III)OAc$ in dimethyl sulfoxide and (B) spectrum of acido(phthalocyanine)Rh(III) complexes: (——) $Rh(Pc)(OH_2)_2{}^+$, ($\cdots$) $Rh(Pc)(CH_3OH)Cl$, (----) $Rh(Pc)(CH_3OH)Br$, in CH_3CN [10]. In (A) the absorption at $\lambda_{max} \sim 1330$ nm has been assigned to a ligand to metal charge transfer transition [9b]. No such absorptions were observed with Rh(III) complexes but spectral differences below 400 nm can be assigned in this series to (axial ligand to metal) charge transfer transitions [10].

The properties described above are those of monomeric phthalocyanines; polymeric species have a different photochemical behavior characterized by intrinsic excited state lifetimes and characteristic photoredox properties. Such new properties of the polymers can be considered the result of a significant overlap between the electronic clouds of molecular neighbors. In this regard the photochemical properties of monomers and dimers are discussed (see below) separately.

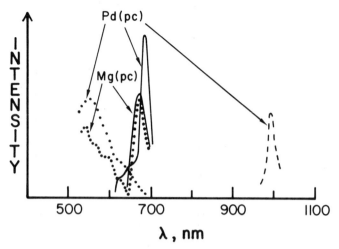

Figure 4 Relationship between absorption and emission of light by phthalocyanines. In this figure, the solid lines correspond to the 0–0 phonon line in absorption, dotted lines show molecular fluorescence from the vibronic excited lowest lying singlet, and the dashed line is the phosphorescence from the lowest lying triplet. Emission experiments in chloronaphthalene at 77 K. Adapted from Refs. [11,12].

The photochemical properties of the phthalocyanines in homogeneous solutions are presented in the following sections together with characteristic redox reactivity in the ground state.

B. LOWEST LYING LIGAND-CENTERED EXCITED STATES (Q-BAND EXCITATION)

The monomeric phthalocyanines exhibit intense absorptions in the 600- to 700-nm region. In a number of compounds, the most intense feature corresponds to a 0–0 line of a vibronic transition involving the ground state and a $\pi\pi^*$ ligand-centered excited state which can also be the lowest lying singlet excited state (Fig. 4) [11]. Fluorescence, observed between 600 and 700 nm, with half lives shorter than tens of nanoseconds (Table 1) has been observed with closed shell metal ions [11–18]. In phthalocyanines where spin–orbit coupling is weak, e.g., MgPc or PcH$_2$, the dominant radiationless relaxation mode of the lowest lying excited singlet is internal conversion. Phthalocyanines with a metal-induced strong spin–orbit coupling, e.g., PdPc or RhPc, show the intersystem crossing from the lowest singlet to the triplet as a dominant deactivation mode [13]. In these complexes, the metal ion has a

Table 1 Fluorescence and Phosphorescence from the Lowest Lying Ligand-Centered Singlet and Triplet States in Typical Metallophthalocyanines

Compound	$S(t)$ (ns)	ϕ_F	$\tilde{\nu}(F)$ (cm^{-1})	$T(t)$ (μs)	$\phi_P \times 10^3$	$\tilde{\nu}(P)$ (cm^{-1})	Conditions[a]
Al(Pc)Cl	6.8	0.58	14706				298 K
					0.5		298 K, 9:1 ethanol–methanol
					0.8		77 K, 9:1 ethanol–methanol
Al(tsPc)	5.0	0.6	14706	330			298 K, pH 3–12
Cd(Pc)		$3–8 \times 10^{-2}$	14440		~0.3		300 K, quinoline
				350	~0.3	9120	77 K, quinoline
				350	~0.4	9120	2 K, quinoline
Cd(prz)				0.38			In 9:1 DMF–H$_2$O mixtures
Co(tsPc)$^{3-}$					0.5		298 K, ethanol
					1.6		77 K, ethanol
Cu(Pc)		$<10^{-4}$			0.1		300 K
				<3	1	9240	77 K
				<10	0.1	9390	2 K
					and	9280 and	
Cu(tsPc)$^{4-}$				~0.02	0.035		298 K, ethanol
					0.04		77 K, ethanol
					0.022		298 K, H$_2$O
					0.050		298 K, Me$_2$SO
Cu(tsPc-NR)					0.050		298 K, CHCl$_3$
					0.040		298 K, 10% 2-propanol
						9400	In CHCl$_3$
Ga(Pc)Cl	3.8	0.31	14706		0.001		298 K
Ga(dsPc)$^-$				278			298 K, aqueous solution, pH 11 [b]
Ga(tsPc)$^{4-}$				125			298 K, aqueous solution, pH 11 [b]

(continued)

Table 1 Fluorescence and Phosphorescence from the Lowest Lying Ligand-Centered Singlet and Triplet States in Typical Metallophthalocyanines (*continued*)

Compound	$S(t)$ (ns)	ϕ_F	$\tilde{\nu}(F)$ (cm^{-1})	$T(t)$ (μs)	$\phi_P \times 10^3$	$\tilde{\nu}(P)$ (cm^{-1})	Conditions[a]
H$_2$(Pc)		0.7	14310				77 K
H$_2$				0.28			In 9:1 DMF–H$_2$O mixtures
In(Pc)Cl	0.37	0.031	14706				298 K
Ir(Pc)Cl				3.5	8	10449	77 K
Mg(Pc)	1.0	0.6	14640	1000	~0.005	9000	77 K
Mg(prz)				0.2			In 9:1 DMF–H$_2$O mixtures
Pd(Pc)					0.5	9935	300 K
	10	5×10^{-6}	15070	25	3	10100	77 K
				44	40	10100	4 K
Pt(Pc)					4	10430	300 K
				7	10	10590	77 K
				15	0.7	10590	4 K
					and	and	
					0.7	10290	
					0.04	10590	2 K
					and	and	
				15	0.7	10290	
					<0.01	10590	1.4 K
					and	and	
					0.7	10290	
Rh(Pc)Cl	2	1×10^{-3}	15060	15	2	10101	77 K
		1×10^{-3}	15060	16	2	10101	4 K
		1×10^{-3}	15060	6,20	2	10101	2 K
Rh(Pc)(CH$_3$OH)Cl[c]				6,20	1.4		298 K, CH$_3$CN
				1.3,2.5	1.3		225 K, CH$_3$CN
				6.7,3.0	1.2		298 K, 9:1 ethanol–methanol
					3.2		77 K, 9:1 ethanol–methanol
					1.3		298 K, CH$_2$Cl$_2$

Compound						Conditions
Rh(Pc)(CH$_3$OH)Br^c					1.4	298 K, methanol
				0.2, 2.5	1.2	298 K, CH$_3$CN
				0.33, 2.7		282 K, CH$_3$CN
				0.39, 2.7		273 K, CH$_3$CN
				0.51, 2.9		263 K, CH$_3$CN
				0.63, 2.85		253 K, CH$_3$CN
				1.1, 2.64		243 K, CH$_3$CN
				1.26, 2.70		235 K, CH$_3$CN
				1.33, 2.20		225 K, CH$_3$CN
Rh(Pc)(CH$_3$OH)I^c				0.036	1.3	298 K, 9:1 ethanol–methanol
				1.8	3.1	77 K, 9:1 ethanol–methanol
					1.4	298 K, CH$_3$CN
Ru(Pc)(4Bu^t-py)$_2$				0.079	1.3	298 K, 9:1 ethanol–methanol
					3.6	77 K, 9:1 ethanol–methanol
				0.090		298 K, CH$_2$Cl$_2$
Ru(Pc)(4Me-py)$_2$				0.135		298 K, CH$_2$Cl$_2$
Ru(Pc)(py)$_2$					0.77	298 K, 9:1 ethanol–methanol
					8.5	77 K, 9:1 ethanol–ethanol
				0.146		298 K, CH$_2$Cl$_2$
Ru(Pc)(dmf)$_2$				4.0		298 K, CH$_2$Cl$_2$
Ru(Pc)(dmso)$_2$				4.2		298 K, CH$_2$Cl$_2$
Ru(Pc)(py)(CO)				4.5		298 K, CH$_2$Cl$_2$
Ru(Pc)(dmf)(CO)	100	0.3	14640	1100	< 0.03	300 K
Zn(Pc)			9150	1500	0.1	77 K
			9160	100	0.4	2 K
Zn(prz)					0.13	In 9:1 DMF–H$_2$O mixtures

^{a} In chloronaphthalene unless specially stated. Values from Refs. [1–18]. Abbreviations: $S(t)$ ≡ fluorescence lifetime, $\bar{\nu}(F)$ ≡ fluorescence peak position, $T(t)$ ≡ phosphorescence lifetime, ϕ_F ≡ fluorescence quantum yield, ϕ_P ≡ phosphorescence quantum yield, $\bar{\nu}(P)$ ≡ phosphorescence peak position.

^{b} Two components in the phosphorescent emission.

^{c} Considerable dimerization under these conditions.

Table 2 Intersystem Crossing between Lowest Lying Singlet and Triplet States

Compound	$k_{isc}/10^{9a}$ (s^{-1})	Conditions
Al(Pc)Cl	0.059	298 K, chloronaphthalene
Ga(Pc)Cl	0.18	298 K, chloronaphthalene
In(Pc)Cl	2.4	298 K, chloronaphthalene
Pd(Pc)	670	77 K, chloronaphthalene
Rh(Pc)	10000	77 K, chloronaphthalene

[a] Values from Refs. [11,14].

large effect in determining the rate of conversion of the excited singlet to the triplet (heavy atom effect). For example, the intersystem crossing rate constants 5.9×10^7, 1.8×10^8, and 2.4×10^9 s^{-1} have been measured for Al(tsPc)Cl^{4-}, Ga(tsPc)Cl^{4-}, and In(tsPc)Cl^{4-}, respectively [19]. The value of the rate constant increases in these series along the direction of an increasing spin–orbit coupling. With open shell metal ion phthalocyanines, one observes that the yield of phosphorescence and lifetime of the lowest lying triplet are metal ion dependent, e.g., ca. 20 ns for Cu(tsPc)$^{4-}$ in aqueous solutions and ca. 20 μs for Rh(Pc)Cl in acetonitrile [12,15,16,20]. Such trends show that the intersystem crossing from the lowest excited singlet to the triplet plays a significant role in increasing the phosphorescence yield at the expense of the fluorescence (Tables 1 and 2) [11–13]. It is possible to interpret such a dependence on the metal center as a consequence of metal-induced spin–orbit coupling and a significant mixing of the metal–ligand electronic clouds, effects that are evident in the ESR of Cu(Pc) [21]. Moreover, studies on the phosphorescence of phthalocyanine complexes of the Pt metals between room temperature and 1.6 K have shown significant splitting of the triplet sublevels in a manner which depends on the metal ion [13]. The perturbations determining such a splitting have been defined as crystal field and spin–orbit coupling of unspecified origins. Moreover, departures from a simple first-order kinetics in the rates of relaxation of ($^3\pi\pi^*$)Rh(Pc)X, where X = Cl, Br, I, have been related to the presence of such sublevels [22].

In addition to the $\pi\pi^*$ excited states considered above, evidence has been provided about the existence of $n\pi^*$ levels in a number of metallophthalocyanines (Mg, Zn, Ru, Pd, Pt) and metal-free phthalocyanine [1]. The $^1n\pi^*$, placed ca. 2 kJ above the lowest lying $^1\pi\pi^*$, undergoes insignificant intersystem crossing to the triplet manifold in relationship to the internal conversion populating the lowest lying $^1(\pi\pi^*)$ [11,22,23]. Such a rapid deactivation could be the reason for the lack of a reactivity that is characteristic of $n\pi^*$ excited states, namely, hydrogen abstraction, when phthalocyanines are excited at wavelengths of the Q band.

The photoredox behavior of the lowest lying singlet state has been investigated with a mixture of the di-, tri-, and tetrasulfonated phthalocyanines [24,25]. Insofar as the ground and excited state properties are determined to a large extent by the degree of substitution in the macrocycle, such results must be viewed as qualitative in nature. In this context, the singlet species undergoes oxidation in electron transfer quenching with methylviologen or anthraquinone-2,6-sulfonate with a rate close to the diffusion-controlled limit, i.e., $k_q = 2.0 \times 10^{10} \, M^{-1} \, s^{-1}$ for the viologen and $k_q = 4.5 \times 10^9 \, M^{-1} \, s^{-1}$ for the quinone quenching. The fact that these reactions did not generate the phthalocyanine radicals (see below) with lifetimes in a microsecond time domain was seen as a consequence of a fast back electron transfer process within the solvent cage, i.e., a process which must be much faster than the one corresponding to the escape of the products from the cage. However, reductive electron transfer quenching of the singlet with *p*-benzohydroquinone generates an intermediate detected within a ns time domain and assigned as a geminate ion pair between the reduced phthalocyanine and semiquinone radical ion [24]. Whether the assignment of this transient is correct or not, a point in need of further investigation, does not detract from the fact that the singlet state shows a considerably larger reactivity than the triplet state. Such a reactivity must be a consequence of the different reduction potentials for the singlet, for example,

$$\epsilon^0_{*\,Al(tsPc)^{2-}/Al(tsPc^{\cdot})^-} = -0.63 \text{ V, and the triplet state,}$$

$$\epsilon^0_{*\,Al(tsPc)^{2-}/Al(tsPc^{\cdot})^-} = -0.02 \text{ V, vs NHE, respectively.}$$

In addition to the detection of the lowest lying triplet state from its characteristic emission centered at 800–900 nm, its absorption spectrum has been determined by flash photolysis [20,26]. In most of the metallophthalocyanines the spectrum of the $^3\pi\pi^*$ state has a maximum at 450 nm which has been assigned to a triplet–triplet transition. The kinetics of the lowest $^3\pi\pi^*$ relaxation has been found to be dependent on the nature of the axial ligand [15,17]. For Rh(III) complexes, Rh(Pc)X with X = Cl, Br, I, the decay of the triplet excited state is largely nonexponential and departures from a simple first-order rate law increase along the series Cl < Br < I (Fig. 5) [15]. Such a behavior could be related to either or both perturbations of the phthalocyanine electronic levels and to the axial ligand perturbation of vibrations that determine the rates of relaxation from triplet sublevels. There is also a good correlation between the electron-donor tendencies of the axial ligand and ground or excited state properties in Ru(II) phthalocyanines (Table 3) [15–19,27,28]. The rate of decay of the $^3\pi\pi^*$ state decreases and the ground state becomes more oxidant with an increasing nephelauxetic character of the axial ligand. In addition, the maxima in the spectra of the $^3\pi\pi^*$ state in Ru(II) phthalocyanines, placed at 500 ± 30 nm, is also dependent on the axial ligand in the manner to be expected if axially coordinated species perturb the electronic structure of the phthalocyanine ligand in the complex. Moreover,

Table 3 Ground and Excited State Redox Properties of Metallophthalocyanines

Compound[a]	E (V) (Pc^-/Pc^{2-})	E (V) $(Pc^-/*Pc^{2-})$	Z/R	$k_{ex}/10^7$ $(M^{-1}\,s^{-1})$ $(Pc^-/*Pc^{2-})$	Quencher[b]
Al(Pc)Cl	1.115	0.01	0.044		
Cd(Pc)	0.777	−0.35	0.018		$Fe(CN)_6^{3-}$
Co(tsPc)$^{3-}$	1.327	0.09		0.30	
Cu(tsPc)$^{4-}$	1.109	−0.04	0.040		
Ga(Pc)Cl	1.102	−0.02	0.032		
In(Pc)Cl	1.067	−0.06	0.023		
Mg(Pc)	0.887	−0.22			
Rh(Pc)(CH$_3$OH)Cl	1.277	0.10		0.22	$Fe(CN)_6^{3-}$
Ru(Pc)(4Bu^t-py)$_2$	1.295	0.12		1.33	c
Ru(pc)(4Me-py)$_2$	1.276	0.10		3.65	c
Ru(Pc)(py)$_2$	1.24	0.07		1.9	d
				0.81	e
				0.59	f
				0.158	Fe^{3+}(aq)
				0.282	$Fe(CN)_6^{3-}$
Ru(Pc)(dmf)$_2$	1.285	0.11	9.58	0.81	c
					d

	$(Pc^{2-}/Pc^{\cdot 3-})$	$(*Pc^{2-}/Pc^{\cdot 3-})$				
Ru(Pc)(py)(CO)	1.387	0.21		1.9		^{e} Fe^{3+}(aq)
				0.164		Fe(CN)$_6^{3-f}$
				0.128		Fe^{3+}(aq)
				2.17		Fe(CN)$_6^{3-f}$
				12.4		
Zn(Pc)	0.922	-0.23	0.023			

	$(Pc^{2-}/Pc^{\cdot 3-})$	$(*Pc^{2-}/Pc^{\cdot 3-})$	$(*Pc^{2-}/Pc^{\cdot 3-})$	
			0.5	Fe(CN)$_6^{4-}$ in CH$_3$CN
			27.	Fe(CN)$_6^{4-}$ in CH$_3$CN
Co(tsPc)$^{3-}$		0.70^g		
Rh(Pc)(CH$_3$OH)Cl	-0.43	0.72		
Ru(Pc)(py)CO	-0.45	0.70		
Zn(Pc)	-0.6	0.71		
Zn(Pc[CN]$_8$)	1.0	1.18		
Zn(prz)	-0.3	1.0		

^{a} For excited state lifetimes see Table 1. Values from Refs. [16–18,27,28,32]. Abbreviations: $E \equiv$ redox potential vs NHE, $Z/R \equiv$ ionic potential, $k_{ex} \equiv$ self-exchange electron transfer rate constant.
^{b} Unless specially stated, reactions with nitroaromatic compounds were studied in CH$_2$Cl$_2$ and reactions with Fe^{3+}(aq) and Fe(CN)$_6^{3-}$ in 9:1 CH$_3$CN–H$_2$O mixed solvent ($I = 0.01$ M NaClO$_4$).
^{c} Reactions with nitroaromatic compounds.
^{d} Reactions with nitroaromatic compounds studied in CH$_3$CN.
^{e} Reactions with Paraquat derivative studied in CH$_3$CN.
^{f} In 7:3 CH$_3$CN–H$_2$O mixed solvent.
^{g} Value estimated from spectroscopic properties of Co(tsPc)$^{3-}$ for the hypothetical couple $^3(\pi\pi^*)$Co(tsPc)$^{3-}$/Co(tsPc$^\cdot$)$^{4-}$.

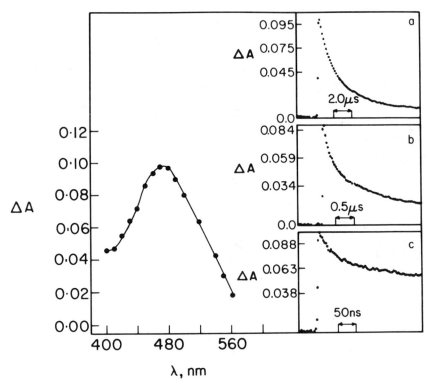

Figure 5 Difference spectrum and traces for the decay of the lowest lying $^3\pi\pi^*$ state in 640-nm flash irradiations of (a) $Rh(Pc)(CH_3OH)Cl$, (b) $Rh(Pc)(CH_3OH)Br$, and (c) $Rh(Pc)(CH_3OH)I$, in deaerated CH_3CN at 298 K. Reproduced with permission from Ref. [15].

the increase of the $^3\pi\pi^*$ lifetime with increase of axial ligand electron-withdrawing character must be the synergistic manifestation of changes in the energy gap between excited and ground state and in the vibrational modes that determine the rate of nonradiative relaxation.

Although no unimolecular redox reactivity of the lowest lying singlet and triplet $\pi\pi^*$ excited states has yet been reported, there has been a considerable amount of work on their electron transfer quenching [17,24,27,29–33]. Insofar as the positions of the 0–0 transitions can be estimated from the absorption and emission spectra of the phthalocyanines, for the excited states redox potentials have been estimated by correcting the ground state redox potentials with these energies. Such potentials permit one to calculate self-exchange rate constants for the excited state [Eq. (1)]:

$$M(\dot{P}c)^{(Z-1)-} + (^3\pi\pi^*)M(Pc)^{Z-} \rightarrow (^3\pi\pi^*)M(Pc)^{Z-} + M(\dot{P}c)^{(Z-1)} \qquad (1)$$

from the rates of electron transfer reactions [Eq. (2)] involving the phthalocyanine's excited state and various organic and inorganic oxidants, O, with well-known redox potentials and self-exchange rate constants [17,27].

$$(^3\pi\pi^*)M(Pc) + O \rightarrow M(\dot{P}c)^+ + O^{\overline{\cdot}} \tag{2}$$

The values determined for the self-exchange rate constants (Table 3) suggest that the major contribution to the Franck–Condon reorganization energy, $\lambda_{reorg} \sim 50$ kJ/mol, comes from the outer sphere reorganization energy (λ_o) [Eq. (3)] [34]:

$$\lambda_{reorg} = \lambda_o + \lambda_i,$$

$$\lambda_o = e^2 \left(\frac{1}{2r_a} + \frac{1}{2r_b} - \frac{1}{r^{\ddagger}} \right) \left(\frac{1}{n^2} - \frac{1}{D_s} \right) \tag{3}$$

(where r_a and r_b are the reactants radius, $r^{\ddagger}$ is the activated complex radius, D_s is the medium dielectric constant, and n is the refractive index) and minor contributions come from the metal-dependent inner sphere reorganizational energy, λ_i. This fact correlates very well with the expected similarity between the nuclear configurations of the radical and the excited state and demonstrates that the reorganization energy is significant in reducing the value of the self-exchange rate constant well below the one expected for a diffusion-controlled process. In this regard, conclusions about the mechanism of redox quenching, based on the arbitrary assumption of diffusion-controlled self-exchange electron transfers, must be considered in error [30].

Inspection of the redox potentials in Table 3 reveals that the $^3\pi\pi^*$ state is a reductant capable of reducing substrates such as methyl viologen ($\epsilon^0_{MV^{2+}/MV^+} = -0.44$ V) and is also a poor oxidant unable to react with weak reducing agents, e.g., amines. Substituents in the macrocyclic ligand induce dramatic changes in the ground and excited state redox potentials. For example, a reduction potential, $\epsilon^0 = -0.6$ V vs NHE, has been reported for a number of M^{2+}-metal ion phthalocyanines while this potential is $\epsilon^0 = -0.3$ V and 1.0 V vs NHE for the porphyrazines and octacyanophthalocyanine, respectively [18]. This point is also demonstrated in the reductive quenching of pyridinoporphyrazines by EDTAH$_2^{2-}$ or cysteine, [Eq. (4)] [18].

$$(^3\pi\pi^*)Zn(pyPc)^{4+} + Q \rightarrow Zn(pyPc^{\cdot})^{3-} + Q^{\overline{\cdot}} \tag{4}$$

$$(Q = EDTA, \text{ cysteine})$$

Nitroaromatic compounds, φ, quench the excited state not only by outer sphere electron transfer; the formation of exciplexes, [Eq. (5)]

$$(^3\pi\pi^*)Ru(Pc)(py)CO + \varphi \rightarrow {}^*[Ru(Pc)(py)CO - \varphi]$$

$$\rightarrow Ru(Pc)(py)CO + \varphi \tag{5}$$

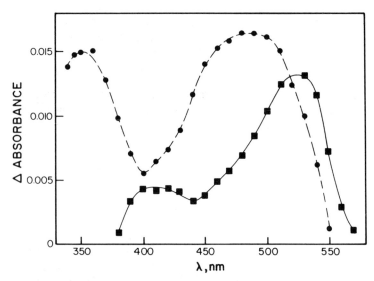

Figure 6 Difference spectra of $(^3\pi\pi^*)$Ru(Pc)(Me$_2$SO)$_2$ (----) and the exciplex generated when this excited state is quenched by 0.1 M p-dinitrobenzene. Spectra recorded 5 μs after the flash irradiation. Reproduced with permission from Ref. [16].

has been followed by flash photolysis (Fig. 6 and Table 4) [16]. Decay of the exciplex by forming excited states of the quencher is largely endoergic and is, therefore, an implausible way of degrading the electronic energy. Since products characteristic of the redox dissociation, e.g., phthalocyanine radicals, have not been observed in time resolved experiments, such a decay must be regarded as a radiative and/or radiationless return to the ground state.

The spectral properties of these exciplexes suggest that their stability is associated with weak charge transfer interactions, a proposition that is also in agreement with the excited state tendency to form excimers by triplet–triplet annihilation processes [Eq. (6)] [3,4].

Table 4 Overall Rate Constant for the Disappearance of Phthalocyanine Exciplexes

Compound	Quencher	$k^a(\text{s}^{-1})$	q^b
Ru(Pc)(Me$_2$SO)$_2$	p-Dinitrobenzene	5.3×10^5	0.22
Ru(Pc)(Me$_2$SO)$_2$	Paraquat	5.5×10^5	0.23
Ru(Pc)(Me$_2$SO)$_2$	Diaquat	5.7×10^5	0.24
Ru(Pc)(dmf)CO	p-Dinitrobenzene	1.27×10^6	0.55
Ru(Pc)(py)CO	p-Dinitrobenzene	1.30×10^6	0.55
Rh(Pc)(CH$_3$OH)Cl	p-Dinitrobenzene	2.6×10^6	0.84

[a] Rate constants determined in CH$_3$CN. Values from Ref. [32].

[b] Ratio of the rate constant for the exciplex decay to the excited state decay rate constant.

$$(^3\pi\pi^*)Al(Pc)Cl + (^3\pi\pi^*)Al(Pc)Cl \rightarrow {}^*[Al(Pc)Cl]_2 \qquad (6)$$

$$\longrightarrow Al(Pc)Cl^{\ddagger} + Al(Pc)Cl^{\bar{\cdot}} \qquad (7)$$

$$^*[Al(Pc)Cl]_2 \Big\langle$$

$$\longrightarrow Al(Pc)Cl + (^1\pi\pi^*)Al(Pc)Cl \qquad (8)$$

Dissociative reactions lead to the formation of radicals in solvents with large dielectric constants [Eq. (7)] or the lowest lying singlet excited state in solvents with low dielectric constants [Eq. (8)].

Another interesting feature in the formation of phthalocyanine exciplexes is that the formation rate experiences a significant deceleration when the reaction is carried out under intense magnetic fields (Fig. 7) [36,37]. Moreover, the study of the phosphorescence in matrices at low temperature shows that these fields induce splitting of the lowest lying triplet, i.e., a state that under D_{4h} symmetry is labeled ^3E [38]. Therefore, it is possible that the splitting of the excited state by the effect of the field leads to forbidden crossings with the excimer's potential surface and a deceleration of the reaction.

Inasmuch as quenching of the lowest lying triplet of the phthalocyanines by $^3O_2$ produces the reactive $^1O_2$ with a nearly diffusion-controlled rate [39], and the excited state of the macrocycle can be generated with red light (600–700 nm), such a sensitization reaction of the phthalocyanines has been applied to cancer phototherapies [39–41]. Indeed a number of these compounds exhibit antitumor activities which have been associated with their ability to generate singlet oxygen [40,41]. However, time-resolved studies have shown that, in addition to this mechanism, the direct oxidation of substrates, namely tryptophane, by excited phthalocyanines is a possible route which competes with oxidations by singlet oxygen [39].

In addition to the photoreactivity described above, photolabilization reactions [Eq. (9)] have been observed in photolyses of hemoglobin, phthalocyanine, and other Fe(II) macrocycles complexed with CO or alkyl isocyanides [42].

$$FeN_4(L)X \xrightarrow[L]{h\nu} FeN_4(L)_2 + X \qquad (9)$$

(N_4 = macrocyclic ligand, L = CH_3CN or methylimidazole)

The products of some photodissociations revert back to the parent complexes in the dark, i.e., they exhibit a photochromic behavior. The use of appropriate concentrations of methylimidazole or acetonitrile leads, however,

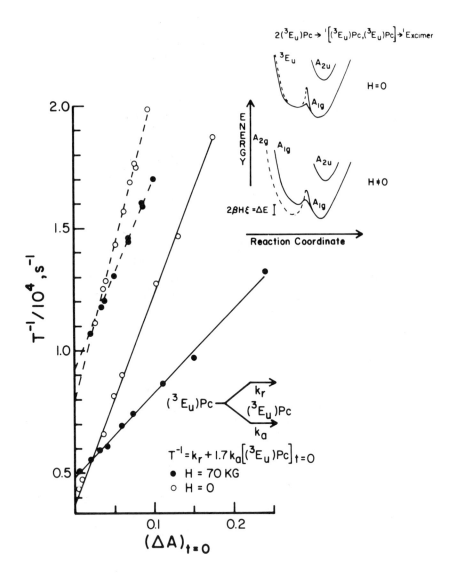

Figure 7 Dependence of the ^3Eu half-life, T, on the initial concentration of exicted state in the absence (○) and in the presence (●) of the magnetic field. $(\Delta A)_{t=0}$ is the prompt absorbance change. The half-life is given in terms of the rate constant for the unimolecular relaxation of the excited state, k_r, and the rate constant for the associative process, k_a, as indicated at the bottom of the figure. The experiments were carried out with PcH$_2$ (dashed line) and Si(Pc)(OEt)$_2$ (solid line), respectively. The inset (top of the figure) shows the hypothetical potential curves along a reaction coordinate for the formation of the excimer from an encounter complex (see equation at the top of the figure) in the absence ($H = 0$) and presence ($H \neq 0$) of the magnetic field. Reproduced with permission from Ref. [36].

to irreversible photosubstitutions and quantum yields have been measured under such conditions. These yields are macrocycle dependent, namely they follow the order Pc < DMG < TIM, and reflect (at least in the case of the phthalocyanines) the different nature of the band irradiated. Indeed, the ligand-centered, lowest lying $^1(\pi\pi^*)$ state is the one populated in irradiations of the phthalocyanines instead of the charge transfer states populated in the case of the DMG or TIM complexes [43]. Nevertheless, similarities between the photochemical reactivities of all these Fe(II) macrocycles signal that the reactive excited state has the same nature in all of them and most of the researchers have assigned it as a low lying metal to ligand charge transfer state [42].

C. LIGAND VERSUS METAL-CENTERED REDOX PROCESSES

Either the redox quenching of the lowest lying excited state or outer sphere electron transfer reactions of the ground state lead to products which have one added or removed electron with respect to the parent phthalocyanines. Such a behavior is, to a certain point, related to electrochemical results showing the stepwise oxidation or reduction of these compounds. In cyclic voltammetry, reduction or oxidation of the metal center can be observed together with redox processes involving the phthalocyanine ligand. For example, the oxidation of the metal center can take place at lower potentials than the oxidation of the ligand in Co(II) tetrasulfonated phthalocyanines while the situation is reversed in the reduction of Mn(II) phthalocyanine (Fig.

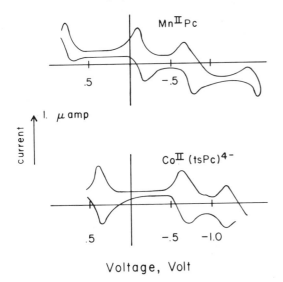

Figure 8 Typical cyclic voltammograms of phthalocyanines. Top, curve recorded with Mn(Pc) dissolved in dimethylformamide containing tetraethylammonium perchlorate as supporting electrolyte. Bottom, curve obtained with an aqueous solution of Co(tsPc)$^{4-}$ containing NaClO$_4$ as supporting electrolyte. From Potentials vs. NHE. B. Van Vlierberge and G. Ferraudi, unpublished observations.

8) though note that the potentials of such processes are solvent dependent and their relative order changes. Although the overall transformation in thermal or photochemical electron transfer reactions is in accord with the thermochemical driving force, the mechanism of the transformation is complex involving several steps which in a number of cases have been investigated by pulsed techniques. Oxidation of Co(II) tetrasulfonated phthalocyanines with a number of oxidants produces phthalocyanine radicals [Eqs. (10) and (11)], a species conveniently characterized by means of the optical and ESR spectra (Table 5) [10,45–48]. This reaction is followed by an intramolecular electron transfer from the Co(II) center to the phthalocyanine radical [Eq. (11)] a process forming the Co(III) tetrasulfonated phthalocyanine as expected from a thermochemical standpoint [47].

$$Co^{II}(tsPc)^{4-} + R \underset{k_{11}}{\overset{k_9}{\Bigg\langle}}$$

$$\xrightarrow{k_9} Co^{II}(tsPc^{\cdot})^{3-} + P \tag{10}$$

$$\Big\downarrow k_{10}$$

$$Co^{III}(tsPc)^{3-} \tag{11}$$

$$\xrightarrow{k_{11}} Co^{III}(tsPc)^{3-} + P \tag{12}$$

$$[R = Ce(IV), Cl_2^-, Br_2^- \text{ and } P = Ce(III), Cl^-, Br^-]$$

The reason for a two-stepped-oxidation instead of the one-stepped electron transfer [Eq. (12)] is clear when one considers these outer sphere oxidations from the standpoint of Marcus theory [34,37]. The rate constants for Co(III)/Co(II) self-exchange electron transfers in macrocyclic complexes are in general much smaller than the rate constants determined for the radical/ligand electron exchange, $M^z(Pc^{\cdot})/M^z(Pc^{2-})$. For example, the self-exchange electron transfer between Rh(III) phthalocyanine and the corresponding Rh(III) ligand radical is $10^6 \ M^{-1} \ s^{-1}$ vs. constants smaller than $10^2 \ M^{-1} \ s^{-1}$ for the Co(III)/Co(II) exchange in macrocyclic complexes. This difference in self-exchange rate constants favors the oxidation of the phthalocyanine and cannot be compensated by the larger driving force of the corresponding metal center oxidation. In terms of the Marcus theory, the relationship between the rate constants for these processes is,

$$\frac{k_9}{k_{11}} = \left(\frac{k_{L/L} \cdot K_{L/L} \cdot f_L}{k_{M/M^+} K_{M/M^+} + f_M} \right)^{1/2} \quad \text{with } \log f_L = \frac{(\log[K_{L/L} / K_{R/O}])^2}{4 \log[k_{L/L} \cdot k_{R/O} / Z^2]}$$

$$\log f_M = \frac{(\log[K_{M/M^+} / K_{R/O}])^2}{4 \log[k_{M/M^+} k_{R/O} / Z^2]} ; \quad K_{M/M^+} = \exp\frac{\mathcal{F} \Delta \epsilon^0_{M/M^+}}{RT}$$

Table 5 Optical and ESR Spectra of Selected Phthalocyanine Radicals

Radical[a]	λ_{max} (nm)	g	Linewidth (Gauss)
$Al^{III}(Pc^{\cdot})Cl^{+\,d}$	525	2.0024[b]	6[b]
$Co^{II}(tsPc^{\cdot})^{3-\,d}$	500	2.002	22
		2.0024[c]	6[c]
$Co^{III}(tsPc^{\cdot})^{2-}$	520[e,f]	2.002[d]	10[d]
$Cu^{II}(tsPc^{\cdot})^{3-}$	515	2.002[d]	
		2.002[c]	85[c]
$Ni^{II}(tsPc^{\cdot})^{3-}$	520[e]		
$[Ni^{II}Ni^{III}(tsPc)(tsPc^{\cdot})]^{4-}$	~515[e]	2.002	19
		2.002, 2.08[b,d]	13, ~50[b,d]
$Rh^{III}(Pc^{\cdot})(CH_3OH)Cl^+$	530[e]	2.002[b,d]	20[b,d]
		2.0037	6.6
$Rh^{III}(Pc^{\cdot})(CH_3OH)Br^+$	530[e]	2.002[b,d]	15[b,d]
		2.0010	9.36
$Rh^{III}(Pc^{\cdot})(CH_3OH)I^+$		2.002[b,d]	10[b,d]

[a] The data in this table were obtained in homogeneous solutions at room temperature unless specified. Some of the radicals can be dimeric species, i.e., associated with one molecule of the parent phthalocyanine.
[b] Data obtained at 77 K from glassy solutions.
[c] Refs. [44,45].
[d] Ref. [46].
[e] Ref. [10].
[f] Ref. [47]

and

$$K_{R/O} = \exp \frac{\mathcal{F}\Delta\epsilon^0}{Rt} \, R/O$$

By using these expressions it is easy to demonstrate that the rate constants satisfy the condition $k_9/k_{11} > 1$. A graphic description of the chemical transformations can be provided by potential surfaces (Fig. 9) showing the activation barriers for the one- and two-stepped process, respectively.

The mechanism discussed [Eqs. (10)–(12)] in connection with the oxidation of Co(II) tetrasulfonated phthalocyanine can be generalized for any appropriate metallophthalocyanine [Eqs. (13) and (14)].

$$[M^z(tsPc)]^{z-4}+R \quad\begin{array}{c} \xrightarrow{k_{12}} [M^{z\pm1}(tsPc)]^{\{z-4\}\pm1}+P \\ \\ \xrightarrow{k_{13}} [M^z(tsPc^{\cdot})]^{\{z-4\}\pm1}+P \end{array} \qquad (13)$$

$$\xleftarrow{\quad k_{13''}\quad}$$

$$\overset{k_{13'}}{[M^{z\pm1}(tsPc)]^{\{z-4\}\pm1}} \qquad (14)$$

(R = oxidant or reductant, P = product)

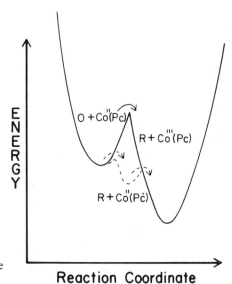

Figure 9 Potential curves describing the oxidation of Co^{II}Pc in a single step and involving the formation of an intermediate ligand radical, Co^{II}Pc.

Reaction Coordinate

Such a mechanism describes the oxidation or reduction of a metallophthalocyanine by one [Eq. (13)] and by two steps [Eq. (14)], respectively. In a mechanism where the radical formation takes place with $k_{13} > k_{13''}$, the rate constants k_{12} and k_{13} determine the significance of the corresponding paths in the overall redox process. Observation of the radical, for example with pulsed techniques, is limited by the inequality $k_{13}\,[[M^z(tsPc)]^{z-4}] > k_{13'}$, in experiments where the phthalocyanine concentration exceeds the concentration of the other reactant. Simple pseudo-first-order and first-order rate laws are, respectively, obeyed by the rates of radical formation and disappearance. Moreover, for radical formation rates, $k_{13}[[M^z(tsPc)]^{z-4}]\,[R]$, comparable to the reverse rates, $k_{13''}[[M^z(tsPc^{\cdot})]^{\{z-4\}+1}]\,[P]$, and a rapid equilibration of the phthalocyanine and radical concentrations, the mediation of the radical is significant when $k_{13'} > k_{12}$. In these conditions, the formation of the product obeys a complex rate law,

$$\frac{d[[M^{z\pm1}(tsPc)]^{\{z-4\}\pm1}]}{dt}$$

$$= k_{13'}(-p + \sqrt{p^2 + K[[M^{z\pm1}(tsPc)]^{\{z-4\}}]} \cdot (C_0 - [[M^{z\pm1}(tsPc)]^{\{z-4\}\pm1}]))$$

where

$$p = (K[[M^z(tsPc)]^{\{z-4\}}] + [[M^{z\pm1}(tsPc)]^{\{z-4\}\pm1}])/2$$

and C_0 is the initial concentration of phthalocyanine.

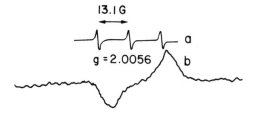

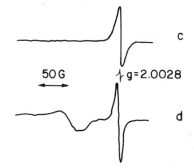

Figure 10 Aqueous solution ESR spectra recorded during the oxidation of $Co^{II}(tsPc)^{4-}$ by Ce(IV) (a), Fremy's salt used as reference and (b), phthalocyanine radical in a similar reaction between $Ni^{II}(tsPc)^{4-}$ and Ce(IV) at (c) room temperature and (d) 77 K. For other experimental conditions see Table 5. Adapted from Ref. [46].

The fate of the phthalocyanine radicals is not always determined by intramolecular electron transfers which regenerate the phthalocyanine ligand. In the oxidation of the Ni(II) tetrasulfophthalocyanine, Ni(II)–ligand radicals undergo associative disproportionation [Eqs. (15) and (16)] into a dimeric species whose ESR spectrum reveals that one of the metal centers should be described as a Ni(III) (Fig. 10).

$$2\,Ni^{II}(tsPc^{\cdot})^{3-} \rightleftharpoons [Ni^{II}Ni^{III}(tsPc)(tsPc^{\cdot})]^{6-} \tag{15}$$

$$[Ni^{II}Ni^{III}(tsPc)(tsPc^{\cdot})]^{6-} \rightarrow Ni(aq)^{2+} + Ni^{II}(tsPc)^{4-}$$
$$+\,\text{ligand degradation products} \tag{16}$$

The dimeric species has been identified as an intermediate of the decomposition process which forms one equivalent of $Ni(aq)^{2+}$ and induces the disappearance of one equivalent of phthalocyanine, [Eq. (16)]. Although similar dimeric species have not been detected in the oxidation of Co(III) phthalocyanines, they have also been postulated as reaction intermediates [Eq. (17)].

$$Co^{III}(tsPc^{\cdot})^{2-} \rightleftharpoons Co^{II}(\text{macrocycle})^{2-} \rightarrow Co^{2+}(aq)$$
$$+\,\text{ligand degradation products} \tag{17}$$

Table 6 Rate Constants for the Disappearance of Phthalocyanine Radicals

Radical	k_I (s^{-1})	k_{II} $(M^{-1} s^{-1})$
$Co^{III}(tsPc^·)^{2-}$	1.2×10^{-1a}	1.8×10^{3a}
$Cu^{II}(tsPc^·)^{3-}$	1.0×10^{4a}	6.0×10^{2a}
$Ni^{II}(tsPc^·)^{3-}$	1.9×10^{5a}	3.4^a
$Rh^{III}(Pc^·)(CH_3OH)Br^+$	$t_{1/2} > 24$ h	
$Co^{II}(tsPc^·)^{5-}$	1.7×10^{3b}	
$Co^{III}(tsPc^·)^{4-}$	1.6×10^{2c}	
$Rh^{III}(Pc\text{-}H^·)(CH_3OH)Br$	1.4×10^{7c}	

^{a} Rate constants for decomposition, k_I, and disproportionation, k_{II}, of the radicals determined in $[H_2SO_4] = 0.05$ M containing 0.15 M $NaClO_4$ unless stated. Values from Refs. [10, 45–47].
^{b} Intramolecular oxidation of the metal center.
^{c} Intramolecular reduction of the metal center.

The decay of the radical derived from copper(II) tetrasulfonated phthalocyanine is slower than the one observed with the corresponding Ni(II) radical (Table 6) but leads to the decomposition of phthalocyanine with the same overall stoichiometry of the Ni(II) reaction [Eqs. (15) and (16)]. Dimeric intermediates have passed undetected either for the reasons already discussed in relationship with the reaction of $Co(tsPc)^{3-}$ or because formation of a dimeric species with a Cu(III) metal center requires too much energy.

The reactions of the phthalocyanine radicals discussed above show that they are very reactive species. There are, however, a number of Rh(III) and Ru(II) phthalocyanine radicals, stable for periods of many hours, whose photochemical reactivities have been investigated by conventional, or sequential biphotonic techniques [46,47,49,50]. The photoinduced regeneration of the phthalocyanine by the reduction of the radical [Eqs. (18)–(23)] has been characterized as a major photoprocess, $\phi \sim 10^{-2}$ in the 254- to 350-nm region and $\phi < 10^{-5}$ at 530 nm, in the UV or VIS irradiations of the Rh(III) radicals [46].

$$Rh(Pc^·)(CH_3OH)X^+ \xrightarrow{h\nu} A \qquad (X = Cl \text{ or } Br) \qquad (18)$$

$$A + CH_3CN \rightarrow B + CH_2CN^· \qquad (19)$$

$$A + (CH_3)_2CHOH \rightarrow B + (CH_3)_2COH^· \qquad (20)$$

$$2B \rightarrow 2\ Rh(Pc)(CH_3OH)X \qquad (21)$$

$$2\ CH_2CN^· \rightarrow (CH_2CN)_2 \qquad (22)$$

$$2\ (CH_3)_2COH^· \rightarrow (CH_3)_2CO + (CH_3)_2CHOH \qquad (23)$$

Time-resolved studies of the reaction mechanism reveal that two intermediates, A and B in Eqs. (18)–(23), participate in the regeneration of the phthalocyanine. Reducing species, e.g., $(CH_3)_2CHOH$, compete with CH_3CN [Eqs. (19) and (20)] for the reduction of the intermediate A and increase its rate of disappearance (Fig. 11). In this regard, the short-lived species A has a redox reactivity that is in better accord with the properties of an excited state than those of an isomer of the parent radical. A $^4n\pi^*$ excited state of the radicals correlates with the reactive $n\pi^*$ of the phthalocyanine and is likely the observed short-lived intermediate. Insofar as the reduction of the excited radical [Eq. (19)] must proceed through hydrogen abstraction, the reaction product is expected to have a protonated Pc ring as is indicated in Eq. (24) [46].

$$\left\{ \begin{array}{c} \bullet\!\!\diagdown\!\!\diagup\! N \end{array} \right\}^{\!*} + \quad SH_2 \quad \longrightarrow \quad \diagdown\! N\!\!-\!\!H \quad + \quad SH\cdot \quad (24)$$

phthalocyanine
radicals

Such a protonation disrupts the cyclic π-system of the macrocycle and removes, therefore, $\pi^* \leftarrow \pi$ transitions from the spectrum of B. Restoration of the original π-system of the phthalocyanine ligand, i.e., the recovery of the phthalocyanine spectrum in time-resolved experiments, obeys a second-order rate law which is characteristic of dimer-assisted deprotonations in metallophthalocyanines (see below).

Irradiation of the radicals at $\lambda_{excit} \sim 525$ nm induces decomposition with much higher yields, $\phi \sim 10^{-2}$, than the regeneration of the phthalocyanine, $\phi < 10^{-4}$, described above. The study of the photodecomposition by using sequential biphotonic excitations with a number of metallophthalocyanine radicals has shown that the mechanism involves several steps, [Eq. (25)–(28)] [50].

$$M(Pc^{\cdot})^+ \xrightarrow{h\nu} {}^*M(Pc^{\cdot})^+ \rightarrow ML \qquad (25)$$

$$(M = Rh, Zn, Al)$$

$$ML \rightarrow M(Pc^{\cdot})^+ \qquad (26)$$

$$ML \rightarrow \text{decomposition products} \qquad (27)$$

$$ML + ML \rightarrow M(Pc^{\cdot})^+ + \text{decomposition products} \quad (28)$$

ML is an intermediate assigned as an isomer of the parent phthalocyanine radical, $M(Pc^{\cdot})^+$, which decays by regenerating the radical [Eq. (26)] or by decomposing in processes with first- and second-order rate laws [Eqs. (27) and (28)]. The rate constants (Table 7) for the three processes that determine the lifetime of ML, [Eqs. (26)–(28)] reveal a metal-dependent

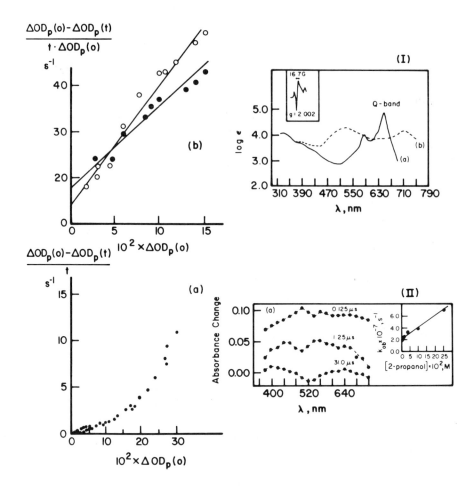

Figure 11 Typical plots (a and b) for the study of the transient kinetics as a function of the photolyzed phthalocyanine radical [5]. In the ordinate, $\Delta OD_p(0)$ and $\Delta OD_p(t)$ represent the change in the optical density induced by the photolyzing pulse at two different instants, 0 and t, of the reaction. The deviation from linearity in (a) signals departures from a first-order kinetics while linearized plots (b) give information about the rate constants for a first-order (intersect) and a second-order (slope) kinetics. These results were collected in 520-nm photolyses of (a) $Zn(Pc^{\cdot})^+$ and (b) $Al(Pc^{\cdot})Cl^+$ (○) and $Rh(Pc^{\cdot})Br^+$ (●) in deaerated Me_2SO. Adapted from Ref. [50]. The insets in the right side of the figure show the spectral properties of $Rh(Pc^{\cdot})(CH_3OH)Cl^+$ (I) and the difference transient spectra of an intermediate generated in 337-nm photolyses of such a radical in CH_3CN [46]. In (I), (a) is the spectrum of $Rh(Pc)(CH_3OH)Cl$, (b) is the spectrum of $Rh(Pc^{\cdot})(CH_3OH)Cl^+$, and the ESR spectrum of the radical is shown in the box. In (II), the plot shows the increase of the rate constant (for the decay of the intermediate) with 2-propanol concentration. Adpated from Refs. [45,50].

Table 7 Rate Constants for Decomposition and Regeneration of Phthalocyanine Radicals[a]

Radical	$k_{26}/10$ (s^{-1})	$k_{27}/10$ (s^{-1})	$k_{28}/10^8$ $(M^{-1} s^{-1})$
$Rh(Pc^{\cdot})(CH_3OH)Br^+$	1.4	7.1	5.2
$Al(Pc^{\cdot})Cl^+$	1.8	9.0	4.6
$Zn(Pc^{\cdot})^+$	1.3	2.5	4.1

[a] Reactions, Eqs. (26–28), investigated in deaerated Me_2SO. Values from Ref. [50].

reactivity. Decomposition of ML with a first-order rate law [Eq. (27)] and the unimolecular regeneration of the phthalocyanine radical [Eq. (27)] show that the Al(III) and Rh(III) species have neary the same reactivity and that this is larger than the reactivity exhibited by Zn(II) species. The regeneration of the phthalocyanine radical [Eq. (28)] via a second-order process suggests that association of two ML molecules, in accordance with the tendency to dimerize of the phthalocyanines, might facilitate structural reorganizations required for the product formation. Therefore, it is possible that such metal-dependent reactivity reflects the effect of the metal charge on the macrocycle, i.e., in a manner that resembles the dependence of the phthalocyanine redox potential in the ionic potential of the metal center [9a].

Decomposition quantum yields (Table 7) have been used as a measure of the excited state ability to produce ML [Eq. (25)]. Although quantum yields reveal a dependence on medium conditions that one would expect for an excited state with a medium dependent lifetime, they are not dependent on the metal ion. In this regard, the reactive excited state must have little if any charge transfer character and must be considered as a ligand centered excited state.

D. UV PHOTOCHEMISTRY OF PHTHALOCYANINE DIMERS

A number of metallophthalocyanines exhibit a pronounced tendency to form polymers [51–60]. Properties of the dimers have been investigated in aqueous and nonaqueous solvents. For example, the differences in the optical electronic spectra of dimers and monomers (Fig. 12) signal that electronic interactions between the components of the dimer are strong. In this context, it has been proposed that the interaction should be described as an overlap between π-electronic clouds. The solvent plays, however, a significant role in stabilizing these species and a given phthaloycanine can be monomeric in a given solvent, e.g., $M(tsPc)^{4-}$ ($M = Cu^{II}, Ni^{II}, Co^{II}$) in DMSO [6] and dimeric in another solvent, e.g., H_2O [51]. Protonation also affects the degree of

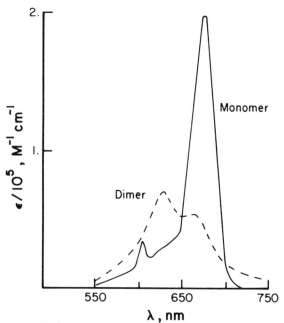

Figure 12 Typical differences between the spectra of monomeric an dimeric sulfophthalocyanines in the spectral region of the Q-band. In the figure, the spectra corresponds to the monomeric and dimeric Cu(II) complexes, $Cu(tsPc)^{4-}$ and $[Cu(tsPc)^{4-}]_2$, respectively. Adapted from Ref. [51].

association. In consequence, the simple dimerization equilibrium [61]

$$2 \; M(tsPc)^{4-} \rightleftharpoons [M(tsPc)^{4-}]_2$$

describes the chemical behavior of the phthalocyanines over a narrow set of conditions which are limited by the formation of larger polymers and the dependence of the equilibrium constant on medium conditions (Table 8).

Although irradiation of the phthalocyanine dimers at wavelengths of the Q-band produce the same low-lying excited state of the monomers, irradiations at about the Soret band or shorter wavelengths induce photoredox dissociations [Eqs. (29) and (30)] [47,61,62].

$$\rightarrow M^{I}(tsPc)^{5-} + M^{III}(tsPc)^{3-} \tag{29}$$

$$[M(tsPc)^{4-}]_2 \xrightarrow{+h\nu} \qquad [M = Co(II)]$$

$$\rightarrow M^{II}(tsPc^{\cdot})^{5-} + M^{II}(tsPc^{\cdot})^{3-} \tag{30}$$

$$[M = Cu(II), \; Co(III), \; Fe(II)]$$

Table 8 Typical Equilibrium Constants for the Association of Phthalocyanines

Compound	log K	Conditions
H$_2$tsPc	7	6.2°C, H$_2$O[a]
Co(tsPc)$^{4-}$	5.3	58°C, H$_2$O[a]
	5.9	38°C, H$_2$O[b]
	5.7	48°C, H$_2$O[b]
	5.3	58°C, H$_2$O[b]
	5.6	20°C, H$_2$O[c]
	4.2	20°C, 25% MeOH in H$_2$O[c]
	2.8	20°C, 50% MeOH in H$_2$O[c]
Cu(Pc)	4.2	Room temp., benzene[d]
	2.2	Room temp., tetrahydrofuran[d]
Cu(tsPc)$^{4-}$	7.2	61°C, H$_2$O[a]
	4.8	25°C, 20% EtOH in H$_2$O[e]
	6.0	25°C, 20% EtOH in 10^{2-} M NaCl[f]
	5.5	25°C, 20% MeOH in H$_2$O[f]
Cu(tsPc-NR)	4.2	22°C, Benzene[g]
	6.5	22°C, CCl$_4$[g]
Zn(tsPc)$^{4-}$	6	58°C, H$_2$O[a]

[a] Ref. [52].
[b] Concentrations of phthalocyanine below 10^{-7} M. See Ref. [57].
[c] Ref. [53].
[d] Concentrations in the 10^{-6}–10^{-3} M range. See Ref. [60].
[e] Value of the association constant for concentrations of phthalocyanine below 5×10^{-4} M. See Ref. [55].
[f] Ref. [54].
[g] Concentrations in the 10^{-6}–10^{-4} M range. See Ref. [59].

Photoredox dissociations leading to phthalocyanines with oxidized and reduced metal centers [Eq. (29)] or ligand radical species [Eq. (30)] have been detected in flash photolysis of the Cu(II), Co(II), Co(III), and Fe(II) tetra-sulfonated phthalocyanines (Fig. 13). The fate of these products depends on the properties of the phthalocyanine radicals discussed above and the rate of recombination into the dimeric species. For example, the photodissociation of the Cu(II) complex into radicals [Eq. (30)] is followed by the regeneration of the dimer with a second order kinetics [Eq. (31)] in the absence of radical scavengers [61].

$$Cu^{II}(tsPc^{.})^{3-} + Cu^{II}(tsPc^{.})^{5-} \rightarrow [Cu^{II}(tsPc)^{4-}]_2 \qquad (31)$$

However, interception of the oxidized species with alcohols leaves the reduced phthalocyanine as a stable product and a transient product associated with the abstraction of hydrogen [Eq. (32)].

$$Cu^{II}(tsPc^{.})^{3-} + (CH_3)_2CHOH \rightarrow Cu^{II}(tsPcH)^{3-} + (CH_3)_2COH^{.} \qquad (32)$$

$$Cu^{II}(tsPcH)^{3-} \rightarrow Cu^{II}(tsPc)^{4-} + H^+ \qquad (33)$$

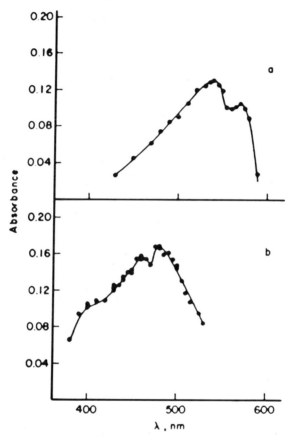

Figure 13 Transient spectra generated in flash photolysis of (a) $[Cu(tsPc)^{4-}]_2$ and (b) $[Co(tsPc)^{4-}]_2$ in 10^{-1} M HClO$_4$. The spectra have been assigned to phthalocyanine radicals in photolyses of the Cu(II) complex and to the equimolar mixture of Co(I) and Co(III) phthalocyanines in photolyses of the Co complex, respectively. Reproduced with permission from Ref. [47].

$$2\ Cu^{II}(tsPcH)^{3-} \rightarrow [Cu^{II}(tsPc)^{4-}]_2 + 2\ H^+ \qquad (34)$$

$$Cu^{II}(tsPcH)^{3-} + Cu^{II}(tsPc)^{4-} \rightarrow [Cu^{II}(tsPc)^{4-}]_2 + H^+ \qquad (35)$$

Recovery of the dimeric species takes place by a process with a pH-independent rate that exhibits a second-order dependence on the concentration of $Cu^{II}(tsPcH)^{3-}$ (Fig. 14). These observations suggest that the equilibrium [Eq. (34)] must be totally displaced toward the protonated species at the acid concentrations used in such experiments. It is therefore possible to give an upper limit, $K \sim 10^{-5}$, for the equilibrium constant of the acid–base equilibrium. From a structural standpoint, the transient, $Cu^{II}(tsPcH)^{3-}$, must

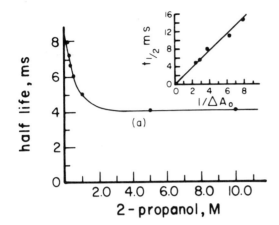

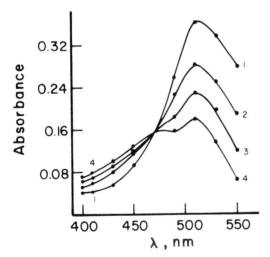

Figure 14 Scavenging of the radicals generated in flash photolyses of $[Cu(tsPc)^{4-}]_2$ (see Fig. 13). The top figure shows the dependence of the radicals half-life on 2-propanol concentration and (insert) on the reciprocal of the initial absorption [48]. The bottom figure shows the transient spectra generated in aerated solutions (scavenging by O_2) at (1) 0 ms, (2) 2 ms, (3) 4 ms, and (4) 6 ms. Adapted from Ref. [47].

be regarded as a protonated isomer of the monomeric phthalocyanine, $Cu^{II}(tsPc)^{4-}$, with a weak dissociable proton–macrocycle bond. Such a weak acidity could be the result of an azomethine group amination, namely one of the bridging nitrogens, in the abstraction of hydrogen [Eq. (32)]. A similar reaction has been discussed above in connection with the photochemistry of phthalocyanine radicals [Eq. (24)] [46].

Interception of the reduced Cu(II) phthalocyanine photodissociation product with oxygen does not prevent the regeneration of the dimer despite the formation of an adduct or peroxo compound, [Eqs. (36) and (37) and Fig 14].

$$Cu^{II}(tsPc^{\cdot})^{5-} + O_2 \rightarrow Cu^{II}(tsPc{-}O_2^{\cdot})^{5-} \qquad (36)$$

$$Cu^{II}(tsPc\text{—}O_2^{\cdot})^{5-} + Cu^{II}(tsPc^{\cdot})^{3-} \rightarrow [Cu^{II}(tsPc)^{4-}]_2 + O_2 \tag{37}$$

Although these reactions [Eqs. (36) and (37)] describe the photochemistry in aqueous solutions, the photobehavior of the Cu(II) complex is markedly different in nonaqueous media [16]. Derivatization of the tetrasulfonated phthalocyanine's sulfonic groups with amines, e.g., to N-octadecylsulfamoyl, results in phthalocyanines with large solubilities in halocarbon solvents. The photoreactivity of the Cu(II) dimer in such solvents must be described in terms of the primary formation of Cu(I) and Cu(III) species instead of radicals [Eqs. (38)–(40)] [16].

$$[Cu^{II}(tsPc\text{—}NR)]_2 \rightarrow {}^*[Cu^{II}(tsPc\text{—}NR)]_2$$
$$\rightarrow Cu^{I}(tsPc\text{—}NR)^- + Cu^{III}(tsPc\text{—}NR)^+ \tag{38}$$

$$Cu^{I}(tsPc\text{—}NR)^- + CHCl_3 \rightarrow Cu(tsPc\text{—}NR)(CHCl_2) + Cl^- \tag{39}$$

$$Cu(tsPc\text{—}NR)(CHCl_2) \xrightarrow{H^+} Cu^{III}(tsPc\text{—}NR)^+ + CH_2Cl_2 \tag{40}$$

The spectral transformations observed in continuous wave photolysis signal the decomposition of the Cu(III) complex and it has been proposed that the mechanism of such a decomposition [16] [Eqs. (41)–(43)] involves equilibration with an oxidized radical similar to that of the Co(III) radical phthalocyanine [Eq. (17)].

$$Cu^{III}(tsPc\text{—}NR)^+ \rightarrow Cu^{II}(tsPc\text{—}NR^{\cdot})^+ \tag{41}$$

$$Cu^{II}(tsPc\text{—}NR^{\cdot})^+ \begin{cases} \xrightarrow{SH} Cu^{II}(tsPc\text{—}NR) + S^{\cdot} + H^+ \tag{42} \\\\ \longrightarrow \text{degradation products} \tag{43} \end{cases}$$

The photochemical generation of metal-oxidized and metal-reduced species instead of the corresponding Cu(II)-phthalocyanine radicals suggests that the nonaqueous medium has either or both stabilized the mono- and tripositive oxidation states of the metal and change the nature of the excited state, e.g., increasing the charge transfer character.

The photodissociation of the Co(II) dimer [Eq. (29)] into Co(I) and Co(III) species is interesting because interception of the Co(I) product with O_2 forms $Co^{III}(tsPc)^{3-}$ in a fast process [Eqs. (44)–(46)] that very likely involves peroxo and superoxo complexes [47]. By contrast, the thermal oxidation of the Co(II) phthalocyanines with O_2 can be an extremely slow reaction with a pronounced dependence in the presence of species that will act as axial ligands in the Co(III) product [47,61].

$$Co^I(tsPc)^{5-} + O_2 \rightarrow Co(tsPc)O_2^{5-} \qquad (44)$$

$$Co(tsPc)O_2^{5-} + Co^{II}(tsPc)^{4-} \rightarrow [Co(tsPc)]_2O_2^{9-} \qquad (45)$$

$$[Co(tsPc)]_2O_2^{9-} \xrightarrow{2H^+} Co^{II}(tsPc)^{4-} + Co^{III}(tsPc)^{3-} + H_2O_2 \qquad (46)$$

Evidence about the formation of the peroxo species in Eqs. (44)–(46) has been found in continuous photolyses of the dimeric complex in oxygen saturated solutions [47]. Under this experimental condition, the irradiation develops new spectral features, in particular a new absorption band with λ_{max} = 660 nm, which evolve toward the spectrum of the Co(III) complex (Fig. 15) in a postirradiation process.

The photochemical behavior of the Co(III) and Fe(II) phthalocyanine dimers is similar to the species described above [Eqs. (29)–(31)]. It must be noticed, however, that the primary product, i.e., and Fe(II)-reduced phthalocyanine radical, is converted by a slow reaction to Fe(I) phthalocyanine [Eq. (47)].

$$Fe^{II}(tsPc^{\cdot})^{5-} \rightarrow Fe^I(tsPc)^{5-} \qquad (47)$$

The photodissociation quantum yields of Cu(II), Co(III), and Fe(II) phthalocyanine dimers increase with photonic energies (Table 9). However, the quantum yield for the photodissociation of the Co(III) dimer at 225 nm is smaller than the one determined at 254 nm, a result interpreted in terms of the population of the reactive state in competition with other states. The nature of the reactive state has not been definitely established and two possible cases have been considered: a charge transfer state involving units of the dimer or a ligand-centered state which induces the redox dissociation in the manner already described for the redox quenching of excited monomeric phthalocyanines [47,61]. Neither the charge transfer state nor the ligand-centered state is required to form products which are the most stable species from a thermochemical stand point. Kinetic constraints may determine the formation of other oxidized or reduced phthalocyanines as discussed above for the ground state redox reactions, see Section B. In the case of a ligand-centered state mediating the redox process, it is possible to compare the energy required for these reactions with those of given electronic transitions in the dimer, e.g., $\pi\pi^*$ transitions within the Soret and Q-bands. Such energies can be easily calculated by using reported thermochemical quantities (Table 3) and

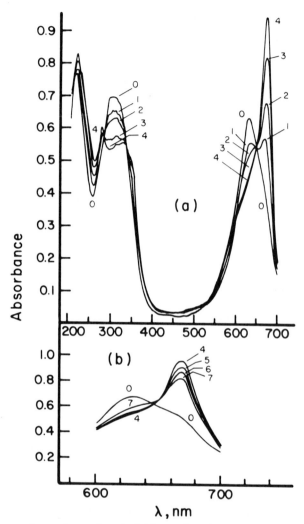

Figure 15 Spectral transformations in 254-nm photolyses of aerated solutions $[Co(tsPc)^{4-}]_2$ in 10^{-1} M $HClO_4$. (a) Spectra recorded after (0) 0 min, (1) 5 min, (2) 10 min, (3) 15 min, and (4) 20 min irradiations with $I_0 = 8.0 \times 10^{-3}$ Einstein liter^{-1} min^{-1}. (b) Disappearance of a metastable photolysis product, i.e., a peroxo or superoxo Co(III) complex, generated by a 20 min irradiation (see a). The spectra were recorded at (5) 20 min, (6) 40 min, and (7) 720 min after the irradiation. Reproduced with permission from Ref. [61].

appropriate cycles (Fig. 16). For example, the dimerization of the $Cu^{II}(tsPc)^{4-}$ is exoergonic in 41.2 kJ/mol and the photodissociation [Eq. (30)] is endoergonic in 195.6 kJ/mol, an energy that is between the electronic transition energies to the lowest lying singlet and triplet states of the monomer (Q-band). Therefore, the energies associated with excitation at wavelengths of the Soret band are in great excess with regard to those required for the redox

Table 9 Typical Quantum Yields for the Photochemical Redox Dissociation of Dimeric Phthalocyanines[a]

λ_{exc} (nm)	$10^4 \times I_{ab}$ (Einstein/liters/min)	ϕ_d	Conditions[a,b]
		$[Cu(tsPc)^{4-}]_2$	
254	8.3	$2.0 \times 10^{3-}$	0.1 M HClO$_4$, 2.7 M 2-propanol
	1.0	2.4×10^{-3}	0.1 M HClO$_4$, 2.7 M 2-propanol
225	0.30	9.3×10^{-2}	0.1 M HClO$_4$, 2.7 M 2-propanol
		$[Co(tsPc)^{4-}]_2$	
254	5.0	2.3×10^{-3}	0.1 M HClO$_4$, O$_2$ saturated
225	0.30	1.3×10^{-1}	0.1 M HClO$_4$, O$_2$ saturated
		$[Co(tsPc)(OH_2)_2^{3-}]_2$	
345	0.64	$<6 \times 10^{-4}$	10.6 M 2-propanol
280	0.58	2.5×10^{-3}	10.6 M 2-propanol
254	2.2	7.5×10^{-3}	10.6 M 2-propanol
225	0.30	5.0×10^{-5}	10.6 M 2-propanol
		$[Fe(tsPc)(OH_2)_2^{4-}]_2$	
345	0.64	$<10^{-4}$	3.3 M 2-propanol
280	5.8	4.8×10^{-3}	3.3 M 2-propanol
254	8.0	2.5×10^{-2}	3.3 M 2-propanol
225	0.30	3.2×10^{-2}	3.3 M 2-propanol

[a] Data from Refs. [48,62].
[b] Solutions deaerated with streams of N$_2$ unless stated otherwise.

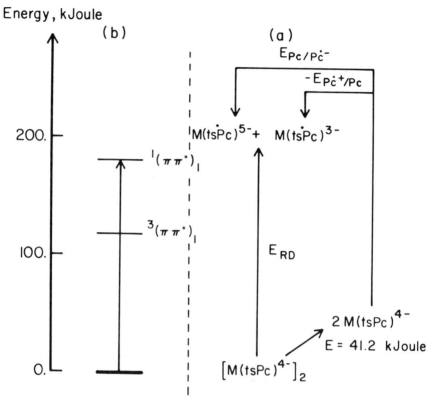

Figure 16 Semiquantitative diagram relating the relative energies of several species involved in the photoredox dissociation of dimeric phthalocyanines (a) to the position of the lowest lying triplet and singlet, $\pi\pi^*$, electronic states (b). In the figure, E_{Pc/Pc^-} and $E_{Pc^+/Pc}$ stand for the standard reduction and oxidation energies of the macrocycle in the monomeric phthalocyanines and E_{RD} is the threshold energy for the photoredox dissociation.

photodissociation. Moreover, the action spectrum, based on the photodissociation quantum yield, gives a threshold energy for the photoreaction, $E_{th} \sim 390$ kJ/mol, that is above the threshold for absorption in the Soret band. The first mechanism considered above, i.e., the population of a specific charge transfer state, seems to be in good accord with such a thermochemistry.

Photoredox dissociations have also been studied with mixed dimers, namely species formed when $Cr(bipy)_3^{3+}$ or $Ru(bipy)_3^{2+}$ form adducts with various metallosulfophthalocyanines [62]. Irradiations at wavelengths corresponding to the population of either the lowest lying states of $Cr(bipy)_3^{3+}$ or $Ru(bipy)_3^{2+}$ monomeric complexes, respectively, cause photoredox dissociations described in the corresponding photochemical steps of the cycles in Fig. 17. Investigation of such phototransformations by flash photolysis reveals that back electron transfer reactions between photochemical products close these cycles, i.e., the overall processes exhibit photoreversibility. The Cr(II) and

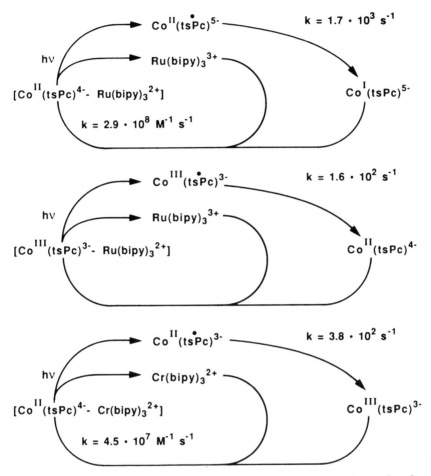

Figure 17 Cycles describing the flash photochemically investigated photoredox dissociations of mixed dimers [61]. The rate constants for intramolecular electron transfers and back electron transfers are indicated within the cycles. Reproduced with permission from Ref. [61].

Ru(III) intermediates in these cycles are the same species observed in the redox quenching of the corresponding Cr(III) and Ru(II) polypyridine complexes.

E. THE UV PHOTOCHEMISTRY OF METALLOPHTHALOCYANINE MONOMERS

Monomeric phthalocyanines of various metal ions, i.e., Mn(III), Co(II), Co(III), Fe(II), Rh(III), Cu(II), Al(III), Si(IV), exhibit a rich, wavelength-dependent, UV photochemistry [10,15,16,37,48,62,63]. In all these com-

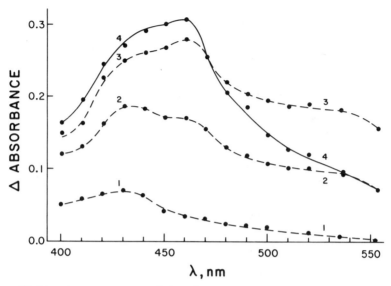

Figure 18 Transient spectra recorded in flash photolysis of Co(tsPc)$^{4-}$ in deaerated 3 M 2-propanol solutions. The spectra for the intermediates were determined at (1) 0 μs, (2) 200 μs, (3) 1 ms, and (4) 10 s after the irradiation. Reproduced with permission from Refs. [47,61].

pounds, excitations at wavelengths shorter than 300 nm induce redox reactions that have been characterized as abstractions of hydrogen from the solvent or appropriate hydrogen donors. The evidence about the nature of these reactions has been obtained from flash photochemical experiments where instead of detecting the transient products of electron transfer reactions (see Section B), the time-resolved spectra (Fig. 18) show the formation of radicals associated with hydrogen abstractions [Eq. (48)].

$$M(Pc) \underset{\longleftarrow\!\!\sim\!\!\sim}{\overset{h\nu}{\longrightarrow}} {}^*M(Pc) \xrightarrow{SH_2} M(Pc\!-\!H^{\cdot}) + SH^{\cdot} \tag{48}$$

The fate of the radical with a partially hydrogenated ring depends very much on the metal ion. For example, Co(II) tetrasulfonated phthalocyanine abstracts hydrogen from 2-propanol [Eq. (49)] and the radical produced in such a reaction undergoes dimerization [Eq. (50)] before the formation of the Co(I) product [Eq. (51)].

$$Co^{II}(tsPc)^{4-} \underset{\longleftarrow\!\!\sim\!\!\sim}{\overset{h\nu}{\longrightarrow}} {}^*Co^{II}(tsPc)^{4-} \xrightarrow[-SH^{\cdot}]{+SH_2} Co^{II}(tsPc\!-\!H^{\cdot})^{4-} + (CH_3)_2COH^{\cdot}$$

$$\tag{49}$$

$$2\,Co^{II}(tsPc{-}H^{\cdot})^{4-} \rightarrow [Co(tsPc^{\cdot})]_2^{6-} + 2\,H^+ \tag{50}$$

$$[Co(tsPc^{\cdot})]_2^{6-} \rightarrow 2\,Co^{I}(tsPc)^{3-} \tag{51}$$

The photochemical behavior of the monomeric copper(II) tetrasulfonated phthalocyanine in aqueous solutions is similar to the one described in Eq. (49) for the related Co(II) complex [48]. However, the large stability of the reduced radical [compared with the Cu(II)/Cu(I) reduction potential] prevents further transformation into the Cu(I) species. In nonaqueous media, e.g., solutions in $CHCl_3$, the hydrogen abstraction leading to the formation of Cu(II)–ligand radicals [Eq. (52)] has smaller yields than the oxidation of electron donors [Eq. (53)] [16].

$$Cu^{II}(tsPc{-}NR)^{4-} \xrightarrow{\;h\nu\;} {}^*Cu^{II}(tsPc{-}NR)^{4-} \xrightarrow{\;SH_2\;}
\begin{cases}
\rightarrow Cu^{II}(tsPc{-}NRH^{\cdot})^{4-} + SH^{\cdot} & \tag{52}\\[2em]
\rightarrow Cu^{I}(tsPc{-}NR)^{5-} + SH_2{+} & \tag{53}
\end{cases}$$

A change in photochemistry with respect to aqueous solutions that (in accord with the photochemistry of the dimers described in the previous section) is in accord with a gain in the stability of the Cu(I) species vs. the corresponding Cu(II)–ligand radical.

These hydrogen abstractions have been related to the population of a reactive $n\pi^*$ state by the absorption of UV light, i.e., light with photonic energies above the frequencies corresponding to the threshold for absorption in the Soret bands. Insofar as the phthalocyanine ligand has several unoccupied π^* orbitals, there are a number of $n\pi^*$ excited states with different energies (Figs. 1 and 2) which could be engaged in such reactions. Theoretical calculations, for example, have placed the energies of $n\pi^*$ excited states close to those of the $\pi^* \leftarrow \pi$ optical transitions which correspond to the absorption of light in the Soret and Q-band spectral regions. Experimental support for the existence of such electronic states comes from features on the absorption spectra of the phthalocyanines, e.g., the size of bandwidths has been attributed to $\pi^* \leftarrow n$ transitions under the intense $\pi^* \leftarrow \pi$ transitions that make for most of the Soret and Q-bands intensities. Moreover, the study of the fluorescence in a number of phthalocyanines suggests the existence of an $n\pi^*$ state above the lowest lying ${}^1(\pi\pi^*)$. Photophysical data suggest that in a number of phthalocyanines, this ${}^1(n\pi^*)$ undergoes rapid and efficient relaxation toward the lowest lying $^1(\pi\pi^*)$ [15]. Insofar as such an $n\pi^*$ state is photochemically unreactive despite having an energy in excess of the one required for driving the observed hydrogen abstractions, it is possible to assign the lack of reactivity to a lifetime that is too short for an efficient participation in such reactions. By contrast, a longer lifetime and/or faster rates of hydrogen abstraction must be characteristic of upper $n\pi^*$ states in order for them to show the described photoreactivity. The observation of emission from an upper state, $\lambda_{em} \sim 420$ nm, in a number of metallophthalocyanines has provided information about the upper excited states. In acido(phthalocyaninato)rhodium(III) complexes, $Rh(Pc)(CH_3OH)X$ with $X = Cl$, Br, and I, a comparison between excitation and action spectra for the hydrogen abstraction (Fig. 19) strongly suggests that the reactive $n\pi^*$ state has a poor communication with an upper emissive $\pi\pi^*$ state and that these states achieve population by different paths.

Although the reactive $n\pi^*$ state of the metallophthalocyanines is populated by excitation in UV bands, its characteristic redox reactivity has been detected in sequential biphotonic photolyses, namely in experiments carried out by irradiating with high intensities at wavelengths of the Q-band [16,64,65]. The dependence of the product and lowest lying, $\pi\pi^*$, yields on light intensity (Fig. 20) is in accord with a mechanism where the lowest lying triplet state absorbs a second photon of visible light and populates the reactive $n\pi^*$ state [Eq. (54)].

$$M(Pc) \underset{\longleftarrow}{\overset{h\nu}{\longrightarrow}} {}^3\pi\pi^* \xrightarrow{h\nu} {}^{1,3}n\pi^* \xrightarrow{SH_2} M(Pc\text{—}H^{\cdot}) + SH^{\cdot} \qquad (54)$$

One important characteristic of the photoprocesses induced with VIS light in sequential biphotonic absorptions is that quantum yields of product formation are larger than those determined with UV light in single photon excitations. It has been proposed that such a difference in quantum yields reflects the inability of states populated by the absorption of UV light, namely $^1\pi\pi^*$ and/or charge transfer states, to populate the reactive $n\pi^*$ states [63,64].

The redox photochemistry of monomeric metallophthalocyanines is not limited to the reactions induced in $n\pi^*$ excited states; a number of compounds exhibit the characteristic photoreactivity of ligand to metal charge transfer states, CTTM. Flash photochemical studies have shown, for example, the formation of radicals characteristic of the axial ligand oxidation [Eq. (55)], when acidocobalt(III) sulfophthalocyanines, $Co^{III}(tsPc)(OH_2)_{2-n}X_n^{(3+n)-}$ with $X = Cl$, Br, I, SCN, are irradiated at in the near UV [62].

$$Co^{III}(tsPc)(OH_2)_{2-n}X_n^{(3+n)-} \underset{\longleftarrow}{\overset{h\nu}{\longrightarrow}} CTTM$$

$$\rightarrow Co^{II}(tsPc)^{4-} + X^{\cdot} \qquad (55)$$

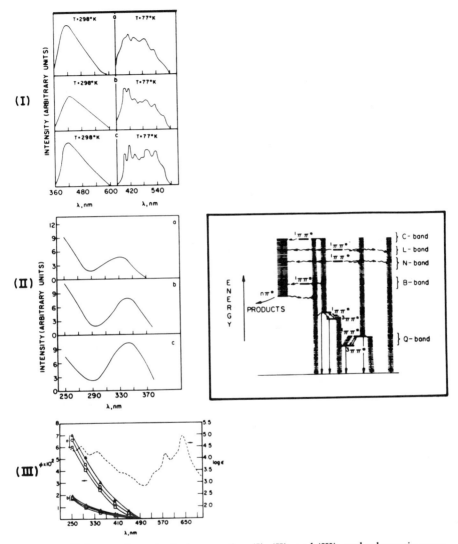

Figure 19 Relevant photophysical properties, (I), (II), and (III), and schematic representation of the principal radiative and nonradiative transformations between excited states of Rh(III) phthalocyanines. The emission spectra for UV excitations of the Rh(III) complexes, (a) Rh(Pc)(CH$_3$OH)Cl, (b) Rh(Pc)(CH$_3$OH)Br, and (c) Rh(Pc)-(CH$_3$OH)I, in deaerated CH$_3$CN (298 K) or 9:1 ethanol–methanol (77 K) are shown in (I). The excitation spectra (λ_{ob} = 420 nm) for the same compounds in CH$_3$CN at 298 K are shown in (II). In (III), the figure shows the action spectra (solid lines) from the dependence of the photoreaction quantum yield on excitation wavelength. The photoreaction was investigated in (a) 2.5 M 2-propanol in CH$_3$CN and (b) 0.5 M 2-propanol in CH$_3$CN with ($\triangle$) Rh(Pc)(CH$_3$OH)Cl, ($\bigcirc$) Rh(Pc)-(CH$_3$OH)Br, and ($\square$) Rh(Pc)(CH$_3$OH)I. The absorption spectrum of Rh(Pc)-(CH$_3$OH)Cl (dashed line) has been superimposed on the action spectra. Adapted from Ref. [15].

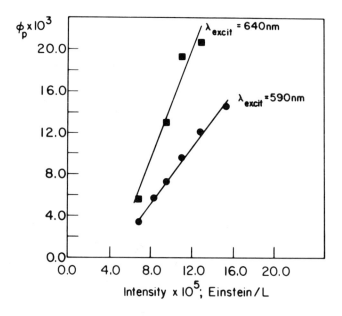

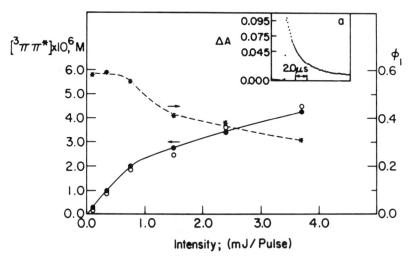

Figure 20 Typical photochemical behavior of Rh(Pc)(CH$_3$OH)Cl under a sequential biphotonic excitation regime. The dependence of the product yields on light intensity is shown in the top. The influence of the secondary photolysis and ground state depletion on the yield and concentration of the lowest lying $^3\pi\pi^*$ as a function of the laser pulse intensity is shown at the bottom. The insert is a typical trace for the decay of the excited state under a sequential biphotonic regime. The decay of the excited state under such conditions exhibited the same lifetime measured under a monophotonic regime. Adapted from Ref. [63].

In Co(III) phthalocyanines, threshold energies for charge transfer photochemistry are smaller than those required for the hydrogen abstractions by $n\pi^*$ states, a result that one would expect for a CTTM state that is underlying the ligand-centered state. This is not, however, a general property of the metallophthalocyanines since the position of charge transfer states is highly dependent on the redox properties of the metal and the ligand. In this regard, phthalocyanines of Rh(III) which are isoelectronic with similar Co(III) compounds show no charge transfer photochemistry because the energy required for the reduction of the metal center, larger than for Co(III), is expected to shift the CTTM states to a position above the ligand centered states [66]. Low-lying CTTM states are available in Mn(III) phthalocyanines, where two primary photoprocesses [Eqs. (56) and (57)],

$$\mathrm{Mn^{III}(tsPc)(OH_2)(OH)^{4-}} \xrightarrow{h\nu}
\begin{cases}
\xrightarrow{\phi_1} \mathrm{CTTM} \to \mathrm{Mn^{II}(tsPc)^{4-}} + \mathrm{OH^{\cdot}} & (56)\\[2ex]
\xrightarrow{\phi_2} n\pi^* \to \mathrm{Mn^{II}(tsPc)^{4-}} + \mathrm{SH^{\cdot}} + \mathrm{H^+} & (57)
\end{cases}$$

result in the reduction of the parent compound and oxidations of the solvent and axial ligand respectively [63,67]. The quantum yields of the two photoprocesses vary with excitation energy, from $\phi_1 / \phi_2 \sim 1$ at $E_{\mathrm{excit}} \sim 320$ kJ to $\phi_1 / \phi_2 \sim 0.1$ at $E_{\mathrm{excit}} \sim 430$ kJ, in a manner that reflects the conversion of the Franck–Condon or thermalized states into the reactive CTTM and $n\pi^*$ excited states. In the μ-oxo dimer, $(\mathrm{OH_2})(\mathrm{tsPc})\mathrm{Mn^{III}}\text{–O–}\mathrm{Mn^{III}}(\mathrm{tsPc})(\mathrm{OH})^{3-}$, the primary process leading to the μ-oxo ligand photooxidation [Eq. (58)]

$$(\mathrm{OH_2})(\mathrm{tsPc})\mathrm{Mn^{III}}\text{–O–}\mathrm{Mn^{III}}(\mathrm{tsPc})(\mathrm{OH})^9 \xrightarrow{h\nu} \mathrm{CTTM} \to \mathrm{Mn^{II}(tsPc)^{4-}}$$

$$+ \mathrm{Mn^{III}(tsPc)(O^{\cdot})^{4-}} \tag{58}$$

is similar to the photooxidation of the hydroxyl ligands in Eq. (55). Therefore, it is possible, considering the photochemistry described above, that Mn(III) phthalocyanines having axial ligands which are better reductants than hydroxide or the μ-oxo Mn(III) ligand will manifest charge transfer photochemistry at photonic energies below 320 kJ.

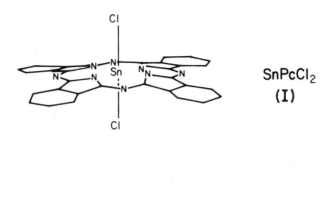

$$SnPcCl_2$$
(I)

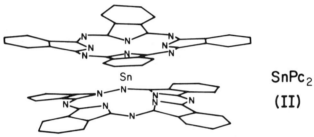

$$SnPc_2$$
(II)

Figure 21 Structures determined for the dichlorophthalocyaninate complex (top) and the bisphthalocyaninate of Sn(IV) (bottom).

F. BISPHTHALOCYANINATE COMPLEXES

Although there are a number of well-characterized bisphthalocyanines (Fig. 21) the photochemical and thermal reactivity have not been extensively investigated [67]. A recent study has shown that coordination in bisphthalocyaninate Sn(IV), $Sn(Pc)_2$, must induce little mixing between the electronic clouds of the macrocycles [68]. Indeed, the crystallographic structure reveals that the macrocycle–macrocycle distance is too long for such an interaction, a fact that is also in agreement with the spectroscopy and electrochemistry of the compound. For example, the presence of typical phthalocyanine spectral features, e.g., Q- and Soret bands, in the bisphthalocyaninate spectrum gives support to the idea of weak mixing while the appearance of a broad and intense band at $\lambda_{max} \sim 750$ nm suggests the existence of charge transfer transitions at low photonic energies. In flash photochemical experiments, the irradiation of $Sn(Pc)_2$ at wavelengths of the Q-band led to the observation of an intermediate assigned as the lowest lying ligand-centered state, $^3\pi\pi^*$. The decay of such a state takes place with a lifetime ($\tau \sim 5$ μs) shorter than the $^3\pi\pi^*$ in $Sn(Pc)Cl_2(\tau \sim 40$ μs) and leads to transient species associated with homolytic and heterolytic dissociations of the bisphthalocyaninate [Eq. (59)– (63)].

$$Sn^{IV}(Pc)_2 \xrightarrow{h\nu} (^3\pi\pi^*)Sn(Pc)_2 \qquad (59)$$

$$(^3\pi\pi^*)Sn(Pc)_2 \rightarrow (CT)Sn(Pc)_2 \qquad (60)$$

$$(CT)Sn(Pc)_2 \begin{cases} \longrightarrow Sn(Pc)_2 & (61) \\ \longrightarrow Sn^{II}(Pc) + Pc^0 & (62) \\ \longrightarrow Sn^{IV}(Pc)^{2+} + Pc^{2-} & (63) \end{cases}$$

In this mechanism, the short lifetime of the excited bisphthalocyaninate and the appearance of photoreactivity at lower energies than in monomeric phthalocyanines are attributed to the presence of the lowest lying charge transfer state [Eqs. (59) and (60)]. Indeed, it is expected according to the preceding discussion that an excited state with a strong $^3\pi\pi^*$ ligand-centered character will not engage in unimolecular redox reactions as is the case with $(^3\pi\pi^*)Sn(Pc)Cl_2$ and other metallophthalocyanines. By contrast, the photodissociations [Eqs. (62) and (63)] are in better accord with the reactivity expected in a state with a pronounced charge transfer character.

G. CONCLUSIONS

Although the studies about the photophysical and photochemical properties of the phthalocyanines that we have reviewed in previous sections provide extensive information about these systems, a number of problems still remain to be resolved. Little is known, for example, about the communication between different states, i.e., charge transfer, metal-centered, and ligand-centered states. The Mn(III) phthalocyanine photochemistry provides an example of a compound with photoactive CTTM and $n\pi^*$ states, a behavior that one can expect in a number of metallophthalocyanines, e.g., compounds of Co(III), Sn(IV), and Pt(IV). Action spectra based on quantum yields of photochemical and photophysical processes can be useful in establishing the role of these different excited states. The reported change in the Cu(Pc) photoreactivity with medium, i.e., from protic to aprotic solvents, suggests that medium effects and the nature of the substituents attached to the phthalocyanine macrocycle determine the nature of the photochemical transformations. In this regard, studies with an homologous series, where the main difference between members is in ligand substituents, and in solvents which allow one to vary the viscosity, dielectric constant, and other medium properties in a controlled manner are yet to be carried out. Moreover, it will be useful to learn about the role of ligand-centered and charge transfer states in compounds related to the phthalocyanines, e.g., porphyrazines, bisphthalocyaninates, and superphthalocyanines. Results of this research can be useful in the design of new compounds with an enhanced tumoricidal activity in phototherapies or for specific applications in photocatalysis. Investigations about the nature of the reaction intermediates in thermal and photochemical reactions of phthalocyanines have been largely limited to the products of electron transfer and hydrogen abstraction. Despite the fact that a number of reactants, e.g., Cl_2, are known to form adducts with the phthalocyanine macrocycle, the mechanism of formation and/or the role of such adducts in a number of chemical reactions have not yet been investigated by time-resolved techniques.

REFERENCES

1. A. M. Schaffer, M. Gouterman and E. R. Davidson, *Theor. Chim. Acta (Berlin)*, 30 (1973) 9.
2. A. Enrikson, B. Roos and M. Sundbom, *Theor. Chim. Acta (Berlin)*, 27 (1972) 303.
3. L. Edwards and M. Gouterman, *J. Mol. Spectrosc.*, 33 (1970) 292.
4. M. J. Stillman and A. J. Thomson, *Trans. Faraday Soc. II*, 70 (1974) 790.
5. A. J. McHugh, M. Gouterman and C. Weiss, *Theor. Chim. Acta (Berlin)*, 24 (1972) 346.
6a. L. D. Rollman and R. T. Iwamoto, *J. Am. Chem. Soc.*, 90 (1968) 1455.
6b. C. Grant Birch and R. T. Iwamoto, *Inorg. Chem.*, 12 (1973) 66.
7. R. M. Hochstrasser and C. Marzzacco, *J. Chem. Phys.*, 49 (1968) 971.
8. A. N. Sevchenko, S. F. Shkirman, V. A. Mashenkev and K. N. Solov'ev, *Soviet Phys.—Dokl*, 12 (1968) 710.
9a. A. B. P. Lever and P. C. Minor, *Inorg. Chem.*, 20 (1981) 4015.
9b. A. B. P. Lever, S. R. Pickens, P. C. Minor, S. Licoccia, B. S. Ramaswamy and K. Magnell, *J. Am. Chem. Soc.*, 103 (1981) 6800.
10. S. Muralidharan, G. Ferraudi and K. Schmatz, *Inorg. Chem.*, 21 (1982) 2961.
11. E. R. Menzel, K. E. Rieckoff and E. M. Voigt, *Chem. Phys. Letters*, 13 (1972) 604.
12. P. S. Vincent, E. M. Voigt and K. E. Rieckhoff, *J. Chem. Phys.*, 55 (1971) 4131.
13. E. R. Menzel, K. E. Rieckhoff and E. M. Voigt, *J. Chem. Phys.*, 58 (1973) 5726.
14. J. H. Brannon and D. Magde, *J. Am. Chem. Soc.*, 102 (1980) 62.
15. G. Ferraudi and S. Muralidharan, *Inorg. Chem.*, 22 (1983) 1369.
16. D. R. Prasad and G. Ferraudi, *Inorg. Chem.*, 21 (1982) 2967.
17. G. Ferraudi and D. R. Prasad, *J. Chem. Soc. Dalton Trans.*, (1984) 2137.
18. D. Wohrle, J. Gitzel, I. Okura and S. Aono, *J. Chem. Soc., Perkin Trans. II*, (1985) 1171.
19. J. McVie, R. S. Sinclair and T. G. Truscott, *J. Chem. Soc. Faraday Trans. 2*, 74 (1978) 1870.
20. S. E. Harrison and J. M. Assour, *J. Chem. Phys.*, 40 (1964) 365.
21. T. Huang, K. E. Rieckhoff, E. Voigt and E. R. Menzel, *Chem. Phys.*, 19 (1977) 25.
22. T. Huang, K. E. Rieckhoff and E. Voigt, *J. Phys. Chem.*, 85 (1981) 3322.

23. J. R. Darwent, J. McCubbin and D. Phillips, *J. Chem. Soc. Faraday Trans. 2*, 78 (1982) 347.
24. J. Darwent, *J. Chem. Soc. Chem. Commun.*, 17 (1980) 805.
25. P. Jacques and A. M. Braun, *Helv. Chim. Acta*, 64 (1981) 169.
26. D. R. Prasad and G. J. Ferraudi, *J. Phys. Chem.*, 86 (1982) 4037.
27. A. B. P. Lever, S. Licoccia, B. S. Ramaswamy, S. A. Kandil and D. V. Stynes, *Inorg. Chim. Acta*, 51 (1981) 169.
28. P. Mallard, S. Gaspard, P. Krausz and C. Giannoti, *J. Organomet. Chem.*, 212 (1981) 185.
29. T. Ohno and S. Kato, *J. Phys. Chem.*, 88 (1984) 1670.
30. T. Nyokong, Z. Gasyna and M. Stillman, *A.C.S. Symp. Ser.*, 321 (1986) 309.
31. Z. Gasyna, W. Browett and M. Stillman, *A.C.S. Symp. Ser.*, 321 (1986) 298.
32. D. R. Prasad and G. Ferraudi, *Inorg. Chem.*, 22 (1983) 1672.
33. R. Marcus, *Discuss. Faraday Soc.*, 20 (1960) 21.
34. N. S. Hush, *Trans. Faraday Soc.*, 57 (1961) 557.
35. M. Frink and G. Ferraudi, *Chem. Phys. Lett.*, 124 (1986) 576.
36. G. Ferraudi, M. Frink and D. K. Geiger, *J. Phys. Chem.*, 90 (1986) 1924.
37. W. Chen, K. E. Reickhoff and M. E. Voigt, *Chem. Phys.*, 95 (1985) 123.
38. G. Ferraudi, G. A. Arguello, H. Ali and J. E. Van Lier, *Photochem. Photobiol.*, 47 (1988) 657.
39. N. Brasseur, H. Ali, D. Autenrieth, R. Langlois and J. E. van Lier, *Photochem. Photobiol.*, 42 (1985) 515.
40. E. Ben-Hur and I. Rosenthal, *Radiat. Res.*, 103 (1985) 403.
41. D. Brault, C. Vever-Bizet and M. Dellinger, *Biochimie*, 68 (1986) 913.
42. C. Irwin and D. V. Stynes, *Inorg. Chem.*, 17 (1978) 2683.
43. N. Sanders and P. Day, *J. Chem. Soc. A* (1969) 2303.
44. P. George, D. J. E. Ingram and J. E. Bennett, *J. Am. Chem. Soc.*, 79 (1957) 1870.
45. C. Ferraudi, S. Oishi and S. Muralidahran, *J. Phys. Chem.*, 88 (1983) 5261.
46. D. K. Geiger, G. Ferraudi, K. Madden, J. Granifo and D. P. Rillema, *J. Phys. Chem.*, 89 (1985) 3890.
47. G. Ferraudi and E. V. Srisankar, *Inorg. Chem.*, 17 (1978) 3164.
48. T. B. Nyokong, Z. Gasyna and M. Stillman, *Inorg. Chem.*, 26 (1987) 1087.
49. D. Dolphin, B. James, A. J. Murray and J. R. Thornback, *Can. J. Chem.*, 58 (1980) 1125.
50. B. Van Vlierberge and G. Ferraudi, *Inorg. Chem.*, 26 (1987) 337.
51. K. Bernauer and S. Fallab, *Helv. Chim. Acta*, 44 (1961) 1287.
52. E. W. Abel, J. M. Pratt and R. Wheland, *J. Chem. Soc. Dalton* (1976) 509.
53. R. J. Blackgrove and L. C. Gruen, *Aust. J. Chem.*, 26 (1973) 258.
54. R. J. Blackgrove and L. C. Gruen, *Aust. J. Chem.*, 25 (1972) 2553.
55. Z. A. Schelly, R. D. Farina and E. M. Eyring, *J. Phys. Chem.*, 74 (1970) 617.
56. Z. A. Schelly, D. J. Harward, P. Hemmes and E. M. Eyring, *J. Phys. Chem.*, 74 (1970) 3040.
57. R. D. Farina, D. J. Halko and J. W. Swinehart, *J. Phys. Chem.*, 76 (1972) 2343.
58. A. R. Monahan, J. A. Brado and A. DeLuca, *J. Phys. Chem.*, 76 (1972) 446.
59. M. Abkowitz and A. R. Monahan, *J. Chem. Phys.*, 58 (1973) 2281.
60. J. A. De Bolfo, T. Smith, J. F. Boas and J. Pilbrow, *Trans. Faraday Soc.*, 48 (1976) 1172.
61. G. Ferraudi, *Inorg. Chem.*, 18 (1979) 1005.
62. G. Ferraudi and J. Granifo, *J. Phys. Chem.*, 89 (1985) 1206.
63. S. Muralidahran and G. Ferraudi, *J. Phys. Chem.*, 87 (1983) 4877.

64. G. Ferraudi, *J. Phys. Chem.*, 88 (1984) 3938.

65. G. Engelsma, A. Yamamoto, E. Markham and M. Calvin, *J. Phys. Chem.*, 66 (1962) 2517.

66. A. B. P. Lever, B. Licoccia and B. S. Ramaswamy, *Inorg. Chim. Acta*, 64 (1982) L87.

67. K. Kasuga and M. Tsutsui, *Coord. Chem. Rev.*, 32 (1980) 67.

68. W. E. Bennett, D. E. Broberg and N. C. Baenziger, *Inorg. Chem.*, 12 (1973) 930.

69. B. Kraut and G. Ferraudi, *Inorg. Chim. Acta*, 149 (1988) 273.

5

Phthalocyanine Films in Chemical Sensors

Arthur W. Snow and
William R. Barger

343

A. INTRODUCTION

The observation that the semiconducting properties of phthalocyanines are modulated by the absorption and desorption of gases has led to significant efforts toward their incorporation in chemical sensors. Phthalocyanines possess other properties that are also favorable for the sensing application. These properties include manipulation as microelectronic device compatible thin films, good chemical and thermal stability toward many environments, and good potential for development of gas specificity. It is hoped that the gas specificity may be developed by making appropriate substitutions of metals in the cavity and organic substituents at the periphery of the phthalocyanine structure.

In its simplest form a phthalocyanine sensor consists of a planar interdigital electrode coated with a thin phthalocyanine film (Fig. 1). If a particular vapor is absorbed by the film and affects the conductivity, its presence may be detected as a conductivity change. Microelectronics technology has exerted an influence on phthalocyanine sensor development. Phthalocyanines are generally weakly conducting materials (e.g., 10^{-12} S/cm), and voltages of 10–700 V are necessary to measure conductivity changes with conventional electrodes. In addition to this power requirement, there is concern about such high voltages causing electrochemical degradation at the phthalocyanine–electrode interface. With microelectronic substrates having an electrode spacing ranging from 1 to 25 μm it is possible to make these measurements at potentials of less than 1 V. In addition to a microelectrode substrate, planar silicon microelectronic device technology has made available a variety of other sensing substrates for organic semiconductor films [1]. This occurs with a large reduction in size and cost.

Several research groups are exploring various aspects of phthalocyanines in sensor research with efforts ranging from the fundamental chemistry of phthalocyanine–gas interactions to performance characteristics of sensing devices. In attempting to prepare an integrated summary of this research, we

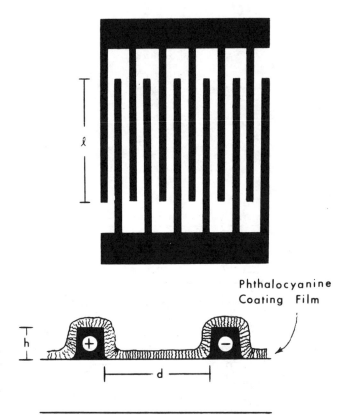

Phthalocyanine
Coating Film

Figure 1 Top and side views of a chemiresistor interdigital electrode sensor substrate illustrating the parameters of electrode spacing, d, height, h, and overlap length, l. In this sketch the coating film thickness is less than the electrode height, and the thickness would be used for the value of h in calculation of a bulk conductivity.

have tried to present concepts and examples that will provide a general picture of this subject.

B. PHTHALOCYANINE–GAS INTERACTION AND CONDUCTIVITY MODULATION

In general form, a microsensor consists of two components: a microelectronic substrate and a chemically sensitive coating. The substrate electronically monitors the coating for a chemically stimulated property change in response to the presence (or absence) of a particular gas. It is the function of the phthalocyanine coating to absorb or form a weak bond with that particular gas and to undergo a large change in conductivity as a result of this interaction. This is thought to occur as a donor–acceptor complex formation with the equilibrium driven by the partial pressure of the gaseous component. It has been observed that acceptor gases such as oxygen, nitric oxide, and nitrogen dioxide cause an increase in phthalocyanine conductivities, a decrease in the activation energies, and the appearance of charge transfer bands in the visible spectra [2]. On the other hand, a donor gas such as ammonia causes a conductivity decease in most phthalocyanines and reverses the effects of the acceptor gases. On the basis of the general response to donor and acceptor gases, phthalocyanines have been classified as p-type semiconductors.

There are many factors that have an influence on the charge transfer interaction and the consequent conductivity change. These factors will be considered in the remainder of this section.

The effect of oxygen was recognized when phthalocyanines were discovered to be semiconductors in 1948 [3]. This effect was studied quantitatively by Heilmeier and Harrison [4] using copper phthalocyanine single crystals in vacuum, air, oxygen, and hydrogen atmospheres. Admission of oxygen was found to increase conductivity and lower the activation energy. This was determined to be a bulk rather than a surface effect. Admission of hydrogen reversed the oxygen effect to yield a lower conductivity and higher activation energy, which remained constant in subsequent evacuations.

The effects of oxygen, nitric oxide, nitrogen dioxide, and ammonia on sublimed films of metal-free, Fe-, Co-, Ni-, Cu-, and Zn-phthalocyanine were studied by Kaufhold and Hauffe [2]. The nitrogen oxides, NO, and NO_2, resulted in large increases in conductivity and reductions in activation energies. The effect was similar but much weaker for oxygen. Also, charge transfer interactions were evident in the visible spectra of these films. Ammonia reversed the electrical effects of the oxidizing gases, but appeared in some way to react with the films. A charge transfer interaction was proposed.

The interaction between oxygen and phthalocyanines has been described as a weak charge transfer equilibrium producing paramagnetic superoxide and oxidized phthalocyanine species [5].

$$MPc + O_2 \rightleftharpoons MPc^+ + O_2^-$$

While oxygen has a demonstrated effect on the conductance of phthalocyanine crystals and films, it is quite small. It corresponds roughly to an increase in conductance from 10^{-14} to $10^{-13} \, \Omega^{-1}$ for a single crystal [4] and from 10^{-14} to $10^{-12} \, \Omega^{-1}$ for a sublimed film [6]. This is not far removed from what the effect would be on a blank quartz substrate [6]. Indeed, if a glass substrate is used, the oxygen effect is not observed since the substrate conductivity level is comparable to that induced by the oxygen [7]. This effect does not interfere with other gas sensing measurements [7]. It has been

Table 1 Conductivity Activation Energies of Phthalocyanine Single Crystals in Vacuum and Saturated in NO_2 Vapor[a]

Phthalocyanine	E_{act} (eV) vacuum	E_{act} (eV) NO_2 saturated
MnPc	0.37	0.10
CoPc	0.60	0.22
NiPc	0.74	0.10
CuPc	0.79	0.17
ZnPc	0.69	0.17
PbPc	0.58	0.28
H_2Pc	0.85	0.15

[a] From ref. [10].

suggested that absorbed oxygen may cause slower response times, since incoming gases would have to displace it from an absorption site before the interaction of the incoming gas could be sensed [8,9]. Operation of sensors at elevated temperatures has improved their response times.

The interaction of nitrogen dioxide gas with phthalocyanine crystals and films has one of the largest effects on conductivity and has received more study than other gas interactions. Many of the observations and results give useful insight into the mechanism by which the interactions promote a conductivity change and will be summarized. Nitrogen dioxide itself is a somewhat complicated gas in that two species, monomeric NO_2 and dimeric N_2O_4, exist in a strongly temperature-dependent equilibrium.

Nitrogen dioxide was reported by early workers [2,6] to have large effects (6 to 8 orders of magnitude) on the conductivity of a variety of sublimed phthalocyanine films. This was accompanied by large changes in the activation energy for conductivity, and charge-transfer interactions were suggested [2]. Later this was quantitatively studied using single crystals of metal-free, Mn-, Co-, Ni-, Cu-, Zn-, and Pb-phthalocyanines [10]. The conductivity change induced by nitrogen dioxide exposure is a surface and not a bulk effect as indicated by the use of a guard ring. All of the phthalocyanine crystals initially responded with a large conductivity increase from $> 10^{-15} \, \Omega^{-1}$/square (except Mn and PbPc which have higher conductivity) to 10^{-7} to $10^{-8} \, \Omega^{-1}$/square at saturation. The changes in activation energies for conductivity in vacuum and saturated with NO_2 (0.5 atm) were quantitatively reported as presented in Table 1. On evacuation, little change was observed in the conductivities of Pb-, Mn-, and Co-phthalocyanines, while the conductivity of the other phthalocyanines decreased by factors of 50 to 1000. Heating under vacuum at 150°C returned metal-free, Ni-, Cu-, and Zn-phthalocyanine conductivities to within a factor of 5 of the initial level, while the Pb-, Mn-, and Co-phthalocyanines required heating at 250°C for 12 h to achieve the effect. Exposures to a small pressure of ammonia restored the initial conductivity in all cases. The Seeback coefficient was positive and became larger with increasing NO_2 induced conductivity.

These observations are interpreted as follows [10]. A charge-transfer complex is formed between a phthalocyanine donor and NO_2 acceptor, and the charge carriers are the holes produced in the phthalocyanine matrix. Ammonia is a competing electron donor and can displace phthalocyanine in the interaction. The Pb-, Mn-, and Co-phthalocyanines, which are less reversible to the NO_2 interaction, form stronger complexes by being able to adopt oxidation states greater than $+2$, thereby enhancing the donor power of the phthalocyanine. It is also pointed out that the nature of the acceptor is important for the conductivity-enhancing effect. Nitrogen dioxide is a π-electron acceptor, and the accepted electron would be delocalized over the planar NO_2 and N_2O_4 structure. Since the hole is also delocalized over the phthalocyanine structure, the coulombic force between the opposite charges is weakened and charge carrier movement is facilitated. This phenomenon is contrasted with the effect induced by exposure to boron trifluoride which is a strong σ-electron acceptor. Copper and nickel phthalocyanine display conductivity increases, but significantly less than those observed with NO_2. This BF_3 effect was not reversible by evacuation, but exposure to a small pressure of ammonia reversed the effect. Boron trifluoride forms a stronger charge transfer complex, but the transferred electron is more localized in a σ orbital. The resulting coulombic forces more strongly retard charge carrier movement of the positive holes in the phthalocyanine matrix.

Spectroscopic studies of sublimed copper and metal-free phthalocyanine films exposed to NO_2 have been reported and offer additional insight into the chemistry of this interaction [11,12]. In the transmission UV-visible spectrum, exposure to NO_2 causes the strong absorptions at 700 and 625 nm to decline in intensity with appearance of a new absorption at 560 nm. These changes are characteristic of formation of the phthalocyanine radical cation. Subsequent heating to 150°C reverses these spectroscopic changes and, in part, restores the original spectrum. In the infrared spectrum, exposure of the phthalocyanine films to NO_2 causes the appearance of bands corresponding either to neutral NO_2 and N_2O_4 or to the NO_2^- ion depending on whether the spectrum is transmission or reflectance. The transmission spectrum is interpreted to indicate that NO_2 and N_2O_4 percolate into the phthalocyanine lattice interstices. The reflectance spectrum displays bands corresponding only to the NO_2^- ion, indicating that charge transfer occurs at the surface.

The conductivity response of sublimed films to NO_2 appears to be somewhat different than that of single crystals. Different phthalocyanine films display different conductivity dependences on the concentrations of NO_2 exposure [13]. Evacuation at room temperature does not completely reverse the NO_2 exposure effect nor is it completely reversed by exposure to donor gases such as ammonia or hydrogen sulfide [11,12]. Sublimed films may have an amorphous or polycrystalline character. Defects and disordering affect gas absorption sites as well as charge carrier transport. Indeed, simple annealing of a sublimed film in the α crystalline form to the β crystalline form changes its electrical conduction and gas response characteristics [14]. It is reported that the surface structure of the sublimed films can be varied by control of substrate temperature and optimized for response to vapors [8,15]. Rigorous control of the sublimation conditions is required for reproducible films.

Another film deposition technique for preparing phthalocyanine films on microelectrodes is based on the Langmuir–Blodgett (L-B) technique of transferring films from a water surface. This technique (see Section D) is very compatible with microelectronic technology and offers excellent control over film thickness and reproducibility. Organic solvent-soluble phthalocyanine compounds are processed as thin films by this technique. This includes phthalocyanines with peripheral or axial substitution and peripherally unsubstituted lithium phthalocyanine [16] which hydrolyzes to the metal-free analogue on water contact, and magnesium phthalocyanine [64]. The phthalocyanine L-B films display significantly different responses to vapors as compared with sublimed films. A comparison of the response to ammonia exposure of an L-B film of a tetracumylphenoxy substituted copper phthalocyanine film with that of a sublimed unsubstituted copper phthalocyanine film is presented in Fig. 2 [17]. The ammonia exposure is a 500 ppm challenge conducted to the sensor from an exponential dilution flask. The signal expected if the sensors tracked the exponential dilution of the 500 ppm initial concentration of the ammonia is shown by the dotted lines. Three features are noteworthy. The signal for the L-B film corresponds to a conductivity increase while that for the sublimed film is for a conductivity decrease. The L-B film displays a greater change in conductivity. The recovery for the L-B film is much faster as shown by a much closer tracking of the signal to the diluted ammonia concentration. Similarly, an asymetrically substituted trismethylene isopropylamine phthalocyanine L-B film exposed to NO_2 showed greatly enhanced response and recovery times in comparison with an L-B film of unsubstituted phthalocyanine [18]. The peripheral group substitution apparently places the phthalocyanine structure in a matrix where absorption and desorption of gases are more facile. The mixed isomer nature of the peripherally substituted phthalocyanines may have the advantage of supressing crystallization and promoting a more gas-permeable matrix. A large variation in sensitivity and specificity is observed for different metal phthalocyanines with tetracumylphenoxy peripheral substitutions [19]. Molecular orbital calculations of Co-, Ni-, Cu-, and Zn-phthalocyanine and of the effect of an axial interaction with ammonia have been performed and, assuming a model of nearly isolated phthalocyanine units, have been correlated with experimental conductivity measurements [20].

These observations indicate that the phthalocyanine film morphology plays an important role in the film's electrical response to gas exposures. The morphology must accommodate both the charge transfer interaction and charge carrier transport. While these mechanisms are not well understood, it is believed that the charge transfer interaction is more facile if the face of a

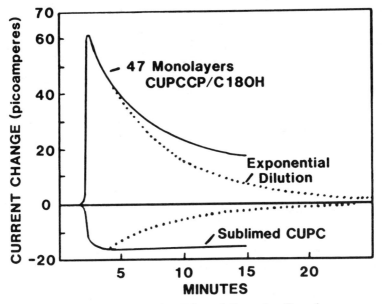

Figure 2 Comparison of response of a 47-layer L-B coating film of tetracu-mylphenoxy copper phthalocyanine with a sublimed coating film of unsubstituted copper phthalocyanine. The signal expected if the response tracked the exponential dilution of the 500 ppm initial concentration of ammonia is shown by the dotted lines. Reproduced from ref. [17] with permission. Copyright 1986, American Chemical Society.

phthalocyanine ring rather than an edge is available for an electron transfer through complex formation. On the other hand charge carrier transport is facilitated by a long-range stacking of cofacially oriented phthalocyanine rings. Unsubstituted phthalocyanine films, when crystalline, are composed of crystallites with a columnar stacking of cofacially oriented phthalocyanine rings. This feature coupled with partial oxidation promotes charge carrier transport through the film [21]. The peripherally substituted phthalocyanines dissolve in a variety of organic solvents and exist as oligomeric aggregates of cofacially oriented molecules [22]. Films of these peripherally substituted phthalocyanines are also believed to consist of these oligomeric aggregates having short-range order instead of crystallites with long-range order. Such a morphology would make available a greater number of phthalocyanine ring faces for charge transfer interactions with gases as well as a less dense molecular packing for more facile gas permeation. However, the long-range order is not present resulting in lower levels of conductivity attainable after charge transfer complex formation. Such an effect has been observed. Tetracumylphenoxy phthalocyanine L-B films were observed to increase conductivity by 4 orders of magnitude on exposure to iodine vapor, while unsubstituted phthalocyanine L-B films will increase conductivity by 10 orders of magnitude by absorption of a similar quantity of iodine [23]. Phthalocyanine film morphology is discussed in greater detail in Section D.

The use of peripheral group substitution to vary vapor sensitivity and selectivity of phthalocyanine films has promising prospects for gas detection. Peripheral groups can modulate the vapor response by morphology effects described above or by an inductive effect on the phthalocyanine ring which modulate the strength of the charge transfer interaction. Some very encouraging work has been reported on varying conjugated macrocyclic ring size and nature of electron donating and withdrawing substituents as an alternative to the phthalocyanine ring system [7,11,12,24].

C. ELECTRONIC SUBSTRATES

Electronic substrates are electrical devices which support coating films (phthalocyanine films in this case) being monitored for a chemically induced property change. The substrates to be discussed are the chemiresistor and the surface acoustic wave (SAW) delay line oscillator. The chemiresistor directly monitors the coating film for an electrical conductivity change. The SAW device responds to changes in the coating's mass, conductivity, and mechanical properties with their relative contributions depending on the SAW device construction and magnitude of the changes. While the chemiresistor is the simpler device, it requires the coating film to undergo both an absorption and a conductivity change, whereas the SAW sensor may respond to absorption of the vapor alone because of the increased mass of the coating. It is possible to have a SAW device complement the chemiresistor by combining the design features of both devices on the same substrate and using it to make simultaneous measurements by both techniques. This combined measurement will also be described.

i. Chemiresistor

Historically, the use of a chemiresistor device to detect conductivity changes in phthalocyanine films can be traced to the work of Tollin, Kearns, and Calvin in 1960 [25]. The basic interdigital electrode design (see Fig. 1) has changed very little, although materials, fabrication techniques, and dimensions have advanced. Reference [26] presents an example with design and measurement details. The device consists of an interdigital array of metal electrodes patterned on an insulating substrate. Two important features are an ohmic contact of the electrodes with the overlying film and an electrode spacing and geometry that make possible the measurement of very small currents. A tabulation of chemiresistor materials, electrode spacing, geometry, . and applied voltages used by various workers is presented in Table 2. A typical

Table 2 Chemiresistor Device Materials and Design

Electrode and substrate	Electrode spacing (mm)	Geometry	Bias (V)	Reference
Gold/quartz	1	Not reported	50	[2]
Gold/quartz	Not reported	Interdigital	10–700	[6]
Gold/quartz	1	Interdigital	Not reported	[14]
Gold/quartz	0.025	Interdigital	Not reported	[17]
Gold/quartz	0.020	Interdigital	<1	[19,20]
Gold/quartz	0.015	Interdigital	<1	[23]
Gold/quartz	0.05	Interdigital	Not reported	[33]
Gold/glass	1	Concentric	100	[36]
Gold/glass	5	Parallel strip	15	[11,12]
Copper glass	Not reported	Interdigital	Not reported	[18]
Aluminum/glass	0.1	Interdigital	100	[25]
Aquadag/glass	0.0007	Interdigital	Not reported	[17]
Niobium/quartz	Not reported	Interdigital	1.5	[34,35]
Platinum/alumina	Not reported	Interdigital	Not reported	[9]
Platinum/alumina				

example for a chemiresistor response is illustrated in Fig. 3 for exposure of a lead phthalocyanine film to 50 ppb of nitrogen dioxide [34].

Ohmic contact has been depicted as an electrical contact where there is an excess or reservoir of free charge carriers in the insulator region of the contact such that the bulk current through the insulator matrix is not dependent on the amount of excess charge at the boundary [27]. If a sufficient excess of charge carriers is not available at the boundary of the insulator, a saturation current is observed at high fields contrary to Ohm's Law. Ohmic contact behavior for gold electrodes on sublimed copper phthalocyanine films [28] and for gold, silver, platinum, and Aquadag (fine particulate graphite in a binder) on pressed, mixed metal-free/copper phthalocyanine pellets [29] has been reported. Aluminum, nickel, tungsten, and molybdenum electrodes were reported to behave differently [29]. In particular, aluminum was reported to form a blocking contact with copper phthalocyanine [30]. It has also been pointed out that gas sensitivity may be affected by a catalytically active metal such as platinum [31]. Many workers have made a general assumption of ohmic contact behavior for gold and silver electrodes on other types of phthalocyanines. Subjectively, this approach appears to be valid but may be checked by obtaining linear I–V curves with electrodes of different spacing and determining if the ratio of slopes is the same as the ratio of electrode spacings.

The electrode spacing and geometry of the chemiresistor are critical to the sensitivity of the current measurement for films of very low conductivity (e.g., phthalocyanine films). Minimizing the electrode spacing and maximizing the number of interdigital electrode finger pairs and electrode finger overlap lengths can greatly increase the quantity of measurable current and reduce the amount of voltage necessary to generate it. This voltage reduction is desirable, since very high voltages can cause electrochemical degradation of the electrode–semiconductor interface. Given the chemiresistor interdigital electrode geometry (Fig. 1), the important parameters are the electrode spacing (d), the overlap length (l) of the electrode finger, the number of electrode finger pairs (n), and the electrode thickness or height (h). Either a bulk or a surface conductivity may be measured depending on the nature of the coating film and its response to a particular vapor. The bulk conductivity of the film (σ) may be calculated as follows:

$$\sigma = \frac{J}{E} = \frac{I/A}{V/d} = \frac{I}{V}\frac{d}{(2n-1)lh}$$

where J is the current density, E is the electric field, I is the measured current, A is the cross-sectional area between electrodes, and V is the bias voltage. If the coating film is thick enough to fill the channel between the electrode fingers, then A may be calculated as the product of the number of channels between the electrode fingers ($2n - 1$), the overlap length of the electrode

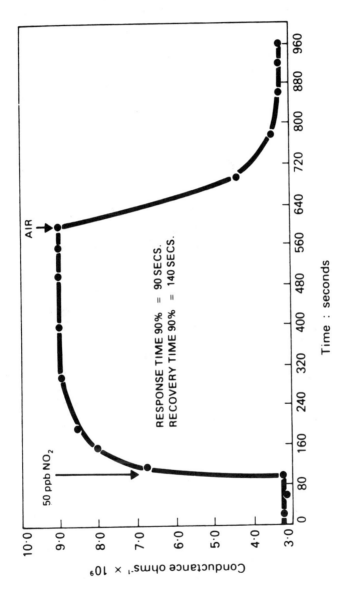

Figure 3 Response and recovery of a chemiresistor coated with a sublimed lead phthalocyanine film and exposed to 50 ppb of nitrogen dioxide in air at 150°C. Reproduced from ref. [34] with permission. Copyright 1986, Elsevier Sequoia.

fingers, and the electrode height. If the film is thinner than the electrode height, then h becomes the film thickness. Electrode thicknesses are typically 500–2000 Å. If the chemistry is such that surface conductivity (σ_s) is being measured, the resistive element is a square with an edge length defined by the electrode spacing (d). The squares spanning the electrode gap are considered to be resistors in parallel so that the resistivity of a square (R_s) is the product of the measured resistance (R) and the number of squares within the electrode array as shown by the following equations.

$$\frac{1}{R} = \sum_{1}^{[(2n-1)l]/d} \frac{1}{R_s} = \frac{(2n-1)l}{d}\frac{1}{R_s}$$

$$\sigma_s = \frac{1}{R_s} = \frac{I}{V}\frac{d}{(2n-1)l}$$

If the conductivity is exclusively either bulk or surface, it may be distinguished by a dependence on film thickness for film thicknesses less than the electrode height. In practice, the current is the measured quantity, and for phthalocyanine films, it is quite small (e.g., on the order of nano- or picoamperes). Since the current is inversely dependent on d and linearly dependent on n and l, sensitivity can be enormously enhanced by using microlithographically produced electrodes which minimize d (25 μm and less) and greatly increase n (50 and greater). With such electrodes it is possible to measure conductivity of monomolecular films [23].

As may be noted from Table 2, a great variation exists in designs of chemiresistor devices. Most were fabricated in the researchers' laboratories by evaporation or sputtering of metal films through a mask onto an inert insulator substrate. As research progresses on phthalocyanine chemiresistor sensors, it becomes apparent that standard device electrodes will be important for coordination of results between laboratories. Currently, through microelectronic technology, high precision microelectrode arrays are commercially available [32], although a designation of standard electrode designs has not been made.

Another issue where great variations exist is in reporting the electrical responses to vapor exposures. The electrical responses have been reported as changes in current, percentage conductivity, surface resistivity and conductivity, conductance, and current density. These require varying degrees of knowledge of the conduction mechanism, the electrode, and the measurement conditions. Efforts should be made for careful documentation of these features so that results will be more widely useful to other researchers.

ii. SAW Device

Use of a SAW device as a vapor sensor component was first reported by Wohltjen and Dessy in 1979 [37], and its mechanism of operation has recently been described [38]. An illustration of a very simple SAW device is presented in

SAW DELAY LINE

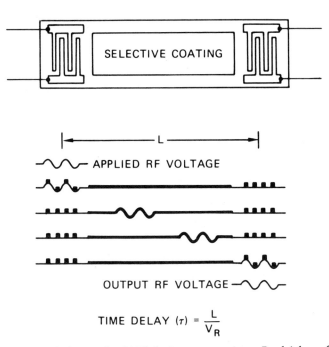

Figure 4 Top and side views of a SAW device propagating a Rayleigh surface wave. The vertical displacement of the wave is exaggerated for clarity. Reproduced from ref. [38] with permission. Copyright 1984, Elsevier Sequoia.

Fig. 4. The device consists of a piezoelectric slab (usually quartz or lithium niobate) on which two sets of interdigital microelectrodes have been fabricated using optical lithography. A radio-frequency (rf) voltage is applied to the transmitter electrode of the transmitter–receiver electrode pair to generate a mechanical Rayleigh surface wave on the piezoelectric substrate. This wave propagates across the surface to the receiver electrode and is converted back into an rf voltage. Connection of this electrode pair through an rf amplifier in a feedback circuit makes the device oscillate at a resonant frequency determined by the interdigital electrode spacing and the Rayleigh wave velocity. Coating the device with a thin film causes a substantial reduction of the Rayleigh wave velocity and a corresponding decrease in the resonant frequency of the device. Vapor absorption further alters the mass, mechanical and electrical properties of the coating, and consequently produces easily measured frequency shifts for the vapor sensing application. Figure 5 presents an example of a lead phthalocyanine coated 110 MHz lithium niobate SAW sensor responding to 10 ppm nitrogen dioxide [39].

A variety of factors associated with the coating may perturb the frequency of SAW devices by a coupling and energy exchange with the surface

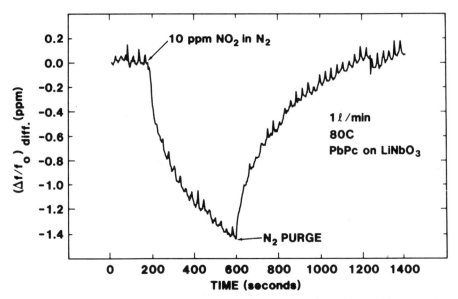

Figure 5 Response and recovery of a 110-MHz lithium niobate SAW device coated with a sublimed lead phthalocyanine film and exposed to 10 ppm nitrogen dioxide in air at 80°C. Reproduced from ref. [39] with permission. Copyright 1985, Elsevier Sequoia.

wave. These factors include a change in mass loading, film modulus, and electrical conductivity. The mechanism of this perturbation may involve a lossy mechanical exchange with the coating mass or a coupling with the electric field component of the Rayleigh surface wave. The SAW frequency shift (Δf) dependence on mass has been derived as a linear relationship as follows [38]:

$$\Delta f = (k_1 + k_2) f_0^2 h\rho$$

where k_1 and k_2 are material constants for the piezoelectric substrate, f_0 is the unperturbed frequency of the SAW device, and the quantity $h\rho$ is the film thickness–density product or the mass per unit area of the coating. The constants k_1 and k_2 represent negative numbers, thus an increase in the coating mass results in a decrease in resonant frequency. The SAW frequency shift dependence on electrical conductivity has been derived as the following relationship [39]:

$$\Delta f = \frac{-K^2}{2} \frac{\sigma_{sh}^2}{\sigma_{sh}^2 + v_0^2 C_s^2} f_0$$

where K^2 is the electromechanical coupling coefficient, σ_{sh} is the sheet conductivity of the film, v_0 is the unperturbed surface wave velocity, C_s is the capacitance per unit length of surface, and f_0 is the unperturbed frequency of the SAW device. Thus, an increase in coating conductivity results in a negative frequency shift. The SAW frequency shift dependence on film modulus has not been treated by itself and is usually considered negligible compared with the mass effect [38].

When a coating film absorbs a vapor, there are usually changes in its mass and electrical conductivity. The relative contributions of these effects to the SAW frequency shift depend on the magnitude of these changes and the SAW device construction. A lithium niobate SAW device has a high electromechanical coupling coefficient, making it well-suited for response to electrical conductivity. The predominance of this contribution has been nicely demonstrated using a lithium niobate SAW device with a lead phthalocyanine-nitrogen dioxide coating-vapor system where the conductivity change is large and mass change is small [39]. On the other hand, the mass effect has been presumed (but not demonstrated) to be the major contributor to the frequency shift for a quartz SAW device with a peripherally substituted phthalocyanine exposed to iodine vapor where the conductivity change is smaller and mass change is quite large [23]. Between these two extremes, a result has recently been reported using a quartz SAW device with a copper phthalocyanine-nitrogen dioxide system where both mass and conductivity effects are claimed to make significant contributions [40]. More research is needed for a better understanding of the mass and conductivity contributions to the SAW frequency shifts.

iii. Combined Chemiresistor–SAW Device

Noting that the interdigital microelectrode is the basic component of both the chemiresistor and SAW device, it is a logical extention to microfabricate both measurement capabilities onto the same substrate. In fact, a dual SAW device is very easily adapted for this purpose as illustrated in Fig. 6 [23]. One pair of electrodes is used for the SAW frequency measurement and either of the remaining two electrodes is used for the conductivity measurement. Figure 7 presents an example of simultaneous SAW and conductivity data for a peripherally substituted copper phthalocyanine Langmuir–Blodgett film exposed to iodine vapor. The iodine exposure causes a conductivity increase of 4 orders of magnitude and a corresponding SAW frequency shift of 10 kHz. The conductivity increase correlates with the iodine content in the film which may be obtained from a knowledge of the mass uptake contribution to the SAW frequency shift. Other vapors of lower molecular weight and weaker electron acceptor strength produce much weaker responses. This dual measurement of coating response to vapor exposure is a new sensing technique with prospects of enhancing detection specificity.

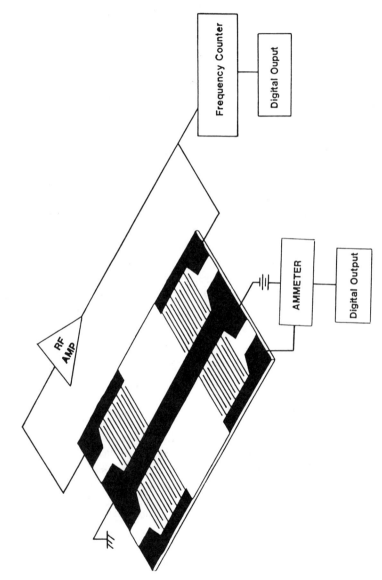

Figure 6 Dual 52-MHz SAW device used for simultaneous mass and conductivity measurements. Reproduced from ref. [23] with permission. Copyright 1986, American Chemical Society.

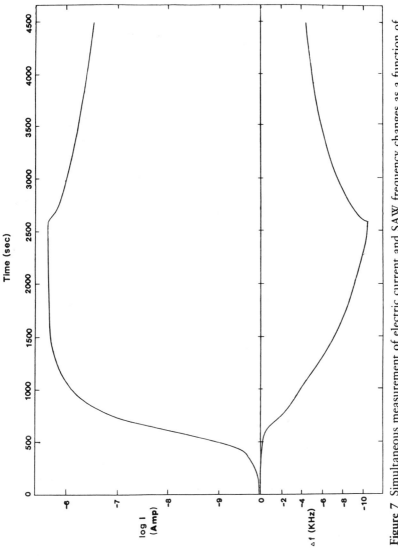

Figure 7 Simultaneous measurement of electric current and SAW frequency changes as a function of time caused by exposure of a tetracumylphenoxy copper phthalocyanine L-B film supported on a 52-MHz dual SAW device to iodine vapor. Reproduced from ref. [23] with permission. Copyright 1986, American Chemical Society.

D. FILM DEPOSITION AND MORPHOLOGY

A critical operation in the construction of a sensor is the physical deposition of the phthalocyanine film onto the electronic substrate. The deposition technique is largely determined by the properties of the phthalocyanine compound, but the process must provide films with reproducible electronic and vapor response characteristics. The film thickness, uniformity, and morphology must be controlled if duplicate sensing devices are to be produced. Currently, there are two phthalocyanine film deposition techniques that are commonly used. Vacuum sublimation is used for most unsubstituted phthalocyanine compounds. The Langmuir–Blodgett film transfer technique may be used for phthalocyanine compounds which are usually substituted at the axial or peripheral positions to make them soluble in volatile organic solvents while remaining insoluble in water. Furthermore, the particular conditions by which a deposition is conducted may have a pronounced effect on the response to vapors since film morphologies are dependent on these conditions. In this section the depositions and morphologies of sublimed and Langmuir–Blodgett films as they pertain to vapor sensors are discussed.

i. Sublimed Films

Phthalocyanine vacuum sublimations are conducted by heating a small quantity of purified material at temperatures ranging from 300 to 500°C at reduced pressure in an inert atmosphere or under vacuum. Electronic substrates are positioned 3–10 cm away from the source and preferably mounted on a temperature-controlled holder. Sublimators are usually custom designed to accommodate the substrates and deposition conditions. A simple sublimator design is illustrated in Fig. 8. The substrate is attached to a metal contact plate which is fixed by spring attachment to the base of the sublimator well. Provision is made for control and monitoring of both the substrate and phthalocyanine reservoir temperatures. Some workers have also used a piezoelectric quartz crystal balance to monitor the rate of film deposition [30]. Obtaining reproducible films is sometimes difficult and may require strict attention to sublimation conditions, phthalocyanine purity, and substrate cleanliness.

The morphology of the phthalocyanine films is variable and strongly dependent on the sublimation conditions. It ranges from amorphous to highly crystalline. Amorphous films are prepared by vacuum sublimation (10^{-6} Torr) onto liquid-nitrogen-cooled (100 K) substrates [41]. These films are stable at room temperature but crystallize to the α phase on annealing over the temperature range 60–140°C followed by an α to β transition at 210°C. If the substrate is held at room temperature, the α phase is obtained at sublimation

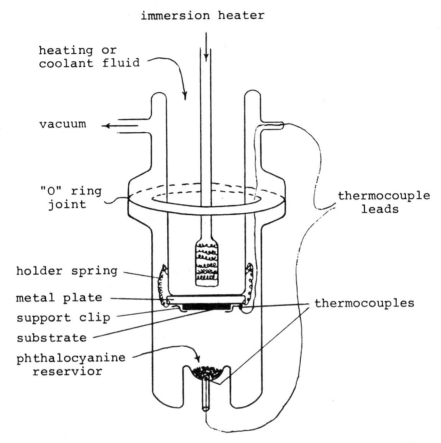

Figure 8 Vacuum sublimator designed for depositing phthalocyanine-coating films on chemiresistor and SAW device substrates.

pressures less than 50 Torr [41]. At higher pressures or at substrate temperatures above 210°C, the β phase is obtained [42]. To avoid decomposition, phthalocyanines should not be heated above 400°C [43]. In appearance, an α phase phthalocyanine film is composed of microscopic granular crystallites as small as 100 Å in diameter while a β phase film consists of larger fibrous crystals [44]. The thermal conversion of an α phase film to a β phase film can result in the formation of fibrous crystallites with a common orientation. A mechanism for this transformation has been studied [44].

This morphology difference is very important for gas sensing applications. The α phase is much more sensitive to the presence of oxygen [45]. While the intrinsic conductivities of the α and β phases are identical, oxygen exposure causes little change in the β phase but a 10^4 conductivity increase in the α phase [46,47,48]. Of additional concern is the observation that exposure

Table 3 Crystallographic Data for α- and β-Polymorphs of Copper Phthalocyanine[a]

	Space group	
	α-form [51] Monoclinic $C2/c$	β-form [50] Monoclinic $P2_1/a$
a (Å)	25.92	19.407
b (Å)	3.790	4.790
c (Å)	23.92	14.628
β (deg)	90.4	120.56
Molecule/unit cell	4	2
Unit cell volume (Å^3)	2350	1166
Density (calc) (g/cm^3)	1.62	1.639
Density (found) (g/cm^3)	1.62	1.63
Tilt angle[b] (deg)	26.5	45.8

[a] From refs. [50,51].
[b] Angle between b axis and normal of phthalocyanine plane.

to some organic solvent vapors can cause an α to β phase transformation [49]. The crystal structures in sublimed films are important determinants of how vapors may enter the lattice and interact with the phthalocyanine units.

Using data for copper phthalocyanine as an example, the molecular arrangements of the phthalocyanine units in the α and β crystal structures are illustrated in Fig. 9, and unit cell characterizations are presented in Table 3 [50,51]. The phthalocyanine units are positioned in columnar stacks with the ring orientation inclined relative to the axis of the stack. The α crystalline form differs from the β form by a smaller tilt angle between the axis of the stack (b axis of the unit cell) and the normal of the phthalocyanine ring, resulting in a considerably shorter metal–metal distance and a longer interstack distance. However, for both forms the interplanar spacing between rings and density calculated from the unit cell data is nearly equal. This indicates that the size of the channels between the columns is also nearly equal, the largest dimensions of which approximate 3.8 Å [50]. Within a stack, the larger tilt angle of the β form positions the pyrrole nitrogen of a phthalocyanine ring above and below the axial coordination sites of a central metal atom in the adjacent phthalocyanine unit. This arrangement approaches a distorted octahedral configuration. In the α form, the axial position is more open. Energetically, the β phase is more stable (by 2.57 kcal/mol for the copper phthalocyanine example [52]). For metal-free phthalocyanine, intermolecular hydrogen bonding of the cavity protons of one phthalocyanine unit with the pyrrole nitrogen of the adjacent phthalocyanine unit has been observed as an 18 cm^{-1} red shift of the N—H stretching in the β phase relative to the α phase [53]. Analytically, the α and β crystalline forms may be distinguished by X-ray diffraction and by infrared and visible spectroscopies [53,54,55]. Also the thermal [41] and organic solvent-induced [49] conversions of the α to the β form may be monitored by these techniques. Phthalocyanines with large metal

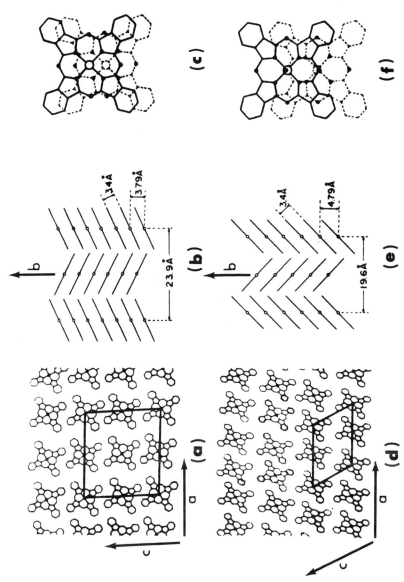

Figure 9 Molecular arrangements in copper phthalocyanine crystals of the α form (a)–(c) and the β form (d)–(f). (a) and (d): projections on (010) plane; (b) and (e): projection on (100) and (001) planes; (c) and (f): superposition of molecules related by translation along the *b* axis. Adopted from ref. [49] with permission. Copyright 1980, American Chemical Society.

atoms (Pt [56] or Pb [57]) or coordinated axial ligands may deviate from one or both of the generalized alpha and beta forms.

For a particular crystalline form, the crystallite size and orientation in a thin film are factors which can affect a sensor's response to vapors. The nature of the substrate surface [58] and the deposition and annealing conditions [44] influence the crystallite size and orientation. Molecular images of these have been observed by high-resolution electron microscopy for a zinc phthalocyanine example [59]. In a carefully controlled experiment directed at the effect of the substrate temperature on sublimed phthalocyanine film morphology, it was discovered that, for a very narrow temperature range, films with exceptionally large and highly oriented crystallites can be obtained [60]. For metal-free phthalocyanine this temperature range occurs at $10 \pm 5°C$ and is conveniently detected by a sharp increase in the Q-band molar absorptivity. The α crystalline form is obtained under these conditions and is very stable toward thermal conversion to the β form. As a coating film for a sensor, this optimization phenomenon was investigated using lead phthalocyanine films prepared at optimized and nonoptimized conditions and exposed to nitrogen dioxide [8]. It was found that the nonoptimized films had a stronger and more rapid response and it was postulated that the nonoptimized film possessed more surface area and defects to serve as absorption sites for the gas.

Finally, in the context of absorption and desorption of vapors, the morphology of the sublimed phthalocyanine film should not be regarded as a fixed matrix. As mentioned above, the α crystalline form may be thermally transformed into the β form with a substantial increase in crystallite size and orientation [44]. Exposure to organic solvent vapors may cause a similar transformation. The mechanism for the conversion of α to β zinc phthalocyanine by exposure to a series of alcohol vapors has been studied [49]. The transformation progresses in two steps. The first involves conversion of the α to an intermediate x form with an increase in crystallite size from 100–200 to 2000 Å and involves a weak charge transfer interaction with the alcohol. The second involves conversion of the x to the β form with characteristic formation of the 10,000- to 20,000-Å long needles and is very dependent on the structure and properties of the alcohol. Sequential electron micrographs of this transformation are presented in Fig. 10. This α to β conversion has also been observed by exposure to other weak nonbonded electron donors (acetone, ether, carbon disulfide), π-electron donors (benzene, toluene, xylene), and π-electron acceptors (nitromethane, nitrobenzene, carbon tetrachloride) [49,61]. With strong nonbonded electron donors (pyidine, piperidine, methylamine, dimethylamine, dimethylsulfoxide, dioxane) and more rigorous exposure conditions, stable stoichiometric complexes with unique crystal structures are formed [61]. These complexes could be thermally decomposed to return the α or β forms. Iodine, a strong electron acceptor, forms parallel chains of triiodide anions in the channels between the phthalocyanine stacks with a considerable alteration in the molecular packing of the phthalocyanine

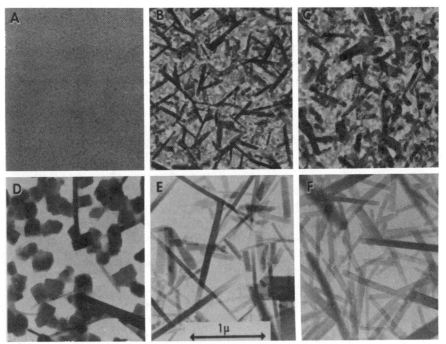

Figure 10 Sequential electron micrographs of zinc phthalocyanine film undergoing α to β transformation in 1-propanol vapor at 55°C: (A) 0 min; (B) 1 min; (C) 5 min; (D) 10 min; (E) 6 h; (F) 70 h. Reproduced from ref. [49] with permission. Copyright 1980, American Chemical Society.

precursor [62]. When considering use of sublimed films for chemical sensors, selection of a polymorph with a reversible vapor absorption is an important factor. Sensors incorporating sublimed films are frequently operated at elevated temperatures to ensure absorption reversibility [34,40].

ii. Langmuir–Blodgett (L-B) Films

The L-B film deposition involves spreading a solution of a phthalocyanine compound onto a water surface, compressing the phthalocyanine molecules into a continuous monolayer film after evaporation or water absorption of the solvent, and transferring the film onto an electronic substrate by a dipping operation. This technique for depositing certain kinds of organic compounds onto solid surfaces one monolayer at a time was developed by Langmuir and Blodgett in the 1930s [63]. Formation of an L-B film from magnesium phthalocyanine was also investigated during that time period [64]. Considerable detail about the L-B film deposition technique may be found in the text by Gaines [65]. For the purpose of this discussion, a phthalocyanine L-

B film is considered to be a mono- or multilayer film transferred by the L-B technique.

By comparison with the sublimation method, the L-B technique may be considered more complex although somewhat complementary. Phthalocyanine compounds which sublime well (i.e., most peripherally unsubstituted phthalocyanines) are not sufficiently soluble to be eligible for the L-B technique. Exceptions are magnesium and lithium phthalocyanine which form unstable [64] or poorly transferable [16] monolayers. A variety of peripherally substituted phthalocyanine compounds have been synthesized and formed into L-B films (Table 4). The sublimation apparatus is much simpler than the L-B film balance. Sublimation requires application of heat, vacuum, and temperature control, while the L-B trough requires scrupulously clean water, carefully synchronized control of film pressure with transfer rate, and multiple dipping operations at constant temperature. The film morphologies are quite different, as are the vapor response characteristics of the deposited films. However, the reproducibility of film depositions and vapor responses appears to be better for the L-B films.

The phthalocyanine compounds used for the L-B film deposition are usually substituted at the ring periphery to promote solubility in organic solvents as illustrated in Table 4. There are a total of 16 substituent positions, but the 8 outer positions are most frequently involved with the substitutions. The tetrasubstituted phthalocyanines are actually isomer mixtures due to positional variation during the synthesis. This mixed isomer character would have the effect of supressing crystallinity but not aggregation, which has been demonstrated by vapor pressure osmometry measurements [22]. With regard to chemical sensor films, the supression of crystallinity is perceived to be an advantage, since the matrix may become more permeable to gases and the morphology may be less complicated by crystalline phase effects compared with what might be observed with isomerically pure substituted phthalocyanines. The asymmetrically substituted phthalocyanines (structures 8 and 11 in Table 4) are particularly interesting in that they incorporate regions of hydrophilicity and hydrophobicity into a phthalocyanine molecule. This design is to achieve a molecular orientation at the air–water interface such that the transferred L-B film will have a similar orientation.

The preparation of an L-B film is conducted on a trough which is equipped with a movable barrier at the water surface for compressing the film, a film pressure or surface tension measuring device (such as a Wilhelmy Plate and electrobalance) and an automated dipping device for transferring the film (Fig. 11). The trough is filled with clean (i.e., triply distilled from quartz, for example) water, and a solution of the phthalocyanine compound (approximately 0.1% in a volatile organic solvent) is spread by dropwise application to the water surface. After evaporation of the solvent or its dissolution into the water, the phthalocyanine film is compressed by the barrier until the film pressure sharply rises, indicating that a continuous surface film has formed.

Table 4 Phthalocyanine Peripheral Group Substituents for L-B Films

Substituent	Number of Groups	Reference
1. CH₃—C(CH₃)(CH₃)—	4	[16,66,67,68,71]
2. CH_3—O—	4	[69]
3. $CH_3(CH_2)_{17}$—O—	4	[70]
4. CH₃—C(CH₃)—CH₂—O—	4	[70]
5. (phenyl)—O—	4	[70]
6. (X,X-phenyl)—O—	4	[69]
7. (biphenyl)—O—	4	[17,19,22,23,70]
8. CH₃—CH(CH₃)—NH—CH₂—	3	[18,71]
9. CH_3—$(CH_2)_{11}$—O—CH_2—	8	[72]
10. CH_3O—, $CH_3(CH_2)_7$— (phthalocyanine polymer)	4 + 4	[73]
11. $CH_3(CH_2)_9$—, HO—C(=O)—$(CH_2)_3$—	6 + 2	[74]

The dependence of the film pressure on the film area is recorded, and transformed into a force–area curve relating the film pressure to the area per molecule in the case of monolayer films. This characterization is related to molecular orientation. Typical data are presented in Fig. 12 and are discussed below. The film transfer is conducted at a constant pressure, usually corresponding to a value in the steep portion of the force–area curve. The substrate is passed through the air–water interface, and the monolayer is

FILM BALANCE

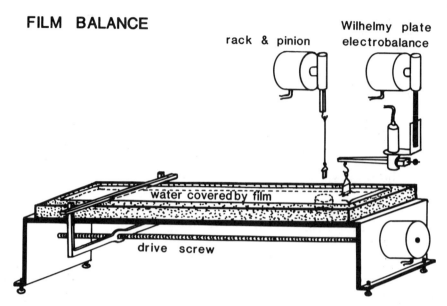

Figure 11 A typical Langmuir–Blodgett film deposition apparatus. Monolayer films adsorbed on water are compressed by a movable barrier and transferred onto a substrate which is passed repeatedly through the air–water interface. Film pressure can be monitored with a Wilhelmy plate suspended from a microbalance.

transferred to its surfaces (Fig. 13). This dipping operation is repeated to build up a multilayer film. The moving barrier maintains constant pressure and is synchronized with the dipping operation by command of the film pressure measurement, usually by way of a computer which automates this process.

Different films sometimes behave differently during the transfer process. Some films transfer on each pass of the substrate through the interface, and some transfer only on the upward or downward pass. When the molecules of the film possess hydrophilic and hydrophobic regions, there is usually a molecular orientation induced by the water surface. An ideal example is stearic acid which forms a compressed film of aligned molecules on the water surface with the acid group oriented toward the water and the hydrocarbon tail directed away from the water. This molecular orientation is preserved during the transfer process. Successive transfers usually result in an alternating molecular orientation where the hydrophilic or hydrophobic surface of the preceding layer directs the orientation of the following layer. Thus, in the ideal case, a layered film morphology with an alternating molecular orientation is built up (Fig. 13).

Phthalocyanine compounds do not appear to behave like ideal molecular components of L-B films. In solution, they aggregate and progress from cofacial dimers in dilute solutions to larger complexes in more concentrated solutions [22,75]. The phthalocyanine π-electronic interaction for dimeriza-

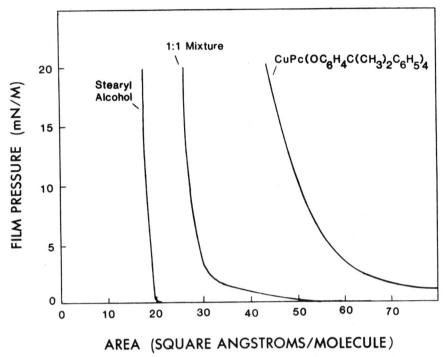

Figure 12 L-B film pressure vs. area isotherms of cumylphenoxy copper phthalo-cyanine, stearyl alcohol, and a 1:1 molar mixture of this phthalocyanine and transfer promoter.

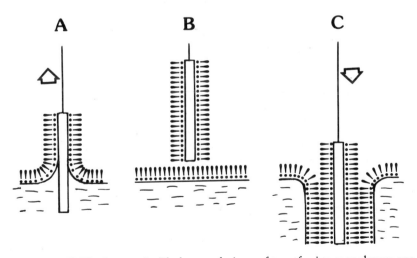

Figure 13 (A–C) The Langmuir–Blodgett technique of transferring monolayers one at a time from the water surface to a substrate to produce a multilayer film with alternating molecular orientations (in the ideal case).

tion is comparable in strength to hydrogen bond formation, as indicated by a dimerization enthalpy of -14 kcal/mol [76]. Phthalocyanines also interact and form weak complexes with solvents, particularly those with π-electron systems [77]. Consequentially, when envisioning the spreading and compression of phthalocyanine-containing L-B films, it is reasonable to anticipate aggregation and solvent retention. It has also been observed that "pure" L-B films of phthalocyanines frequently will not transfer uniformly to substrates, and solvents with special properties (mesitylene [16,66,67]; xylene/THF [71,74]) or transfer promoters (stearyl alcohol [19,23,26]; arachidic acid [67,68]) are used to effect a good transfer. These additives reduce the mechanical rigidity of the phthalocyanine films as well as affect the surface energy for transfer. Their role in promoting the transfer is not completely understood, and their addition can influence the film morphology.

The morphology of the L-B phthalocyanine films has not been studied as extensively as that of the sublimed films. It is closely related to both molecular structure and conditions of L-B film formation and transfer. Several of these aspects have been recently reviewed [79]. General statements that follow should be regarded in the context that the L-B films are composed of phthalocyanine compounds which differ in identity, number, and symmetry of peripheral substituent groups. As indicated above, an L-B film has a layered morphology in the direction of its thickness, and the molecular orientation in one layer may be influenced by an adjacent layer or the substrate surface. Within a molecular layer, the degree of orientation and size of ordered domains are important issues for developing an understanding of charge carrier transport and the coating's response to vapors.

Most work indicates that the ordered domains of transferred phthalocyanine monolayers consist of stacks of cofacially oriented molecules characterized by a dihedral angle between the phthalocyanine ring and monolayer plane and a stack axis orientation (see Fig. 14). The stack axis is usually assumed to be parallel to the monolayer plane and substrate surface. Within the monolayer plane some orientation of the stack axes is reported to be induced by the dipping direction of the substrate. Characterization techniques include film pressure vs. area isotherms, X-ray and electron diffraction, ESR spectroscopy, electronic and infrared polarized light spectroscopies, and electron microscopy. Assembling this body of data and making comparisons between systems are challenging, complex, and controversial problems. However, it may indicate where valid generalizations can be made. Table 5 displays data on systems where diffraction and/or angular orientation data have been reported.

A film pressure vs. area isotherm indicates the average area occupied per molecule in a monolayer film. A value is frequently reported for the molecular area which corresponds to the first steep rise in film pressure (Fig. 12) where two-dimensional solid or liquid films are presumed to become continuous. The area per molecule corresponding to a feature of the isotherm where an

Table 5 Phthalocyanine L-B Film Characterization Data

References	Phthalocyanine compound	Transfer promoter	Spreading solvent [mixture]	Area per molecule (Å^2)	Dihedral angle[a]	d-spacing (Å)	
[66]	CuPc(t-C$_4$H$_9$)$_4$		Xylene	96		3.3	19
[67]	H$_2$Pc(t-C$_4$H$_9$)$_4$	Arachidic acid	[Toluene Chloroform Mesitylene]	60	52°	3.34[b]	17.15[b]
[68]	H$_2$Pc(t-C$_4$H$_9$)$_4$	Arachidic acid	[Xylene Trichloroethylene]		80 ± 10°		
[71]	CuPc[CH$_2$NHCH(CH$_3$)$_2$]$_3$		Chloroform	57		3.29	
[78,80]	H$_2$Pc[OC$_6$H$_4$C(CH$_3$)$_2$C$_6$H$_5$]$_4$	Stearyl alcohol	Chloroform	34	80 ± 10°	3.4[b]	
[73]	+SiPc(OCH$_3$)$_4$(OC$_8$H$_{17}$)$_4$O+ polymer P_w = 15		Chloroform	67		3.4	19.6

[a] Angle between phthalocyanine ring plane and monolayer plane.
[b] X-Ray diffraction data on powdered samples.

Stack Axis
Orientation

Phthalocyanine Ring – Monolayer Plane Dihedral Angle

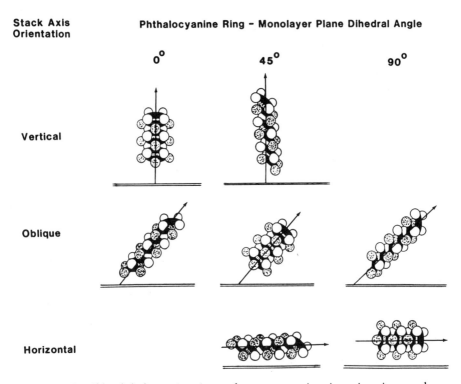

Figure 14 Possible phthalocyanine ring and aggregate axis orientations in monolayers.

orientation effect is presumed to occur (e.g., a shoulder) is also occasionally noted. The reported value should be bracketed by the minimum and maximum cross-sectional areas of the molecule. The breadth and thickness of the phthalocyanine ring are closely approximated by 10 and 3.5 Å, respectively. Additions of peripheral groups require an addition of 2–5 Å to the breadth in correlation with the substituent group size. Thus, the molecular area may range from 42 Å^2 for a very small phthalocyanine standing on edge to 225 Å^2 for a large phthalocyanine lying flat in the monolayer plane. Values from film pressure vs. area isotherms that are less than the minimum calculated molecular area are indicative of molecular layers that are more than one molecule thick. Most phthalocyanine compounds have relatively small molecular areas in Langmuir films (Table 5) indicating that they are standing on edge and/or are in layers more than one molecule thick. The interpretation is complicated by solvent effects and the tendency of phthalocyanines to aggregate. Chloroform as a spreading solvent generally yields films with smaller molecular areas than aromatic solvents such as xylene. For example, the molecular area of CuPc(t-C$_4$H$_9$)$_4$ in a film spread from chloroform is much less than half that in a film spread from xylene [79]. Phthalocyanine aggregation as cofacially oriented stacks is both concentration and solvent dependent. A rapidly evaporating solvent such as chloroform may deposit a collection of rigid aggregates on the water surface, and, after film compression, some of the aggregate stacks may have phthalocyanine molecular units forced above the single monolayer thick plane, causing an apparent reduction of area occupied by the molecules of the Langmuir film. A slowly evaporating aromatic solvent such as xylene may deposit a plasticized flexible film of aggregate stacks on the water. On compression, these aggregates have sufficient molecular mobility to efficiently pack with the stack axis parallel to the monolayer plane and the film uniformly one molecule thick. However, the Langmuir film molecular area measurement may need to be corrected for residual solvent, especially for those that have attractive π-interactions with the phthalocyanine ring. This is a difficult correction since it would require quantification of any retained solvent in the film on the water surface during isotherm measurement instead of determination after the film transfer.

Some very valuable work has been conducted with a siloxane phthalocyanine oligomer (weight average degree of polymerization equal to 15) which may also serve as a model for large nondissociating stacked aggregates [73] (Table 5). Further, for this type of structure the polymer chain is linear (Si—O—Si bond angle of 180°) and the phthalocyanine rings are perpendicular to the siloxane chain [81]. This model may approximate a rigid rod 510 Å long, lying on its side with the phthalocyanine rings perpendicular to the monolayer plane, as would be consistent with the 67 Å^2 molecular area.

At the other extreme, copper octa(dodecoxymethyl) phthalocyanine [72] and copper tetraoctadecyltetrapyridinoporphyrazinium bromide [82] have been reported as forming L-B films with respective molecular areas of 180 and 150 Å^2. The interpretation is that the macrocycle lies flat within the plane of the monolayer. This demonstrates that peripheral groups can exert a profound influence on L-B film morphology.

Diffraction data, although scarce, appear to be very useful as a general diagnostic for aggregation and for correlation with L-B film molecular area measurements. The *d*-spacing data in the left column of Table 5 corresponds to the interplanar spacing of aggregated cofacial phthalocyanine rings. This 3.3- to 3.4-Å interplanar distance appears to be remarkably independent of the substituent groups and corresponds well with that of the siloxane phthalocyanine oligomer as well. The 17- to 19-Å *d*-spacing has been interpreted by analogy to liquid crystal work as an in-plane phthalocyanine length [67] or an interstack distance [66,73]. These spacings have been used as measures of the phthalocyanine thickness and breadth to calculate a minimum molecular area. For the case where the phthalocyanine ring is perpendicular to the monolayer plane, the agreement with the L-B film molecular area is quite good [73]. For the case where the L-B film area is larger, the diffraction data have been used to calculate a tilt angle [67].

Electron spin resonance spectroscopy is a more direct way of determining the dihedral angle and its distribution probability. A spin probe, usually a copper phthalocyanine analog, is dispersed in a diamagnetic phthalocyanine L-B film during its preparation. The anisotropy of the spectrum is measured as a function of the angle between the external magnetic field and the plane of the L-B film. By comparison with copper phthalocyanine single crystal data [68] or by utilizing the single crystal data to simulate the anisotropy of the L-B film ESR spectrum [80], tilt angles have been calculated. It is remarkable that two independent investigations utilizing very different phthalocyanine substituent groups, transfer promoters, and L-B film preparation procedures obtained the same result (Table 5). The 80° dihedral angle is consistent with the small L-B film molecular areas reported. ESR has also been used to confirm that a tetrapyridinoporphyrazine ring lies flat in an L-B monolayer plane as indicated by the film pressure vs. area isotherm [82].

Infrared and electronic polarized light spectroscopies have also yielded information about phthalocyanine orientation in L-B films. The technique of grazing incidence infrared spectroscopy yields spectra whose band intensities are dependent on the vibrational mode's dipole orientation relative to a metal substrate surface [90]. The phthalocyanine L-B film is supported on a flat metal surface, and the infrared radiation is directed at a low grazing angle to the surface with the electronic field polarized perpendicular to the surface. Bands associated with a dipole parallel to the substrate plane orientation should not be observed. The stretching mode of the N—H bond in the plane of the metal-free phthalocyanine ring has been used to demonstrate that the phthalocyanine rings of the cumylphenoxy-substituted system are oriented at an angle to the L-B film plane [78]. Polarized transmission electronic spectroscopy has been used to determine molecular orientation caused by dipping during the L-B film transfer process. Dichroism has been observed where normal incident light polarized perpendicular to the dipping direction is more strongly absorbed than that polarized parallel to the dipping direction. The $\pi-\pi^*$ transition in the visible spectrum is polarized along the plane of the phthalocyanine ring, thus its absorption intensity is maximized when the electronic field of the incident light is parallel to the phthalocyanine plane. The morphological model consistent with the dichroism observations is one with the phthalocyanine rings standing on edge and aggregated in stacks with the stack axis also in the monolayer plane and preferentially oriented parallel to the dipping direction of the substrate. The observations responsible for this model are as follows. An L-B film of $CuPc(CH_2OC_{12}H_{25})_8$ displayed a dichroic ratio (absorption of light polarized perpendicular to the dipping direction to light polarized parallel to the dipping direction) of 1.9 [72]. An L-B film of the oligomer $[—SiPc(OCH_3)_4(OC_8H_{17})_4—]n$ displayed a dichroic ratio of 2.7 which could be increased to 5.8 by annealing at 140°C for 30 min [73].

Electron microscopy has provided a close look at some phthalocyanine L-B films. An image for a monolayer of copper *t*-butylphthalocyanine has been reported and described as small ordered domains embedded in an amorphous

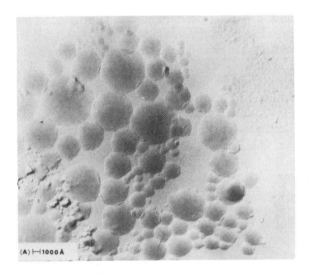

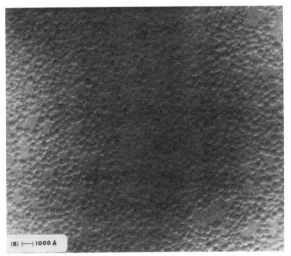

Figure 15 Transmission electron microscopy images of monolayer replicas: (A) cumylphenoxy metal-free phthalocyanine and (B) 1:1 molar mixture of cumylphenoxy metal-free phthalocyanine with stearyl alcohol.

matrix [66]. If compared with a sublimed film image [59], the L-B film appears to be significantly less ordered. An L-B film of tetracumylphenoxy copper phthalocyanine displayed an image consisting of disc-shaped colloidal particles about 5 nm thick and 100–500 nm in diameter [78]. Incorporation of a transfer promoter (e.g., stearyl alcohol) results in a similar but more uniform particle size (Fig. 15). Differential scanning calorimetry has shown that the transfer promoter is phase separated from the disc-shaped colloid [23,78].

Finally, much work is yet to be done concerning the effect of L-B film morphology on absorption and desorption of vapors. A study comparing iodine absorption of a mixed cumylphenoxy phthalocyanine/stearyl alcohol film prepared as an L-B film, a fused L-B film (stearyl alcohol component melted), and a film formed by spraying a solution of the two components has been reported [23]. The sprayed film absorbed more than four times the I_2 absorbed by the L-B film which, in turn, absorbed twice as much as the fused L-B film. However, the conductivity responses were similar. It was suggested that the sprayed film had a looser packing and higher porosity than the L-B film, and that the fused L-B film developed an overcoating from the melted less-dense stearyl alcohol which shielded the phthalocyanine from large amounts of iodine although each film absorbed enough iodine to saturate the conductivity response. To obtain a reasonable understanding of the morphology effect, many more vapors and films need to be studied.

E. RESPONSE OF SENSORS TO GASES AND VAPORS

In this section performance data of phthalocyanine sensors will be summarized. It is not possible to apply strict quantitative parameters for comparison of the studies that appear in the literature due to the different device designs, film application techniques, vapor exposure conditions, and forms of electronic signal measurement. Reflectively, if conventional devices could be identified and conventions for reporting data established, much more quantitative intercomparisons could be made. The performance data will be presented semiquantitatively and qualitatively in terms of (1) sensitivity range, (2) specificity, (3) reversibility, response times, and recovery rates, and (4) stability.

The sensitivity range is the vapor concentration range from which useful detector signals may be obtained. It is tabulated for response to a target vapor as a vapor concentration range and a corresponding detector signal range for various phthalocyanine coatings of a referenced study. These data are not detection limits and will be dependent on the coating film (structure, morphology, thickness, etc.), device design (electrode contact, spacing, cross-sectional area, temperature control), vapor delivery system (sensor housing, dead volume, carrier gas, flow rate), and electronic measurement capability (current measurement sensitivity and precision, bias voltage, frequency measurement).

Specificity relates to the relative magnitude of a response to a particular vapor compared with a second vapor. Its objective is to discriminate between chemically similar vapors or high concentrations of ambient interferent vapors. Response ratios of 1000:1 or higher are desirable. This differentiation is the most challenging problem in chemical sensor chemistry. Specificity is almost totally a function of the coating structure, and much work needs to be done to improve on this aspect. Currently this problem is also being addressed by chemical separation techniques (miniature gas chromatographs and permselective membranes) and computerized pattern recognition schemes for sensor arrays.

Reversibility, response time, and recovery rates are measures of how well the sensor signal corresponds to instantaneous equilibration with the partial pressure of the analyte vapor. In practice, the process of absorption, film diffusion, and desorption may cause a substantially delayed response to vapor concentration changes. This problem centers on coating film morphology, structure, and thickness and has been effectively dealt with by elevated temperature control in the device.

Stability refers to long-term reversibility and reproducibility of sensors with particular attention directed at resistance to the effects of ambients such as light, oxygen, water, ozone, etc. The remarkable thermal oxidative stability of phthalocyanine compounds has been a strong asset for their use in chemical sensors, but at elevated operation temperatures this becomes a concern.

The operation temperature of the sensor also has a large effect on the sensitivity range, selectivity, response times, and stability. The temperature-dependent processes involved in the sensor operation are semiconductor charge transport, vapor absorption, diffusion and desorption, and charge transfer interactions, as well as reaction and degradation of the phthalo-cyanine film. For sublimed films, reversibility and response times attained at elevated temperatures are the primary considerations. The signal sensitivity frequently passes through a maximum with increasing temperature, reflecting an increasing conductivity but a decreasing partitioning of the vapor into the film. Phthalocyanine–vapor reactions or film evaporation place an upper limit on the temperature. The operating sensor temperature is empirically determined, and its effects need to be considered when making comparisons.

i. Sensitivity

Both chemiresistor and SAW phthalocyanine sensors have been studied for NO_2 detection. Signal sensitivity ranges for corresponding NO_2 vapor concentration ranges are tabulated in Table 6. Most data were obtained at elevated temperatures to optimize sensitivity and reduce response times. These data are purposefully displayed in a nonnormalized form (i.e., not in terms of sensitivity per unit vapor concentration, film thickness, or device geometry) because the physical correlation is clearer with directly measured quantities as opposed to calculated sensitivity parameters which have arbitrary correlation with vapor concentrations and may be skewed by selection of an electronic baseline. Also, in many cases, inadequate knowledge of device and film characteristics prevents reduction to an intrinsic fundamental quantity. The

Table 6 Concentration and Signal Range Sensitivity Data[a] of Phthalocyanine Chemiresistor and SAW Device Sensors for Nitrogen Dioxide

Reference	Phthalocyanine	Temperature[b] (°C)	Concentration range (min→max)	Signal range (min→max)
Chemiresistor				
[35]	CoPc		$2.8 \rightarrow 44$ ppb	$1.0 \times 10^{-7} \rightarrow 1.8 \times 10^{-7}\ \Omega^{-1}$
	PbPc		$2.8 \rightarrow 44$ ppb	$3.5 \times 10^{-8} \rightarrow 1.8 \times 10^{-7}\ \Omega^{-1}$
	CuPc	170	$2.8 \rightarrow 44$ ppb	$2.4 \times 10^{-8} \rightarrow 8.4 \times 10^{-8}\ \Omega^{-1}$
	ZnPc		$2.8 \rightarrow 44$ ppb	$8.0 \times 10^{-9} \rightarrow 3.2 \times 10^{-8}\ \Omega^{-1}$
	NiPc		$2.8 \rightarrow 44$ ppb	$4.2 \times 10^{-9} \rightarrow 8.6 \times 10^{-9}\ \Omega^{-1}$
	H$_2$Pc		50 ppb	$10^{-10}\ \Omega^{-1}$
[34]	PbPc	170	$17 \rightarrow 320$ ppb	$4.6 \times 10^{-5} \rightarrow 1.6 \times 10^{-4}$ A
[8]	CuPc	147	$10 \rightarrow 10^{4}$ ppm	$10^{15} \rightarrow 10^{9}\ \Omega/\square$
[11]	CuPc		$10 \rightarrow 100$ ppm	$2.0 \times 10^{-8} \rightarrow 1.6 \times 10^{-7}$ A
[12]	H$_2$Pc		$10 \rightarrow 100$ ppm	$1.7 \times 10^{-8} \rightarrow 2.8 \times 10^{-7}$ A
[18]	CuPc[CH$_2$NHCH(CH$_3$)$_2$]$_3$		$4 \rightarrow 12$ ppm	$2 \times 10^{-10} \rightarrow 6 \times 10^{-10}$ A
[83]	(AlPcF)$_n$	120	$20 \rightarrow 200$ ppm	$0.015 \rightarrow 0.042\ (\Omega\text{-cm})^{-1}$
SAW device				
[40]	CoPc		$40 \rightarrow 160$ ppm	$-1.6\ \rightarrow -3.9$ kHz
	CuPc		$40 \rightarrow 200$ ppm	$-0.96 \rightarrow -2.64$ kHz
	FePc	150	$40 \rightarrow 200$ ppm	$-0.78 \rightarrow -2.08$ kHz
	MgPc		$40 \rightarrow 200$ ppm	$-0.76 \rightarrow -1.52$ kHz
	NiPc		$40 \rightarrow 200$ ppm	$-0.72 \rightarrow -1.08$ kHz
	H$_2$Pc		$40 \rightarrow 200$ ppm	$-0.42 \rightarrow -1.12$ kHz
[39]	PbPc	80	10 ppm	-0.16 kHz

[a] Data are uncorrected for coating film thickness.
[b] If not designated, room temperature is assumed.

noticeable features of Table 6 are that the chemiresistor responds to lower NO_2 concentrations than does the SAW device, and that, within the unsubstituted phthalocyanine series, cobalt, lead, and copper phthalocyanine films have the largest signal sensitivity ranges, while nickel and metal-free phthalocyanine films have the smaller ranges. The concentration dependence of the chemiresistor is linear [18,34,35] while that of the SAW device approaches saturation [40] reflecting the higher NO_2 concentration range sampled. Temperatures of 150–170°C optimize sensitivity for both the chemiresistor [35] and SAW [40] measurements. Note that the sensitivity range difference between copper and metal-free phthalocyanine drops markedly at room temperature [11,12].

Like NO_2, halogens are also strong electron acceptors, and phthalocyanine sensors respond to them similarly. Concentration and signal range sensitivity data are presented in Table 7. The chemiresistor is more sensitive than the SAW measurement and extends into the sub-ppm concentration range. Phthalocyanine metal ion effects on sensitivity are most apparent at low vapor concentrations as perceived in the NO_2 data. The halogen data of Table 7 are not extensive enough to observe this effect. Some relative sensitivity data to 0.25 ppm of fluorine showing a metal ion effect has been reported with copper, lead, and iron phthalocyanine displaying the strongest response [35].

Sensitivity toward ammonia is much less than toward NO_2 or the halogens. Phthalocyanine sensor data for ammonia are displayed in Table 8. The direction of the conductance change for sublimed films is a decrease. Reference [36] describes development of a sensor based on this conductivity decrease for atmospheric ammonia. The conductivity decrease is measured as the change caused by exposure of a vacuum-regenerated copper phthalocyanine film to ammonia in air and is linearly related to the ammonia partial pressure over a 0.02–0.20 Torr range. The conductance change is in the increasing direction for the tetracumylphenoxy copper phthalocyanine Langmuir–Blodgett film and has a larger sensitivity range [17]. There is also a non-linear dependence on ammonia concentration [26]. At a low concentration range (3 ppm), specific metal ion effects for the cumylphenoxy phthalocyanine L-B film series on ammonia sensitivity as well as for other vapors are observed (see Table 9). The response of the SAW device is quite small, and the frequency shift is in a direction opposite to that expected for a mass increase of the film [40]. The nature of this response is not currently understood.

ii. Specificity

Specificity is directed toward strong electron-acceptor gases. Vapors with negligible response include nitrogen, hydrogen, carbon monoxide, carbon dioxide, methane, nitrous oxide, sulfur dioxide, and benzene [7,11,12,19,34,40], which would be expected since they are neither strong electron acceptors nor donors. Moderate electron acceptors (oxygen, nitric

Table 7 Concentration and Signal Range Sensitivity Data[a] of Phthalocyanine Chemiresistor and SAW Device Sensors for Halogens

Reference	Phthalocyanine	Temperature[b] (°C)	Concentration range (min→max)	Signal range (min→max)
Chemiresistor				
[34]	H_2Pc/Cl_2	150	$0.2\rightarrow 0.8$ ppm	$4\times10^{-10}\rightarrow 2\times10^{-9}\ \Omega^{-1}$
[12]	H_2Pc/Cl_2		$10\rightarrow 1000$ ppm	$2.8\times10^{-10}\rightarrow 6.4\times10^{-6}$ A
[11]	$CuPc/I_2$		$0\rightarrow 400$ ppm	$10^{-12}\rightarrow \quad 10^{-4}\ \Omega^{-1}$
[23]	H_2	$MPc[OC_6H_4C(CH_3)_2C_6H_5]_4$ $+I_2$	$0\rightarrow 400$ ppm	$6\times10^{-10}\rightarrow 2\times10^{-6}$ A
	Cu		$0\rightarrow 400$ ppm	$2\times10^{-10}\rightarrow 2\times10^{-6}$ A
	Zn		$0\rightarrow 400$ ppm	$8\times10^{-11}\rightarrow 1\times10^{-6}$ A
	Pt		$0\rightarrow 400$ ppm	$6\times10^{-12}\rightarrow 1\times10^{-6}$ A
	Pd		$0\rightarrow 400$ ppm	$1\times10^{-10}\rightarrow 5\times10^{-7}$ A
	Co		$0\rightarrow 400$ ppm	$3\times10^{-10}\rightarrow 4\times10^{-7}$ A
	Ni		$0\rightarrow 400$ ppm	$3\times10^{-11}\rightarrow 1\times10^{-7}$ A
SAW Device				
[85]	$CuPc/Cl_2$	150	50 ppm	-6.10 kHz
	$CuPc/Br_2$		50 ppm	-0.70 kHz
	$CuPc/I_2$		13 ppm	-0.16 kHz
[23]	H_2	$MPc[OC_6H_4C(CH_3)_2C_6H_5]_4$ $+I_2$	400 ppm	-11.8 kHz
	Cu		400 ppm	-11.3 kHz
	Zn		400 ppm	-12.1 kHz
	Pt		400 ppm	-13.1 kHz
	Pd		400 ppm	-14.4 kHz
	Co		400 ppm	-15.1 kHz
	Ni		400 ppm	-14.4 kHz

[a] Data are uncorrected for coating film thickness.
[b] If not designated, room temperature is assumed.

Table 8 Concentration and Signal Range Sensitivity Data[a] of Phthalocyanine Chemiresistor and SAW Device Sensors for Ammonia

Reference	Phthalocyanine	Temperature[b] (°C)	Concentration range (min→max)	Signal range (min→max)
Chemiresistor				
[6]	H_2Pc	50	$0\to$ 184 Torr	$10^{-12}\to10^{-14}$ A
[36]	CuPc		$26\to$ 260 ppm	$8.5\times10^{-8}\to6.5\times10^{-8}$ A
[17]	CuPc		$0\to \leqslant500$ ppm	$2.7\times10^{-11}\to1.1\times10^{-11}$ A
	$CuPc[OC_6H_4C(CH_3)_2C_6H_5]_4$		$0\to \leqslant500$ ppm	$6.0\times10^{-12}\to6.7\times10^{-11}$ A
SAW device				
[40]	H_2Pc	150	200 ppm	+ 0.00 kHz
	MgPc		200 ppm	+ 0.15 kHz
	FePc		200 ppm	+ 0.52 kHz
	CoPc		200 ppm	+ 0.05 kHz
	NiPc		200 ppm	+ 0.22 kHz
	CuPc		200 ppm	+ 0.34 kHz
	PbPc		200 ppm	+ 1.36 kHz

[a] Data are uncorrected for coating film thickness.
[b] If not designated, room temperature is assumed.

Table 9 Maximum Percentage Change in Conductance of L-B Cumylphenoxy Phthalocyanine Films due to Vapor Exposure[a]

Central atom	NH$_3$ (3 ppm)	(CH$_3$O)$_2$CH$_3$PO (2 ppm)	SO$_2$ (8 ppm)
H$_2$	18	14	26
Co	−1.3	0.5	−4.3
Ni	185	128	13
Pd	63	27	15
Pt	163	69	40
Cu	137	16	0
Zn	−8.1	−2.2	−2.6
Pb	0	0	0

[a] From ref. [19].

oxide, hydrogen chloride) and donors (water, ammonia, hydrogen sulfide, dimethyl methyl phosphonate) cause moderate responses in the sensors.

Because of its ubiquity, oxygen deserves special attention. As described in Sections B and D, oxygen's effect on conductivity is small and morphology dependent. However, it may be strongly absorbed [8]. Indeed, mass spectra of unsubstituted phthalocyanine compounds display a parent ion peak plus 16 amu as evidence of this [84]. A strong argument has been made that absorbed oxygen must be displaced by incoming vapors to observe conductivity changes [8,9]. Thus, the relative amounts of weakly and strongly bound oxygen, sensor temperature, carrier gas, and displacement strength of the incoming vapor are important factors.

Nitric oxide compared with nitrogen dioxide is a weaker electron acceptor. Both chemiresistor and SAW responses are smaller by factors of 1000 and 10, respectively [7,12,85].

Hydrogen chloride is reported to have a weak (<0.1%) conductivity effect on copper and metal-free phthalocyanine films compared with nitrogen dioxide [11,12]. Interestingly, a magnesium phthalocyanine film is reported to have a uniquely strong response to HCl and be almost completely unaffected by other electron-acceptor gases [35].

Water vapor is a ubiquitous, moderately weak electron donor interferent. Its direct effect on phthalocyanine sensors is exhibited as a negligible [34,40] or a very small conductivity decrease [6,17,86]. However, water vapor has been reported to interfere with nitrogen dioxide detection by either reaction with it to form nitric acid or by blocking phthalocyanine film absorption sites [9]. Data have been reported for a lead phthalocyanine film exposed to 10–300 ppb of nitrogen dioxide in an air carrier gas with humidity adjusted to 0, 12, and 33%. The humidity progressively reduced nitrogen dioxide sensitivity, and saturation of the sensor response to nitrogen dioxide above 100 ppb was observed.

Interference from ammonia and hydrogen sulfide in detection of nitrogen dioxide occurs during and subsequent to nitrogen dioxide exposure. Direct exposure effects of ammonia are relatively small compared with those of nitrogen dioxide. Simultaneous and subsequent exposure to ammonia and hydrogen sulfide results in partial compensation of the conductance increase caused by nitrogen dioxide [11,12]. However, the nitrogen dioxide effects are much stronger, and it has been suggested that sublimed metal-free and copper phthalocyanine films may be used to detect 10 ppm levels of nitrogen dioxide in the presence of 100 ppm of ammonia or hydrogen sulfide [7].

Dimethyl methylphosphonate, $(CH_3O)_2CH_3PO$, has been of some interest as an organophosphorus anticholinesterase agent stimulant and may be considered as an example of a moderate electron donor. A study using L-B phthalocyanine films for its detection indicated that ammonia is a strong interferent as indicated in Table 9 [19]. In the context of strong nonbonded electron donors, it was pointed out in Section D that such compounds (pyridine, methylamine, dimethyl sulfoxide, dioxane, etc.) form stoichiometric complexes with phthalocyanines [61]. Thus, it is reasonable to expect general perturbations from strong electron, donor vapors, and these interactions are deserving of more study.

iii. Reversibility and Response Times

The reversibility of a sensor depends on the strength of the interaction between the vapor and the film. Nitrogen dioxide is strongly absorbed by phthalocyanine films, and elevated temperatures (150–170°C) are required for desorption [8,34,35,40]. The nitrogen dioxide interaction is weaker with peripherally substituted phthalocyanines in L-B films, and sensors may be operated at room temperature [18]. For ammonia vapor, a similar situation of sublimed film [36] vs. L-B film [17,19] exists. The degree of reversibility as well as response time parameters may be characterized by modeling the response and recovery process of the film. Empirically, it has been found that a second-order rate law is one mathematical expression that can be used to represent the data for SAW sensor response and recovery [87].

$$\frac{dC}{dt} = kC^2$$

Absorption:

$$C = C_s - \frac{1}{(1/C_s) + k_1 t}$$

Desorption:

$$C = \frac{1}{(1/C_s) + k_2 t}$$

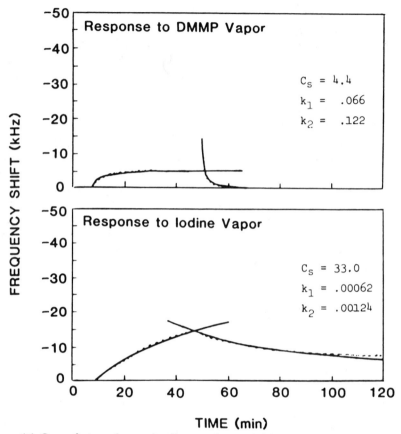

Figure 16 Curve fitting of second-order rate law (solid line) to experimental SAW sensor data (dotted line) of (top) cumylphenoxy nickel phthalocyanine L-B film responding to dimethyl methylphosphonate vapor and of (bottom) cumylphenoxy metal-free phthalocyanine L-B film responding to iodine vapor illustrating good and poor reversibility, respectively.

where

C	= mass of absorbed vapor
C_s	= mass of absorbed vapor at saturation
k_1, k_2	= rate constants
t	= time

Values for C_s, k_1, and k_2 are determined by fitting curves of the above forms to the experimental data by the method of least squares. Curve fitting for the exposure of cumylphenoxy phthalocyanine L-B films to dimethyl methylphosphonate and iodine are presented in Fig. 16 as examples of good and poor reversibility. The rate constants are indices of response and recovery rates, and the difference between them is a measure of the degree of reversibility. It should be emphasized that these particular rate constants are dependent on the measurement conditions (film and device characteristics, dead volume, temperature, vapor concentration, etc.) and the second-order equation is empirical. Response and recovery rates are very temperature and film thickness dependent. For nitrogen dioxide detection an increase in temperature from 120 to 170°C is reported to reduce the 90% response time from 10 to 1.5 min [34], and a reduction of film thickness from 2500 to 300 Å is reported to reduce the response time by about a factor of 4 [83].

iv. Stability

Susceptibility toward irreversible chemical or morphological change in a phthalocyanine film is a concern for the sensor application. Sublimed phthalocyanine films operate at high temperatures (150–170°C) for nitrogen dioxide detection. Evidence has been reported that at these conditions nitrogen oxide species are partially retained by way of aromatic nitration [12]. It has also been reported that halogen exposure has a destructive effect on nitrogen dioxide sensitivity [85]. Irreversible and/or degradative reactions from contact with halogens are known with regard to chlorine [91], bromine [91], and iodine [92] and become more severe with increasing halogen reactivity. Fluorine exposure would be expected to be extremely degradative to the phthalocyanine film. Lead phthalocyanine should be regarded with caution as it is readily demetallated by acids and slowly precipitates from basic organic solvents (DMSO, THF) as metal-free phthalocyanine. Scanning electron micrographs of a lead phthalocyanine used for detecting nitrogen dioxide showed degradation [40]. Six-week prolonged heating experiments on a copper phthalocyanine film showed large fluctuations in sensitivity and response time to nitrogen dioxide [85]. Similar and more accentuated effects would be expected for halogen exposure. Sublimation losses at temperatures higher than 150°C are also a concern [40]. Morphological changes may be caused by vapors that are strong nonbonded electron donors [61] or by iodine [88]. Such changes may alter absorption sites for target vapors or charge transport properties of the film. Amorphous films may offer a way of circumventing crystalline morphology problems.

F. VARIATIONS ON PHTHALOCYANINE STRUCTURE

In the chemiresistor sensor, the phthalocyanine structure functions as a molecular receptor of vapors whose film conductivity is modulated by electron transfer and/or morphological interactions. In this section the phthalocyanine performance is considered for the effect of structural variations and in comparison with other conjugated metallomacrocyclic systems.

As a sublimed film, the peripherally unsubstituted phthalocyanines respond to strong electron acceptors (nitrogen dioxide, chlorine, iodine) with large increases in conductivity and to electron donors (ammonia, hydrogen sulfide, dimethyl methylphosphonate) with small decreases in conductivity. The reversibility is very poor at room temperature, but acceptable at elevated temperatures (150–170°C). The effects of varying the complexed metal ion may be observed at very low vapor concentrations.

Peripheral substitution with alkyl or alkoxy groups and preparation as L-B films result in sensors that have similar baseline conductivities and smaller responses to strong electron acceptor vapors [18,23]. The responses to electron donor vapors are larger [17]. A particularly important difference is rapid reversibility at room temperature. Also, complexed metal ion effects may be observed at low vapor concentration. Largely unexplored and of particularly important consequence are the effects of strongly electron-donating and -withdrawing peripheral substituents. Since electron transfer interactions are involved, these contributions could be important in tuning a sensor's response.

Axial substitution with covalently bound ligands has not been investigated. The fluorine-bridged aluminum phthalocyanine polymer, $(AlPcF)_n$, has been investigated and found to behave similar to the sublimed films with respect to nitrogen dioxide response [83].

For comparison with related conjugated metallomacrocyclic systems, Honeybourne and co-workers have done some very fine work investigating the systems illustrated in Fig. 17 [7,11,12,24]. In addition to the phthalocyanine system (I) shown in the figure are the hemiporphyrazine (II), macrocyclized bis-1,10-phenanthroline (III), dihydrodibenzotetraaza-[14]-annulene (IV), and porphyrin (V) systems. Systems I and V are macroring 18 π-electron systems while systems II, III, and IV have cyclic 6 π-electron subsystems within the conjugated contour. Nitrogen dioxide sensitivity data and comments on reversibility are entered in Table 10 for 100 and 1000 ppm exposures followed by evacuation at room temperature. The phthalocyanine system has a uniquely high response but poor reversibility. Systems II and III displayed no response. System IV had a fast response and good reversibility. Within this system, functionalization with electron-withdrawing and -donating groups was found to trade off sensitivity for reversibility. The nitro substitution reduced sensitivity but enhanced reversibility, while methyl substitution had the contrary effect. The meso-tetraphenyl porphyrin system responds poorly to high nitrogen dioxide partial pressures with poor reversibility. This sytem differs from the others in that the meso-phenyl groups are twisted out of the porphyrin ring plane, thus increasing the intermolecular porphyrin ring spacing and reducing crystallinity. An L-B film of a more planar methyl-substituted porphyrin was reported to increase conductivity by a factor of 10,000 on exposure to 10 ppm nitrogen dioxide [89].

Much structure–property work remains to be done to obtain a better

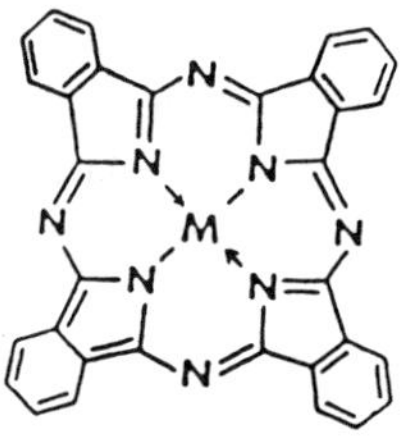

Ia M = H$_2$
Ib M = Cu

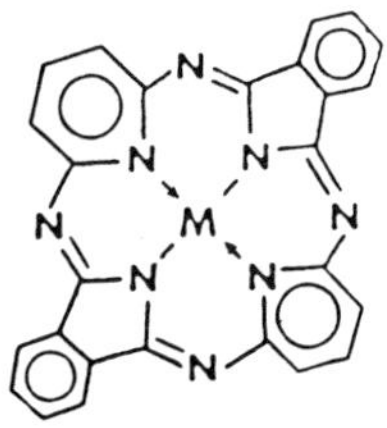

IIa M = H$_2$
IIb M = Cu

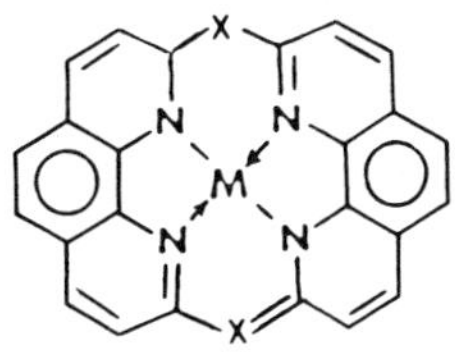

III X = N, CH

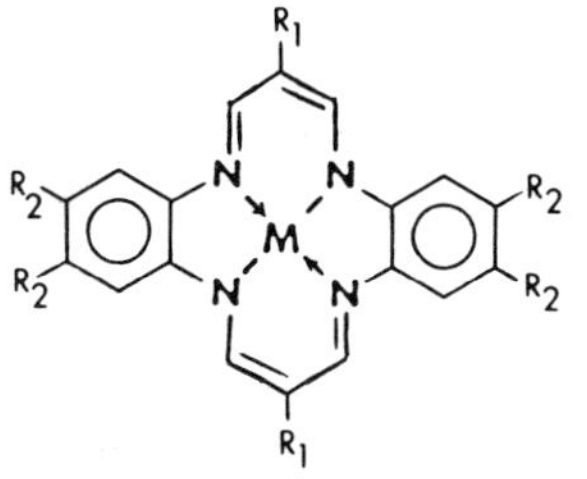

IVa M = H$_2$, R$_1$ = R$_2$ = H
IVb M = H$_2$, R$_1$ = NO$_2$, R$_2$ = CH$_3$
IVc M = H$_2$, R$_1$ = H, R$_2$ = CH$_3$
IVd M = H$_2$, R$_1$ = CO$_2$Et, R$_2$ = CH$_3$
IVe M = Cu, R$_1$ = R$_2$ = H
IVf M = Co, R$_1$ = H, R$_2$ = CH$_3$

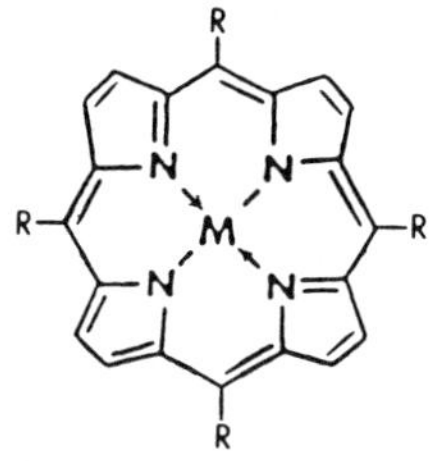

Va M = H$_2$, R = C$_6$H$_5$
Vb M = H$_2$, R = C$_6$H$_4$CH$_3$
Vc M = Pt, R = C$_6$H$_4$OCH$_3$

Figure 17 Structures of comparative conjugated metallomacrocyclic systems investigated as sensor coating films for nitrogen dioxide.

Table 10 Sensitivity and Reversibility of Conjugated Macrocyclic Compound Films to Nitrogen Dioxide[a]

	Response to NO$_2$ (A)		Reversibility by evacuation
Compound	100 ppm	1000 ppm	
Ia	$10^{-6.5}$	$10^{-3.2}$	Very poor
Ib	$10^{-6.3}$	$10^{-3.6}$	Very poor
IIa	No response		
IIb	No response		
III	No response		
IVa	$10^{-10.7}$	$10^{-8.9}$	Good
IVb	10^{-11}	10^{-9}	Good (>IVa)
IVc	$10^{-9.1}$	$10^{-8.7}$	Good (<IVa)
IVd	$10^{-8.3}$		
Va	10^{-10}	$10^{-9.3}$	Poor
Vb	$10^{-10.6}$	$10^{-9.1}$	Poor
Vc	$10^{-10.4}$	10^{-9}	Poor

[a] From refs. [7,11,12,24].

understanding and a better balance of sensitivity and reversibility for phthalocyanines and related systems. The consideration of specificity is a broader problem to be approached by more intricate chemical structures, sophisticated sensor measurements, and multisensor analysis techniques.

REFERENCES

1. H. Wohltjen, *Anal. Chem.*, 56 (1984) 87A.
2. J. Kaufhold and K. Hauffe, *Ber. Bunsenges. Phys. Chem.*, 69 (1965) 168.
3. A. T. Vartanyan, *Zh. Fiz. Khim.*, 22 (1948) 769; and D. D. Eley, *Nature (London)*, 162 (1948) 819.
4. G. H. Heilmeier and S. E. Harrison, *Phys. Rev.*, 132 (1963) 2010.
5. J. Simon and J. J. Andre, *Molecular Semiconductors*, Springer-Verlag, Berlin, 1985, pp. 112–116.
6. G. J. van Oirschot, D. van Leeuwen and J. Medema, *J. Electroanal. Chem.*, 37 (1972) 373.
7. C. A. Honeybourne, R. J. Ewen and C. A. S. Hill, *J. Chem. Soc. Faraday Trans. I*, 80 (1984) 851.
8. J. D. Wright, A. W. Chadwick, B. Meadows and J. J. Miasik, *Mol. Cryst. Liquid Cryst.*, 93 (1983) 315.
9. A. V. Chadwick, P. B. M. Dunning and J. D. Wright, *Mol. Cryst. Liquid Cryst.*, 134 (1986) 137.
10. R. L. van Ewyk, A. W. Chadwick and J. D. Wright, *J. Chem. Soc. Faraday Trans. I.*, 76 (1980) 2194.
11. C. L. Honeybourne and R. J. Ewen, *J. Phys. Chem. Solids*, 44 (1983) 833.
12. C. L. Honeybourne and R. J. Ewen, *J. Phys. Chem. Solids*, 44 (1983) 215.
13. J. J. Miasik, A. Hooper and B. C. Tofield, *J. Chem. Soc. Faraday Trans. I*, 82 (1986) 1117.
14. H. Lars and G. Heiland, *Thin Solid Films*, 149 (1987) 129.
15. P. S. Vincett, Z. D. Popovic and L. McIntyre, *Thin Solid Films*, 82 (1981) 357.
16. S. Baker, M. C. Petty, G. G. Roberts and M. V. Twigg, *Thin Solid Films*, 99 (1983) 53.
17. W. R. Barger, H. Wohltjen, A. W. Snow, J. Lint and N. L. Jarvis, in *Fundamentals and Applications of Chemical Sensors*, D. Schuetzle and R. Hammerle, Eds., American Chemical Society, Washington, D.C., 1986, p. 155.
18. S. Baker, G. G. Roberts and M. C. Petty, *IEEE Proc., Part I: Solid-State Electron Devices*, 130 (1983) 260.
19. W. R. Barger, H. Wohltjen and A. W. Snow, in *Transducers '85, IEEE*, New York, 1985, p. 410 (IEEE No. 85CH2127-9).
20. F. W. Kutzler, W. R. Barger, A. W. Snow and H. Wohltjen, *Thin Solid Films*, 155 (1987) 1.
21. T. J. Marks, *Science*, 227 (1985) 881.
22. A. W. Snow and N. L. Jarvis, *J. Am. Chem. Soc.*, 106 (1984) 4706.
23. A. W. Snow, W. R. Barger, M. Klusty, H. Wohltjen and N. L. Jarvis, *Langmuir*, 2 (1986) 513.
24. C. L. Honeybourne, J. D. Houghton, R. J. Ewen and C. A. S. Hill, *J. Chem. Soc. Faraday Trans. I.*, 82 (1986) 1127.
25. G. Tollin, D. R. Kearns and M. Calvin, *J. Chem. Phys.*, 32 (1960) 1013.
26. H. Wohltjen, W. R. Barger, A. W. Snow and N. L. Jarvis, *IEEE Trans. Electron Devices*, ED-32 (1985) 1170.

27. P. Mark and W. Helfrich, *J. Appl. Phys.*, 33 (1962) 205; F. Gutmann and L. E. Lyons, *Organic Semiconductors*, Wiley, New York, 1967, p. 373.

28. A. Sussman, *J. Appl. Phys.*, 38 (1967) 2738.

29. C. Hamann, *Phys. Status Solidi*, 4 (1964) K97.

30. R. D. Gould, *Thin Solid Films*, 125 (1985) 63.

31. G. Heiland and D. Kohl, *Sensors Actuators*, 8 (1985) 227.

32. (a) Microsensor Systems Inc., Fairfax, Va.; (b) Platfilm Ltd., Bognor Regis, U.K.

33. J. Mizuguchi, *Jpn. J. Appl. Phys.*, 20 (1981) 713.

34. B. Bott and T. A. Jones, *Sensors Actuators*, 5 (1984) 43.

35. T. A. Jones and B. Bott, *Sensors Actuators*, 9 (1986) 27.

36. A. Szczurek and K. Lorenz, *Int. J. Environ. Anal. Chem.*, 23 (1986) 161.

37. H. Wohltjen and R. E. Dessy, *Anal. Chem.*, 51 (1979) 1458.

38. H. Wohltjen, *Sensors Actuators*, 5 (1984) 307.

39. A. J. Ricco, S. J. Martin and T. E. Zipperian, *Sensors Actuators*, 8 (1985) 319.

40. M. S. Nieuwenhuizen, A. J. Nederlof and A. W. Barendsz, *Anal. Chem.*, 60 (1988) 230.

41. M. S. Mindorff and D. E. Brodie, *Can. J. Phys.*, 59 (1981) 249.

42. F. W. Karasek and J. C. Decius, *J. Am. Chem. Soc.*, 74 (1952) 4716.

43. J. F. Boas, P. E. Fielding and A. G. MacKay, *Aust. J. Chem.*, 27 (1974) 7.

44. M. Ashida, N. Uyeda and E. Suito, *J. Cryst. Growth*, 8 (1971) 45.

45. Y. Sadaoka, N. Yamazoe and T. Seiyama, *Denki Kagaku, Oyobi Kogyo Butsuri Kagaku*, 46 (1978) 597. (*Chem. Abstr.*, 90: 113485p).

46. S. E. Harrison and K. H. Ludewig, *J. Chem. Phys.*, 45 (1966) 343.

47. K. J. Beales, D. D. Eley, D. J. Hazeldine and T. F. Palmer, in *Katalyse an Phthalocyaninen*, H. Kropf and F. Steinbach, Eds., Georg Thieme Verlag, Stuttgart, 1973, pp. 1–32.

48. Y. Sakai, Y. Sadaoka and H. Yokouchi, *Bull. Chem. Soc. Jpn.*, 47 (1974) 1886.

49. F. Iwatsu, T. Kobayashi and N. Uyeda, *J. Phys. Chem.*, 84 (1980) 3223.

50. C. J. Brown, *J. Chem. Soc.*, (1968) 2488.

51. M. Ashida, N. Uyeda and E. Suito, *Bull. Chem. Soc. Jpn.*, 39 (1966) 2616.

52. J. H. Beynon and A. R. Humphries, *Trans. Faraday Soc.*, 51 (1955) 1065.

53. J. H. Sharp and M. Lardon, *J. Phys. Chem.*, 72 (1968) 3230.

54. A. A. Ebert, Jr. and H. B. Gottleib, *J. Am. Chem. Soc.*, 74 (1952) 2806.

55. J. M. Assour, *J. Chem. Phys.*, 69 (1965) 2295.

56. C. J. Brown, *J. Chem. Soc.*, (1968) 2494.

57. K. Ukei, *Acta Cryst.*, B29 (1973) 2290.

58. M. Ashida, *Bull. Chem. Soc. Jpn.*, 39 (1966) 2625; 2632.

59. T. Kobayashi, Y. Fujiyoshi, F. Iwatsu and N. Uyeda, *Acta Cryst.*, A37 (1981) 692.

60. P. S. Vincett, Z. D. Popovic and L. McIntyre, *Thin Solid Films*, 82 (1981) 357.

61. T. Kobayashi, N. Uyeda and E. Suito, *J. Phys. Chem.*, 72 (1968) 2446.

62. C. J. Schramm, R. P. Scaringe, D. J. Stojackovic, B. M. Hoffman, J. A. Ibers and T. J. Marks, *J. Am. Chem. Soc.*, 102 (1980) 6702.

63. G. L. Gaines, *Thin Solid Films*, 99 (1983) ix–xii.

64. A. E. Alexander, *J. Chem. Soc.*, (1937) 1813.

65. G. L. Gaines, *Insoluble Monolayers at Gas-Liquid Interfaces*, Interscience, New York, 1966.

66. J. R. Fryer, R. A. Hahn and B. L. Eyres, *Nature (London)*, 313 (1985) 382.

67. G. J. Kovacs, P. S. Vincett and J. H. Sharp, *Can. J. Phys.*, 63 (1985) 346.

68. M. J. Cook, M. F. Daniel, A. J. Dunn, A. A. Gold and A. J. Thompson, *J. Chem. Soc., Chem. Commun.*, (1986) 863.

69. A. W. Snow, W. R. Barger, P. Berg and M. Klusty, *Langmuir*, submitted.

70. W. R. Barger, A. W. Snow, H. Wohltjen and N. L. Jarvis, *Thin Solid Films*, 133 (1985) 197.

71. G. G. Roberts, M. C. Petty, S. Baker, M. T. Fowler and N. J. Thomas, *Thin Solid Films*, 132 (1985) 113.

72. D. W. Kalina and S. W. Crane, *Thin Solid Films*, 134 (1985) 109.

73. E. O. Orthmann and G. Wegner, *Angew. Chem. Int. Ed.*, 25 (1986) 1105.

74. M. J. Cook, M. F. Daniel, K. J. Harrison, N. B. McKeown and A. J. Thompson, *J. Chem. Soc. Chem. Commun.*, (1987) 1148.

75. E. Schnabel, H. Nother and H. Kuhn, in *Chemistry of Natural and Synthetic Colouring Matters*, T. S. Gore, B. S. Joshi, S. V. Sunthanker and B. D. Tilak, Eds., Academic Press, New York, 1962, p. 561.

76. Z. A. Schelly, D. H. Haward, P. Hemmes and E. M. Eyring, *J. Phys. Chem.*, 74 (1970) 3040.

77. R. C. Graham, E. M. Eyring and G. H. Henderson, *J. Chem. Soc. Perkin II*, (1981) 765.

78. W. R. Barger, J. Dote, M. Klusty, R. Mowery, R. Price and A. W. Snow, *Thin Solid Films*, 159 (1988) 369.

79. S. Baker, in *Proc. Intl. Symp. on Future Electron Devices, Bioelectronic and Molecular Electronic Devices*, Research and Development Assn. for Future Electron Devices, Tokyo, 1985, pp. 53–58.

80. M. D. Pace, W. R. Barger and A. W. Snow, *J. Mag. Res.*, 75 (1987) 73.

81. W. J. Kroenke, L. E. Sutton, R. D. Joyner and M. E. Kenney, *Inorg. Chem.*, 2 (1963) 1064; J. R. Mooney, C. K. Choy, K. Knox and M. E. Kenney, *J. Am. Chem. Soc.*, 97 (1975) 3033.

82. S. Palacin, A. Ruaudel-Teixier and A. Barraud, *J. Phys. Chem.*, 90 (1986) 6237.

83. M. Dugay and C. Maleysson, *Synthetic Metals*, 21 (1987) 255.

84. R. B. Freas and J. E. Campana, *Inorg. Chem.*, 23 (1984) 4654.

85. M. S. Nieuwenhuizen and A. J. Nederlof, *Anal. Chem.*, 60 (1988) 236.

86. V. F. Kiselev, V. V. Kurylev and N. L. Levshin, *Phys. Stat. Sol.*, 42 (1977) K61.

87. W. R. Barger, M. A. Klusty, A. W. Snow, J. W. Grate, D. S. Ballantine and H. Wohltjen, *Proc. of the Symposium on Sensor Science and Technology*, Proceedings Volume 87-15, The Electrochemical Society, Inc., Pennington, N.J., 1987, pp. 198–217.

88. T. Kobayashi, K. Yase and N. Uyeda, *Acta Cryst.*, B40 (1984) 263.

89. R. H. Tregold, S. D. Evans, P. Hodge, R. Jones, N. G. Stocks and M. C. J. Young, *Br. Polym. J.*, 19 (1987) 397.

90. J. F. Rabolt, F. C. Burns, N. E. Schlotter and J. D. Swalen, *J. Chem. Phys.*, 78 (1983) 946.

91. P. A. Barrett, E. F. Bradbrook, C. E. Dent and R. P. Linstead, *J. Chem. Soc.*, (1939) 1820.

92. W. A. Orr and S. C. Dahlberg, *J. Am. Chem. Soc.*, 101 (1979) 2875.

6

Phthalocyanines in

Photobiology *

Ionel Rosenthal and
Ehud Ben-Hur

A. INTRODUCTION

Living organisms, as we know them today, adapted through evolution to solar radiation at the Earth's surface, and under normal conditions they are unharmed by visible light. However, the simultaneous presence of an appropriate photosensitizing dye and oxygen converts this innocuous radiation to a potent damaging agent. Modern literature traces the first scientific report of this phenomenon to Raab [120], who demonstrated that *Paramecia* were rapidly killed by visible light in the presence of oxygen and low concentrations of acridine, eosin, or other dyes. Three years later, Jesionek and Tappeiner [70] used eosin and sunlight to treat skin cancer. Soon after, Tappeiner and Jodlbauer [150] found that fluoresceins sensitized the photoinactivation of enzyme preparations and protozoa in the presence of oxygen, and anticipated that this reaction could be much more general than the few cases known at that time. The name "photodynamic action" was coined for this reaction, with the intent to distinguish it from the sensitization of photographic plates by dyes. (Many years later, modern organic photochemistry adopted the term "sensitizer" to describe any physical energy transfer process.) In spite of subsequent objections generated by the indiscriminate use of photodynamic action to describe also anaerobic sensitization such as that of psoralens, today this term is firmly accepted, especially in connection with effects on integral biological systems.

The original interest in this reaction was fueled by the hypothesis that photodynamic action played a crucial role in physiological photobiology due to the presence of endogenous and normally occurring pigments such as chlorophyll, carotenoids, and cytochromes. With time, the lack of supportive evidence in this direction diverted attention to exogenous sensitizers. Many different kinds of dyes are known to be effective, and have been used with a variety of biological systems, from molecular through the cellular and whole organisms [28]. Since this reaction modifies essential biological macromolecules, virtually all kinds of organisms, albeit widely different physiologically, are sensitive to sublethal or lethal photodynamic damage depending on the intensity of the treatment. Outstanding among the plethora of reports on this topic was the description of the severe but transient effects of humans, following systemic administration of hematoporphyrin and exposure to sunlight [90].

Reviews of studies of this process at the chemical and molecular biology levels have been published [57,144]. In spite of the large variety of dyes tested, phthalocyanines (Pcs) were scarcely used as photodynamic sensitizers (in retrospect, this might be explained by the lack of photodynamic activity of the common commercial Pcs which contain Cu or Co). Water-soluble Pc metal complexes (Al, Zn, Ca, Mg, Na, K, Fe^{2+}) were claimed to exhibit bactericidal action on wet cotton fabric exposed to light and air, and a patent registration was requested [116]. Zn-Pc sulfonate was found to be more effective in photokilling of unarmoured dinoflagellate *Ptychodiscus brevis* (Florida red tide) than classical photodynamic sensitizers such as methylene blue, rose bengal, and hematoporphyrin [6].

Photodynamic therapy of cancer is a promising application of photodynamic action. In the 1940s, Auler and Banzer [5] and Figge *et al.* [54] established the affinity of various porphyrins, including hematoporphyrin, to malignant as compared with adjacent normal tissues, by studying the fluorescence produced when the tissues were exposed to ultraviolet light. In attempting to optimize the tumor-localizing ability and photodynamic properties, hematoporphyrin was treated with a mixture of acetic acid–sulfuric acid (19:1), and the acetylated product was dissolved in dilute alkali to improve its solubility [88]. The new product was named hematoporphyrin derivative (HPD) and has been the most frequently used drug for fluorescent visualization and photodynamic induction of tumors' necrosis [48,49,74,98,163].

Photodynamic therapy depends on selective cell injury. First, the selectivity rests on prolonged retention of the photosensitizer in tumor tissue as a result of some incompletely understood biochemical properties of malignant cells (protein binding, membrane changes, etc.) or cancer tissue (increased blood supply, avascular zones, absent lymphatics). Second, additional selectivity may be achieved by spatial localization of the illumination to the target.

The photodynamic therapy is a two-step procedure. HPD is first administered intravenously, and after a delay—usually 1–3 days—to allow for selective retention of the sensitizer in the tumor, the target tissue is irradiated with red light, usually centered at 630 nm. The light may be delivered either through the skin for a relatively superficial tumor, or via optical fibers for treatment of deep-lying tumors which are accessible endoscopically. The volume of the tissue treated is determined by the penetration of the light in the tissue and, as currently practiced, this therapy is suitable mainly for relatively small, localized solid tumors. To date, several thousand patients have been treated by this procedure.

HPD is a complex mixture of porphyrins of a somewhat variable composition, which have been only partially characterized [29,37]. The active component is thought to be dihematoporphyrin ether [50], and/or ester [75]. The uncertain composition and chemical lability may create difficulties in ensuring reproducibility in the use of HPD. Furthermore, although its main absorption is around 400 nm, for therapy the dye is activated by red light (λ = 630 nm) where a minor absorption exists, because of the increased transparency of tissues to red light [114]. This leads to poor efficiency of the overall photoprocess.

A more desirable photosensitizer would be a single compound which has a maximum absorption peak in the red part of the visible spectrum and high photodynamic efficiency, possesses a very low systemic toxicity, and shows a preferential retention/affinity for malignant tumors. The recent suggestion that Pc dyes fit these requirements provided the impetus for the intensive investigation of their photobiology. The use of Pcs as drugs in photomedicine is still in infancy, and much still remains to be revealed. As natural for a new field which generates vivid research activity, a few observations are at present in apparent disagreement. Undoubtedly, more exciting research is to be expected.

B. PHOTOPHYSICS OF PHTHALOCYANINES

Phthalocyanines, as expected from the extensively conjugated aromatic chromophore, possess an intense, more or less Gaussian Q-band about 400 cm^{-1} broad, in the visible range at 650–700 nm ($\epsilon > 10^5$ m^{-1} cm^{-1}) [51]. As with other large, planar molecules, most Pcs substituted with hydrophilic groups form stacked aggregates in a plane perpendicular to the plane of the dye macrocycle. The driving forces for this aggregation are hydrophobic in character, and are the result of the propensity of the Pc skeleton to avoid contact with the water molecules. The existence of these aggregates is reflected spectrophotometrically by the shift of the absorption peak to shorter wavelengths, as well as by hypochromism. When the dye aggregate is excited by light no useful photochemistry can be initiated, since the rate of deactivation by internal conversion to the ground state greatly exceeds that of the monomer [44]. Disaggregation of the stacked dye molecules can be achieved with organic solvents (methanol, dimethyl sulfoxide, pyridine), or with detergents at suitable concentrations [72]. It is noted that Pcs substituted with lipophilic substituents dissolve, and also associate, in organic solvents such as benzene [101]. Equilibrium constants estimated for dimeric aggregation lie within the range 10^5–10^7 M^{-1} and are dependent on the complexing metal. The aggregation follows the order Cu > H > Fe > VO

> Zn > Co ≫ Al. As a matter of fact, ClAl-Pc is in a monomeric form over a wide range of concentrations $(10^{-7}–10^{-4}\ M^{-1})$ [45].

The Pcs occupy an historical place in modern photochemistry. The first demonstration of laser action in an organic dye was made by Sorokin and Lankard [143] using a giant pulsed ruby laser to induce stimulated emission in an ethyl alcohol solution of ClAl-Pc. While still employed as lasing dyes [85], some Pcs show enough bleachable absorption to serve also as repeated passive Q-switching elements for ruby lasers [67].

The nature of the central metal ion also has a significant effect on fluorescence and phosphorescence quantum yields of Pcs. Heavy metal ions enhance the intersystem crossing to the triplet state, but dyes having a central paramagnetic transition ion, such as Cu or Cr, possess very short triplet lifetimes. Redox potentials for some ground and excited state Pcs are also known [44]. Since in all cases when a Pc has been used to photosensitize a net chemical change it is the excited triplet state that serves as the active intermediate, a short life time imposes a very severe limitation on the usefulness of a sensitizer.

C. REACTIONS OF PHOTOEXCITED PHTHALOCYANINES WITH MOLECULAR OXYGEN

There are two major reaction pathways open to an excited sensitizer involved in photooxidations. In the first, the sensitizer reacts directly with another chemical entity, by hydrogen or electron transfer, to yield transient radicals which react further with oxygen. This reaction has been classified as Type I photooxidation. In a second photooxidation, defined as Type II, the reactions' sequence is reversed. The sensitizer triplet interacts with oxygen, most commonly by energy transfer, to produce an electronically excited singlet state of oxygen which can react further with the chemical entity susceptible to oxidation. Less commonly, the electron transfer from sensitizer to oxygen occurs, to generate a superoxide radical anion (O_2^-), and an oxidized form of the sensitizer [57].

Unfortunately, in spite of the desire to assess the relative importance of various possible processes, this assessment is a difficult task, particularly in heterogeneous systems such as membrane or whole cells in which the usual kinetics are not directly applicable, and the exact location of the reagents in the microenvironment is not known. One of the most common diagnostic techniques is the addition of a reagent, such as a furan or a sterically hindered cyclic amine, which reacts with singlet oxygen. Since these reactions are among the least specific for singlet oxygen [58,125], this approach is less than flawless. Similarly, a variety of singlet oxygen physical quenchers is known and can be used to inhibit reactions suspected of involving singlet oxygen. Unfortunately, all the quenchers are systems of low oxidation potential and will almost certainly interact with other oxidants produced in the system. Finally, the use of a deuterated solvent employs the fact that the lifetime of singlet oxygen is longer in D_2O than in H_2O by a factor of 10–15 [96]. Thus, singlet oxygen reactions are expected to be more efficient in D_2O than in H_2O. However, the magnitude of this effect can vary depending on the kinetics of the system; the absence of an isotopic effect is relevant to $^1O_2$ involvement only when the solvent-induced radiationless decay limits the $^1O_2$ lifetime in the system [59]. On the other hand, it is noted that other oxidizing systems in which a solvent isotope effect is observed would also be expected to show a difference in rate in D_2O compared with H_2O. For example, reactions involving superoxide rather than singlet oxygen would also be more efficient in D_2O if they are limited by competing dismutation of O_2^- [26]. In living biological systems, such as cultured cells, the replacement of water with D_2O might interfere with the recovery process and consequently potentiate cell killing; the resulting effect could be mistakenly attributed to $^1O_2$ [13].

In light of these considerations it is concluded that only quantitative kinetic methods can distinguish between different oxidation mechanisms.

The literature has recorded reports related to all possible interactions between photoexcited Pcs and oxygen. Thus, metal-free Pc, and less efficiently, Cu-, Pt-, and Zn-tetra-*t*-butyl Pc, generated singlet oxygen as estimated by oxidation of 2,2,6,6,-tetramethylpiperidine. Co and Ni derivatives were inert. The metal-free and Cu dyes photogenerated O_2^- as well. Zn- and Pt-Pcs removed O_2^- formed by other sensitizers [89]. The ability of some metallophthalocyanines to generate singlet oxygen was evaluated by monitoring the disappearance of dimethylfuran from the reaction mixture. It was found that the central atom in these compounds exerts considerable influence on the sensitization ability. The activity decreases in the order Zn > Mg > Cu > Mn > Co > Fe [165]. Sulfonated Pcs were found to photooxidize L-tryptophan and cholesterol in aqueous and aqueous–organic solutions, respectively. Specific singlet oxygen products and inhibition experiments with sodium azide suggested involvement of singlet oxygen. The rate of L-tryptophan disappearance was the highest for the Ga (100) and Al (65) complexes, followed, in order of decreasing activity, by Zn (20), Cu (11), Mg (8), Mn (7), H_2 (5), and Ce (3). A nonsinglet oxygen pathway was observed in the case of the Mn complex. Substitution with Co, Fe, VO, Ni, or Cr rendered the sulfonated Pc inactive; substitution with carboxylic groups inactivates the ClGa-Pc [86]. In a follow-up, the same group determined the absolute quantum yields of singlet oxygen generation by H_2- (0.12), Zn- (0.62), ClGa- (0.47), and Cu- (0.12) Pc sulfonates. Sulfonation of ClGa-Pc to varying degrees does not alter the kinetics of $^1O_2$ production from the excited dye but affects their tendency to aggregate in solution [160]. In concordance with the previous observation, Mn-Pc was reported to catalyze the oxygenation of 3-methylindole by a nonsinglet oxygen mechanism, mimicking the activity of the enzyme tryptophan-2,3-dioxygenase [156].

Monomeric Zn-Pc sulfonate in aqueous buffer is an efficient photosensitizer for the oxidation of cysteine, histidine, methionine, tryptophan, and tyrosine, as well as guanosine, and for inactivation of lysozyme. These reactions appear to be mediated by singlet oxygen [146]. A correlation between the quantum yields for inactivation of human cells of the line NHIK 3025 and for photodegradation of 1,3-diphenylisobenzofuran by ClAl-Pc sulfonate, suggested a Type II process [100].

The ability of several Pcs to generate singlet oxygen was measured following the bleaching of N,N-dimethyl-4-nitrosoaniline at 440 nm by the transannular imidazole peroxide intermediate [124]. Thus Pcs containing diamagnetic ions such as Zn and Al generate $^1O_2$, with quantum yields of 0.45 and 0.34, respectively, in contradistinction to dyes containing paramagnetic ions (Cu, Fe, VO) which did not photosensitize the oxidation at all. In accord with this finding is the observation that the lifetime of the triplet state transient of Cu-Pc dissolved in chloronaphthalene, as studied by laser flash photolysis, was unaffected by the presence of oxygen [92].

Zn-Pc generates singlet oxygen with a quantum yield of ca. 0.4 in ethanolic solution as determined by time-resolved emission studies at 1270 nm and photooxidation experiments using 1,3-phenylisobenzofuran. The quantum yield of photooxidation slightly increases on incorporation of Zn-Pc into unilamellar liposomes of dipalmitoylphosphatidylchloline [157].

Related to Pcs, the photoproperties of bis(tri-n-hexyl siloxy)silicon naphthalocyanine have been investigated. This compound shows an intense absorption band centered at 776 nm ($\epsilon = 6.5 \times 10^5 \ M^{-1} \ cm^{-1}$), and a fluorescence emission from 750 to 1000 nm. In air-saturated solution, a peak at 1270 nm was superimposed on the general fluorescence background and was attributed to singlet oxygen luminescence ($^1\Delta_g \rightarrow \ ^3\Sigma_g$), although the lifetime of this emission was significantly longer than that accepted for singlet oxygen in benzene which was used as a solvent [55].

The electron transfer reaction from photoexcited dye to oxygen also occurs. Thus, visible light excitation of Pcs dissolved in dimethyl sulfoxide in the presence of oxygen generates a superoxide radical anion which was spin trapped with 5,5-dimethyl-1-pyrroline-1-oxide and identified by electron spin resonance. The quantum yields for O_2^- generation range from 10^{-5} for Zn-Pc to 4.2×10^{-4} for ClGa-Pc [10].

Alternatively, the participation of electronically excited Pcs in direct reactions with organic compounds has also been substantiated in the literature. It has been known that many photoexcited Pcs will readily undergo net electron transfer with suitable redox couples forming the separated ion products. Thus metal Pcs show remarkably high photoactivity in reduction of methyl viologen. This photoreactivity depends strongly on the central metal ion: Mg > Zn > Cd > Cu > Ni. No activity was detected for Fe, Co, or Mn [149]. Different reaction sequences occurred with different Pcs; Mg-Pc and Zn-Pc first undergo a one electron oxidation by methyl viologen [109], while excited ClAl-Pc is initially reduced by EDTA, which is the sacrificial electron transfer agent present in the mixture. Similarly, the photoreduction of methyl viologen by sulfonated Zn-Pc in micellar solutions proceeds by an electron transfer reduction of the excited dye by cysteine [44,46]. The effect of ring substituents on the efficiency of photoreduction of methyl viologen was established for Mg-Pc: tetramethoxy > unsubstituted > tetraphenyl > tetranitro [110]. When a Pc dye sensitizes a photoredox reaction, the major reaction pathway to electron transfer products requires the triplet excited state [44,45,62].

Electron transfer reactions of the photoexcited triplet state of ClAl-Pc were studied by means of nanosecond laser photolysis-kinetic spectroscopy. The triplet state of the dye reacted with 1,4-benzoquinones and with aromatic amines to produce the one electron oxidized radical cation or the neutral radical, respectively [108]. Other examples of one electron transfer reactions are the photosensitized reduction of Fast Red A in the presence of ascorbic acid by a Pc-containing polymer [81] and photogeneration of hydrogen from water in the presence of Zn-Pc sulfonate [38]. The relaxation kinetics of several excited Pcs indicated that in sufficiently concentrated solutions, in the presence of a reducing agent, the triplet–triplet reaction can lead to charge separation [47]. Finally, the photobleaching of ClAl-Pc sulfonate was enhanced markedly by previous complexation with bovine serum albumin, thus suggesting a direct interaction between the sensitizer and substrate [17].

D. PHOTOBIOLOGY OF PHTHALOCYANINES

Several lines of cells in culture were found to be inactivated by the combined action of Pcs and visible light: V79 Chinese hamster fibroblasts by ClAl-Pc [11,12], ClAl-Pc sulfonate [15,32], Zn-Pc [10], ZnPc sulfonate [15], Ce-Pc sulfonate, metal-free Pc sulfonate, Ga-Pc sulfonate [32]; UV-2237 murine fibrosarcoma (a cell line derived from UV light-induced fibrosarcoma in mice) by ClAl-Pc sulfonate [39], Mg-Pc, Zn-Pc, Zr-Pc, ClAl-Pc, Cl$_2$Sn-Pc [40]; NIH/3T3 fibroblasts ("normal" mouse line) by ClAl-Pc sulfonate [39]; and human glioblastoma cell line (U87MG) by metal-free Pc sulfonate and ClAl-Pc sulfonate [1]. The following Pcs had no cytotoxic activity in these assays: Ga-, Co-sulfonate, Cu-sulfonate, Mn-, Fe-, Li-, VO-, Co-, Ni-sulfonate, Alcian blue [10,14], Cu-sulfonate, FCr-sulfonate, VO-sulfonate [32], metal-free-, Cu-, Cu-sulfonate, FCr-, Fe-, Co-, Pd-sulfonate, Ni- and Ni-sulfonate [40].

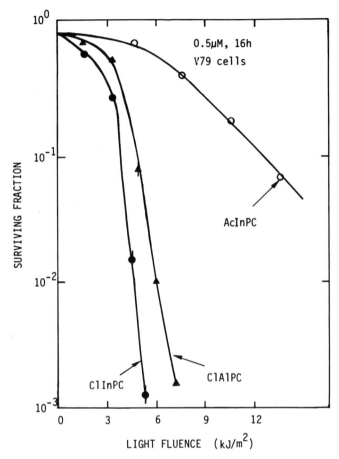

Figure 1 Survival of Chinese hamster cells photosensitized by various Pcs. Cells were incubated in growth medium for 16 h with 0.5 μM ClAl-Pc (▲), ClIn-Pc (●), and acetate In-Pc (○), and then exposed to graded light fluence. Survival was assayed as described by Ben-Hur and Rosenthal [12].

In experiments with human glioblastoma *in vitro*, metal free-Pc sulfonate had only a very weak cell killing effect, while ClAl-Pc sulfonate was a good deal more effective [1]. Metal-free Pc was also inactive in sensitizing the photohemolyses of red blood cells [142].

In general, a paramagnetic metal renders a Pc inactive, as expected from the short lifetime of the triplet state [124]. Conversely, a diamagnetic metal which extends the triplet lifetime, such as In, enhances the phototoxicity. In addition, preliminary measurements indicate that the metal counter ion (Cl^- vs. CH_3COO^-) also affects the phototoxicity (Fig. 1). In this context it is noted that ClAl-Pc sulfonate, which is routinely prepared by direct sulfonation with sulfuric acid to yield a mixture of isomers sulfonated at various extents, is one of the extensively used Pc photodynamic sensitizers. The possibility that this preparation route might replace the chlorine atom by a hydroxyl group [103], apparently has not yet been considered.

ClGa-, ClAl-, and Zn-naphthalocyanines could not photoinactivate V-79 Chinese hamster cells [113].

The action spectrum for photocytotoxicity is expected to follow the chromophore of the Pc sensitizer. This was indeed established for ClAl-Pc [16,115]. The small red shift in the action spectrum is presumably due to the prebinding of the dye to a cell component. In this respect it is noted that the electrostatic interaction of Co-tetracarboxy Pc with polylysine resulted in a bathochromic shift of absorbance from 680 to 695 nm due to an interaction dictated by the polymeric state of the polypeptide [84].

Various factors affect the Pc-mediated phototoxicity in cultured cells:

1. *Kinetics of dye uptake.* In a cellular system, the central metal as well as the peripheral substituents on the Pc macrocycle determine the rate of dye uptake and thus the overall cellular photosensitivity. UO_2-sulfonate and ClAl-sulfonate were taken up at the highest rate followed by Ni-, Zn-, Cu-, Co-, and Cl_2Si-Pc sulfonate. The uptake from the growth medium containing 10% serum, in which only about 15% of the dye is not bound to serum proteins, was 5- to 18-fold slower than in the absence of serum, suggesting that most of the uptake is of free dye. The rate of uptake was unrelated to the state of aggregation of the dye. The rate of uptake was temperature dependent at intervals longer than 1 h. At shorter times, very little temperature dependence was observed. These results suggest that the uptake process takes place in two steps. The first step is passive, involving binding of metallo-Pc sulfonate to a receptor on the cell membrane, while the second is active and involves internalization of the bound dye [19].

While the nature of the central metal affects intrinsically the photochemical activity and also the rate of cellular uptake, the peripheral substituents affect primarily the rate of uptake and indirectly the photobiological activity. Thus, the kinetics of uptake and cell retention are different for ClAl-Pc and ClAl-Pc sulfonate [15] as expected from the different protein-binding capabilities dictated by the different peripheral charges and molecular sizes. Furthermore, for sulfonated ClGa-Pcs the activity was inversely related to the number of sulfonic acid groups for otherwise identical experimental conditions. Large variations in photoreactivity were observed among the four isomeric disulfonated derivatives, with the most hydrophobic isomer exhibiting the highest photoactivity [34]. However, ClAl-Pc monosulfonate was reported to be inactive in sensitizing the HeLa cell photokilling, in contradistinction to the polysulfonated derivatives [147]. The cellular uptake and distribution of ClAl-Pc dyes sulfonated to different degrees [3] were studied in an attempt to correlate hydrophobicity with cell membrane permeability and phototoxicity.

The lower sulfonated derivatives (two sulfonic groups) were 25 times more efficient in photoinactivation of V-79 Chinese hamster cells, than the higher sulfonated dyes (mixed tri- and tetrasulfonated preparations) [112]. Finally, *in vitro*, using V-79 cells, the lower sulfonated Zn-Pc derivatives were the most active with the exception of the poorly water-soluble monosulfonated dye. A mixture of tetrasulfonated isomers obtained by direct sulfonation was 10 times more active than the homogeneous tetrasulfonated derivative prepared by the condensation of sulfophthalic acid. *In vivo*, testing the effect on EMT-6 mammary tumors, the latter dye was completely inactive, whereas the remainder of the sulfonated dyes exhibited a similar structure–activity pattern as observed with V-79 cells, although *in vivo* the variations are less pronounced [35].

In contrast with this trend, the higher sulfonated ClAl-naphthalo-cyanines showed substantial phototoxicity while the lower sulfonated deriva-tives did not succeed in sensitizing cells toward red light [113].

Among substituted Zn-Pcs, the rate of uptake and consequently the phototoxicity deceased in the order H > OH > SO_3Na ≫ neopentoxy [126]. Finally, the carboxylated ClAl-Pc was much less efficient than the sulfonated counterpart [127]. Derived from the different uptake rates was the observation that the photoinactivation of mammalian cells is proportional to the time of incubation of the cells with the dye prior to light exposure, and to the initial dye concentration [15].

While cancer cells in culture do not appear to take up sulfonated ClAl-Pc at different rates as compared with their normal counterparts [39], rates of uptake differ greatly among various cell types. Among the cell lines tested, endothelial cells take up the dye at a faster rate than fibroblasts and are also more photosensitive (Fig. 2). In spite of the lack of discrimination between normal and cancer cells, leukemia seems to be an exception (e.g., compare HPBL and L5178Y-s cells). It is noted that photosensitization by ClAl-Pc sulfonate is capable of causing differential *in vitro* toxicity to residual acute nonlymphoblastic [140] and to clonogenic acute myeloblastic leukemia [141], compared to normal progenitor cells.

2. *Presence of oxygen.* Photocytotoxicity of Pcs requires molecular oxygen [12,32]. The efficiency of photosensitization decreased with oxygen concentrations below a threshold of ca 3.4% O_2. The full oxygen effect is achieved already at ca. 10 mm Hg pO_2, two to three times lower than with HPD [124]. Indeed, there are also other differences between Pcs and HPD. Phototoxicity of HPD is thought to be mediated via singlet oxygen (Type II photooxidation). This is supported by the observations that HPD is more effective in D_2O than in H_2O [99], and that tryptophan, a chemical trap for $^1O_2$, has a very pronounced protective effect against hematoporphyrin phototoxicity [66]. The evidence for $^1O_2$ involvement in photocytotoxicity induced by Pcs is less substantiated. Although these dyes can generate $^1O_2$ with

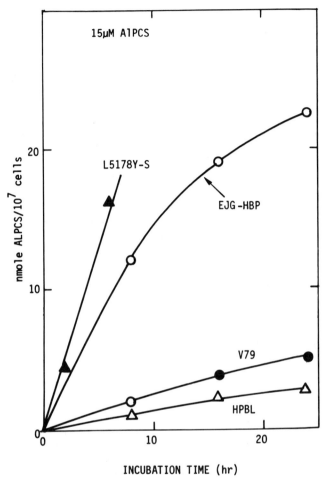

Figure 2 Uptake of ClAl-Pc sulfonate into cultured mammalian cells. Cells were incubated with growth medium containing 15 μM dye and supplemented with 10% fetal calf serum. Dye uptake was measured at intervals as described by Ben-Hur *et al.* [19]. The cells used were Chinese hamster fibroblasts (●), bovine endothelial cell line (○), human peripheral blood lymphocytes (△), and a mouse leukemia line (▲).

reasonable quantum yields [124], and participate in $^1O_2$-specific reactions with biological molecules in solution [86], the situation *in vivo* appears to be different. Thus, phototoxicity is the same in D_2O and H_2O [12,13,15] and tryptophan has only a minor protective effect, with a fluence modifying factor of 1.3 at the 10% survival level (Fig. 3) compared with a factor of 5 for HPD under similar conditions [66]. Cysteamine, a free radical scavenger, has a more pronounced protective effect than tryptophan (Fig. 3). An apparent diminution of $^1O_2$ involvement in Pcs photosensitization may be suspected as the

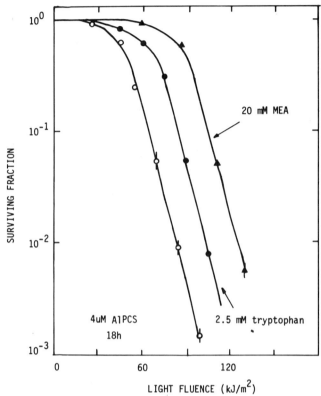

Figure 3 Survival of Chinese hamster fibroblasts. Cells were incubated for 18 h with 4 μM ClAl-Pc sulfonate in growth medium and then exposed to graded light fluence. Prior to light exposure, the cells were incubated with 2.5 mM tryptophan for 2 h (●), with 20 mM cysteamine for 15 min (▲), or without additives (○). Tryptophan and cysteamine were present during irrdiation.

biological system becomes more complex. Using sulfonated ClAl-Pc for photoinactivation of β-hydroxybutyrate dehydrogenase, the D_2O effect in intact mitochondria and submitochondrial particles was 1.7 and 2.2, respectively [123; the values were calculated using the data presented in Fig. 1]. Thus, it appears that in intact cells, Pc phototoxicity is mediated by a Type I photodynamic reaction, although some contribution by Type II cannot be ruled out. A recent flash photolysis study of tetrasulfonated Ga-Pc and ClAl-Pc showed that at low oxygen concentration the Type I mechanism can make significant contributions to some amino acids photooxidations [53]. The photoexcitation of liposomal Zn-Pc in cultured mouse myeloma cells enabled the observation of triplet states but not of singlet oxygen (by infrared luminescence). Certainly, this failure could be explained by the extremely rapid decay of singlet oxygen in cells [56]. In view of the ambiguity of the experimental support in the exclusive favor of one mechanism, it is sensible to question the overall significance of this mechanistic discrimination. Since in the biological milieu the sensitizer is bound to a specific molecular site, a complex dye-biomolecule is actually activated by light. As such, only if the sensitive site whose modification is responsible for the biological damage is remote from the original complex, and a migratory active species is needed, the intermediacy of an excited oxygen is mechanistically relevant.

3. *Light fluence.* As expected, the extent of cell photoinactivation is proportional to the total light fluence [12,15] and also to the light fluence rate; that is, the sensitivity increased with energy density [21]. The latter effect is presumably due to the capacity of the cells to recover from sublethal damage.

4. *The physiology of the cells.* This factor does not affect their intrinsic photosensitivity. Thus, sensitivity is independent of the position in the cell cycle [12]. Plateau-phase cells appear to be more sensitive than log-phase cells, because the former take up more dye [20]. Experiments with split light fluence indicated that log-phase cells can repair sublethal damage induced by Pc-photosensitization, while the repair of such damage in plateau-phase is apparently absent [25].

5. *The pH value of the medium during irradiation.* Photosensitized inactivation of Chinese hamster cells by ClAl-Pc is dependent on the pH value of the medium during irradiation. Thus, in the range of pH values 6–8, the sensitivity was increased at lower values [13]. This effect is displayed to a lesser extent by ClAl-Pc sulfonate.

6. *Hyperthermia treatment.* It has been found that thermal treatment (90 min at 42°C) following light exposure enhanced the ClAl-Pc photosensitization of Chinese hamster cells [13].

The ClAl-Pc sulfonate photosensitized inhibition of mitogenic stimulation of human lymphocytes by plant lectins has also been studied. The resulting inhibition indicates the mitogen's attachment to its receptor as one of the possible targets for photosensitization [82]. The combined action of photosensitization by ClAl-Pc sulfonate and γ-radiation on the response of human lymphocytes or Chinese hamster cells was usually additive regardless of the sequence of application. However, in human lymphocytes irradiated at low temperature, the photosensitization interacted synergistically with the subsequent ionizing radiation as expressed by the yield of micronuclei produced [22].

Regarding the mechanism of phototoxicity at the molecular level not very much information is known. When cultured mammalian cells are photosensitized by ClAl-Pc sulfonate, the macromolecules' biosynthesis and glucose oxidation are inhibited progressively with time. No differences between relative sensitivity of synthesis of protein, RNA, and DNA were found. The inhibition of glucose oxidation was tentatively correlated with cell killing [20]. In spite of the fact that ClAl-Pc does not bind to DNA, it can induce DNA damage after exposure to light. The observation that ClAl-Pc induced DNA damage is potentially lethal is indicated by the enhanced photosensitivity following a preillumination treatment with 5-bromodeoxyuridine. This compound potentiates the cell killing by DNA damaging agents, partly because of interference with the repair process. This, however, does not imply that in its absence, DNA damage is the major lethal lesion. It is conceivable that 5-bromodeoxyuridine interacts synergistically with the dye photosensitization to enhance cell killing. Another synergistic effect was reported for sodium salicylate [13]. The DNA damage is expressed as single-strand breaks under alkaline conditions in the DNA of Chinese hamster cells. However, the frequency of breaks at equitoxic doses is about one-third of that produced by X-rays. During incubation in growth medium after exposure to ClAl-Pc and light, cells rejoined DNA strand breaks at a rate similar to that observed after X-irradiation. Furthermore, Pc photosensitization of Chinese hamster cells does not produce mutation in either the HGPRT locus or the Na^+K^+-ATPase locus (expressed as 6-thioguanine and ouabain resistance, respectively). Since cytotoxic agents that cause DNA damage are also mutagenic, the lack of mutagenicity argues against DNA being a significant target for Pc photosensitization [18]. Similar results were also reported for Ga-sulfonated Pcs. It was also noted that the repair of the DNA damage was not inhibited by aphidicolin, an inhibitor of DNA polymerases α and δ, under conditions in which DNA replicative synthesis was inhibited by more than 90% [69]. Further support for the minor role played by DNA damage was provided by experiments with human fibroblasts derived from patients afflicted by the hereditary disease ataxia telangiectasia. Although these cells are hypersensitive to ionizing radiation [9] and radiomimetic agents, because of a deficiency in their ability to repair radiation-induced DNA damage, the same sensitivity to Pc-sensitization as fibroblasts from normal donors was exhibited (Fig. 4).

The induction and repair rate of DNA single-strand breaks and DNA–protein cross-links induced by ClAl-Pc and visible light in Chinese hamster lung fibroblasts have been determined. The data demonstrate that there are sufficient unrepairable DNA lesions to cause cell death. However, the relative importance of DNA lesions for *in vivo* processes could not be ascertained [121].

In regard to subcellular structures, UO_2-Pc sulfonate appears to localize at the plasma membrane, mitochondrial membranes, and, most prominently, in lysosomes (Fig. 5). A similar behavior was observed for ClAl-Pc sulfonate after subcellular fractionation. Experiments were therefore designed to test the contribution of damage in these subcellular structures to cell killing. In the case of ClAl-Pc sulfonate, plasma membrane damage appears to play only a minor role. This minor effect is inferred from our observation that addition of the

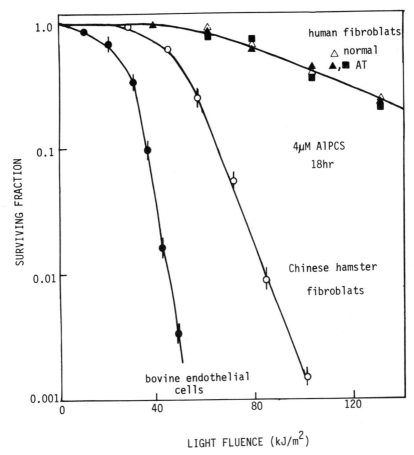

Figure 4 Survival of various cells incubated with 4 μM ClAl-Pc sulfonate for 18 h and then exposed to graded light fluence. The human fibroblasts were described by Ben-Hur *et al.* [9].

detergent cetrimide to the growth medium after light exposure enhances cytotoxicity to only a small extent. Such addition causes a drastic potentiation of the response after treatments that kill cells mostly by plasma membrane damage [105]. It is noted, however, that in certain systems such as hemolysis of red blood cells [14,142], Pc-sensitized membrane damage plays a major role. The role of damage to the lysosomal membranes was studied using hydrocortisone under conditions that cause stabilization of these membranes [68]. It is expected that damage to the lysosomal membranes would cause release of hydrolytic enzymes resulting in cell death. Hydrocortisone should protect against phototoxicity in such a case. As no effect of hydrocortisone was noted on ClAl-Pc sulfonate photosensitization of Chinese hamster cells, it

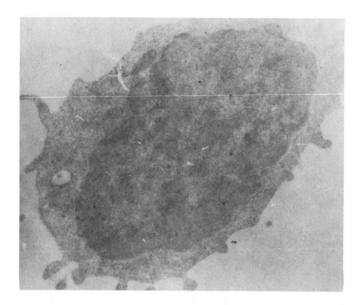

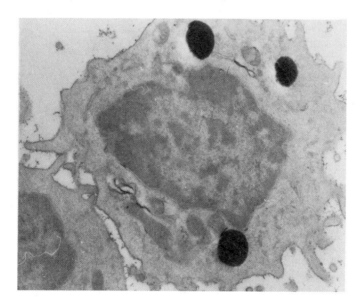

Figure 5 TEM micrographs of lymphocytes from human blood. (A) Control, un-
treated cells, ×20,000. (B) Cells incubated for 16 h with 50 μM UO$_2$-Pc sulfonate,
×15,0000. No further staining with uranyl acetate was done after fixation.

is concluded that damage to lysosomal membranes is not involved in cell killing (our unpublished results).

Ruling out the plasma membrane and the lysosomes leaves mitochondria as major targets for cellular inactivation. Indeed, experimental data support a role for mitochondrial damage in Pc-induced cell killing [20]. These include a pronounced inhibition of energy metabolism and severe structural damage to the mitochondria visualized with the electron microscope. The inhibition of the mitochondrial enzyme β-hydroxybutyrate dehydrogenase by ClAl-Pc sulfonate photosensitization [123] is consistent with this hypothesis.

Related to the effect on membranes, copper containing Pcs were found to induce rapid calcium efflux from actively loaded sarcoplasmic reticulum, in the dark. This effect was explained by the dye oxidation of a pair of neighboring sulfhydryl groups associated with the calcium release channel, to a disulfide [2].

Some data have already been reported on the photobiological response of normal and tumor tissue following Pc administration in experimental animals. Thus, the photonecrotic effect of ClAl-Pc sulfonate was estimated for normal rat liver. As expected, the extent of tissue necrosis depends on the amount of dye and light administered. The period of time from Pc administration to light exposure is also important, since the response is correlated with the levels of dye retained in the tissue. Because of the high absorption coefficient of ClAl-Pc sulfonate at 675 nm, which was the light wavelength employed, saturation of the response occurred at relatively low levels of the dye (5 mg/kg body weight). At higher levels, the response diminished because of the light attenuation resulting from an internal "filter effect." As found in experiments with cultured cells, the effect is oxygen dependent, since the extent of photodynamic necrosis was abolished by occluding the blood flow during therapy. Serum clearance is quite rapid, with a half-life of about 1 h for ClAl-Pc sulfonate as compared with about 3 h for the HPD. A large difference was found in the liver clearance rate, with the HPD disappearing within a week, while ClAl-Pc sulfonate persisted for up to 6 weeks. Finally, less cutaneous photosensitivity was seen with ClAl-Pc sulfonate than with HPD [31].

ClAl-Pc sulfonate was reported to be an effective sensitizer for colonic photodynamic therapy. The lesion healed by regeneration with no reduction in bursting pressure. The potential for colonic therapy with ClAl-Pc seems to be important since it may offer a treatment for polyps without disturbing the collagen architecture and the strength of the colon, and hence perforation would be unlikely [7,8].

Zn-Pc sulfonate showed phototoxicity in two albino mouse transplanted tumor models: (1) pleomorphic sarcoma S180 in Swiss–Webster mice and (2) neuroblastoma, C1300 in A/J mice [148], and ClAl-Pc sulfonate was more effective than HPD for treatment of a subcutaneous transplanted fibrosarcoma in the rat [30]. The response was quantified; at doses below 1 mg/kg, the diameter of necrosis increased with the logarithm of the dose of sensitizer, while at higher doses much smaller increases in necrosis were seen. Peak tumor concentration of ClAl-Pc sulfonate occurred 24–48 h after sensitization, compared with a peak at 3 h in muscle. The peak ratio of tumor–muscle was 2:1 at 24 h [153]. A study of transplantable tumors in rat showed that immediately after light exposure there is a stoppage of blood flow in the treated tumor, followed, 24 h later, by tumor necrosis [136]. We hypothesized that an initial damage to the endothelium was followed by release of clotting factors into the circulation and thrombi formation [23].

Brasseur *et al.* [33] compared the photodynamic activity of various Pcs with that of HPD in treating the EMT-6 mammary tumor model in mice. The Ce complex was the most active sensitizer in terms of dye and light doses required to induce tumor necrosis and cure, but also showed the highest phototoxicity to healthy skin. The metal-free Pc sulfonate was void of photoactivity in *in vivo* conditions. The Al and Ga complexes exhibited tumoricidal activity competitive with the active ingredient of HPD. Syngeneic mice bearing a colorectal carcinoma growing subcutaneously in the flank region received photodynamic therapy 24 h after the i.v. injection of a single dose of ClAl-Pc sulfonate. The tumors removed 5 days after the treatment were substantially reduced in size and showed marked morphological alterations. Histological examination revealed that the treatment induced severe necrosis and cytotoxicity of neoplastic cells with viable tumor limited to a small peripheral margin [41]. BALB/c mice bearing a transplanted MS-2 fibrosarcoma injected with Zn-Pc incorporated into unilamellar liposomes, and exposed to red light, experienced severe photodamage to both neoplastic cells and tumor vasculature. On the other hand, the skin photosensitivity was very limited [97]. The efficiencies of metal-free Pc, ClAl-Pc, and Zn-Pc sulfonates were comparatively tested to several porphyrin sensitizers, using a model system with a C3H mammary carcinoma growing subcutaneously on the dorsal side of mouse feet. Only ClAl-Pc sulfonate sensitized the tumor to any significant extent; metal-free and Zn derivatives were inactive in this assay. ClAl-Pc also sensitized the skin, although less than HPD [52]. In this context, it is also noted that no significant differences in skin phototoxicity were observed among disulfonated Zn-Pc, ClAl-Pc sulfonate (S/Pc ratio 3.4), and ClGa-Pc sulfonate on BALB/c mice bearing EMT-6 mammary tumors [35]. Finally, a preliminary study showed that selective necrosis of malignant glioma cells in a tumor bed in mice can be achieved with ClAl-Pc sulfonate and red light, using treatment parameters that do not damage normal brain under conditions that mimic clinical practice [133].

In recent years, the potential use of PDT has been explored in diseases other than cancer, such as age-related macular degeneration. The rationale of the photochemical treatment is to completely obliterate the subretinal blood vessels and thus stop further extrusion and bleeding. Indeed, the combined action of ClAl-Pc sulfonate and red light efficiently obliterated choroidal vessels in the rabbit eye [24].

The same rationale of controlled occlusion of blood vessels may heal other localized vascular lesions of the skin and subcutaneous tissues such as congenital hemangiomas. For example, portwine stain, which is a flat, red lesion present at birth due to vascular ectasia, has presently no effective treatment. Other hemangiomas, such as "strawberry mark," although usually regress spontaneously, might occasionally require treatment (surgical excision, electrocoagulation, injection of sclerosing solutions, application of dry ice) which leaves scarring. The preliminary results of application of photodynamic therapy for these disorders are positive.

E. AFFINITY OF PHTHALOCYANINES TO BIOLOGICAL TISSUES

The binding of Pcs to normal tissues has been repeatedly substantiated by histological studies [103,104]. Thus basic quaternary substituted Pcs have been used for many years as a differential stain for acidic polysaccharides [87,135], mucosubstances [61], differential staining of gray and white matter in fixed human brain slices [65], and histiocytes and fibroblasts *in vivo* [91]. In general, cationic Pcs have been used to stain and quantify biological polyanions of all kinds [4]. Conversely, negatively charged sulfonated Pc dyes have been shown to stain collagen [132], proteoglycan filaments in rat tail tendons [93], bovine cornea [94], mouse lung alveoli [158], single muscle fibers [77,130], collagen-poor layers at the surface of canine articular cartilage [111], nerve membranes [83,131,139], myelin sheaths and glia fibers in combination with Levafix Red Violet E-2BL [161], and erythrocytes [36,78–80,138]. In the presence of 1 M MgCl$_2$, Cu-Pc sulfonate was found to bind specifically to single-stranded RNA, leaving native DNA, proteins, acid polysaccharides, and phospholipids completely unstained. Under these conditions the dye is complexed by nonelectrostatic bonds with nonstacked purine bases, mainly adenine. Without added cations, the dye stains both DNA and RNA [95,151]. Diazo-Pcs containing Mg, Cu, or Pb were reacted with tissue proteins in aldehyde-fixed material and evaluated as reagents for electron microscopy [152]. Finally, the water-insoluble Cu-Pc has been suggested for vascular labeling of abnormally permeable blood vessels [71]. This straining is achieved by mechanical retention of suspended colored particles of the right size from the colloidal material injected in the "cracks" of an interrupted endothelial surface.

More relevant to photodynamic therapy was the observation that experimental tumors growing subcutaneously and intracranially took up Cu-Pc sulfonate in amounts considerably greater than the surrounding normal tissue [164]. Similarly, healthy brain tissue of laboratory animals rejected the intravenously injected sulfonated uranyl Pc, while the tumor retained it. A dye concentration ratio of 1:50, healthy vs. tumor tissue, could be achieved [60]. These kinds of observations were investigated systematically in more recent studies. Thus tumor uptake and organ distribution were studied in Fischer 344/CRBL female rats bearing the 13762 mammary adenocarcinoma using sulfonated Pcs labeled with radioactive metals. [^{99}Tc]Tetrasulfonated Pc accumulated preferentially in the liver, kidney, and reticuloendothelial system. The dye was also retained by the ovarian follicles and the uterus. Favorable tumor–blood ratio of 5–10 were observed for the brain, muscle, and fat during the 24-h study. The tumor activity also slightly surpassed that of the blood, colon, and intestine. Most of the activity with the tumor was concentrated in the outer cell layers [128]. Substitution of ^{99}Tc by ^{67}Ga in the sulfo Pc complex resulted in a shift in the tissue distribution pattern, with the bulk of material now passing through the hepatobiliary system. The kidneys retain activity at levels similar to those of the spleen, adrenals, and ovaries with overall activity in kidney values at one-third of those of the liver. ^{67}Ga-sulfonated Pc reached both favorable tumor–blood and tumor–muscle ratios. The difference in distribution pattern between sulfonated Ga- and Tc-Pc indicates that the central atom strongly influences the biodistribution, probably because of the stereochemistry of the metal complex [129]. The peripheral ring substituents which dictate the electrical charge and solubility characteristics also affect the tissues' distribution. Thus, the water-soluble sulfonic acid derivative is cleared faster from the blood than the lipophilic nitro Pc while the blood levels of the amino Pc remained intermediate between those of the sulfonated and nitro-Pc. The kinetics of tumor uptake also differ among the various derivatives. Clearance of the water-soluble Pc is mainly via the kidneys, while the least water-soluble analogs are excreted biliarly ([159]; unfortunately no data on solubility or on the medium used for administration are presented). Others reported that tetraamino Pc is insoluble in water [152, and our unpublished results], while in our hands tetranitro-Pc is insoluble in water in spite of the opposite observation [145]). In this context, it is also noted that a colloidal suspension of the insoluble metal Pc was preferred to the water-soluble sulfonated compound for tumor therapy [60].

A study of the uptake and retention of ClAl-Pc sulfonate in a rat colon cancer, a hamster pancreatic cancer, and a mouse glioma showed that all the tumors reached accumulation peaks at 24–48 h after intravenous administration of the dye, compared with peaks at 1–3 h in the normal tissues. The tumors outside the central nervous system reached peak tumor–normal tissue ratios of 2–3:1, and the tumors within the central nervous system reached a far higher ratio of 28:1 [154]. In a comparison of different photosensitizing dyes with respect to uptake by C3H mammary carcinomas and normal tissues in mice, it was found that ClAl-Pc sulfonate was taken up by the tumor more efficiently than was a purified HPD preparation; the uptake of the unsubstituted ClAl-Pc was only half of that of HPD. Both aluminum complexes were taken up in large amounts by kidney, spleen, and liver, and the skin–tumor concentration ratios relative to HPD were higher, a fact which apparently agreed with the levels of skin photosensitivity [100,119]. On the other hand, Bown *et al.* (31), based on their observations on *in vivo* experiments, indicate that one of the major problems of photodynamic therapy, namely, cutaneous photosensitivity, may be alleviated by using sulfonated ClAl-Pc. Indeed, our preliminary studies on the distribution of ClAl-Pc sulfonate in canine tissues show little retention of the dye in skin [166].

Three murine tumors, of different histological type (Colo 26, a colorectal carcinoma, M5076, a reticulum cell sarcoma, and UV-2237, a fibrosarcoma), took up and retain ClAlPc sulfonate to a greater extent than adjacent normal skin and muscle tissue. The specific uptake is, in part, a consequence of uptake by individual neoplastic cells. Maximum uptake in all three tumor types was achieved 24–48 h after dye administration. However, there are differences in uptake between the three murine neoplasms suggesting that Pc affinity is markedly dependent on the nature of the tumor. Photodynamic therapy of the same tumors in mice led to damage correlated with the degree of dye uptake [42].

A pharmacokinetic analysis demonstrated that ClAl-Pc sulfonate administered intravenously in healthy mice is cleared from the blood stream within 1 h. An immediate massive excretion of the dye was noted in urine and a delayed slow release in feces. The dye accumulated mainly in reticuloendothelial rich tissues. The liver is the major site of accumulation. Maximal concentration of the dye in the organs was reached after 24–48 h. Histological examination showed that ClAl-Pc sulfonate accumulated mainly in macrophage-derived and endothelial cells [162]. When the water-insoluble Zn-Pc, incorporated into liposomes, was administered in tumor-bearing mice, it was cleared from the serum via the bile–gut pathway with a half-life of 9 h. After 24 h, a tumor–normal tissue ratio of 7.5 was obtained [122].

At the molecular level, ClAl-Pc has been shown to bind to proteins [137]. Spectroscopic studies of the interaction of Fe- and Co-tetrasulfonated Pc with human serum albumin under conditions of low dye concentration show complex formation between the protein and the Pc derivative at the molar ratio 1:1. The interaction between Co-Pc sulfonate and cyanogen bromide-albumin fragments indicates that there are two high-affinity sites on the protein. The stronger one is located on fragment M and the weaker one on fragment C, at a dye to peptide molar ratio of 1:1 and 1:2, respectively. The reaction of Co-tetrasulfonated Pc with the complex of heme with the M fragment leads to the displacement of heme by the phthalocyanine. This observation implies that the heme and phthalocyanine binding sites on the albumin M fragment are the same. A different result is obtained in the reaction of Co-Pc sulfonate with the heme complex of the C fragment. Displacement of heme occurs to a minimal extent, in spite of Co-Pc sulfonate–C fragment complex formation to a degree comparable to that of the M fragment. This fact suggests that the Pc binding site on the C fragment is not the same as the heme binding site, but it is probably very close to, or overlaps it. In such a situation, Pc binding can produce conformation perturbation of the heme binding center and the release of some heme molecules. Based on structural studies it is assumed that the Co and Fe ions are implicated in the binding to albumin. The predominant binding site of the dye is close to lysine 199. Although both dyes tested are aggregated in aqueous solutions, the complexation with albumin is exclusively with the monomeric form [155]. The same two react with globin to form a green complex at the molar ratio 1:1. These complexes combine reversibly with oxygen [117,118]. Since the photoelectron quantum yield of Pc is two orders of magnitude greater than that of 21 common amino acids and 15 polyamino acids tested in the range 200–240 nm, the feasibility of photoelectron labeling of biological surfaces is suggested [43].

Zn-Pc incorporated into unilamellar liposomes in mainly transported in the serum by specific binding to lipoprotein [97].

The transport, plasma protein binding *in vitro*, and tumor localization *in vivo*, of the Pc-structurally-related, sulfonated derivatives of tetraphenylporphyrin have been examined recently. The number and distribution of sulfonic groups were major factors in accumulation. The dye with two adjacent sulfonates was most efficiently taken up; the presence of more or fewer sulfonate residues led to reduced uptake. Steady-state accumulation of drugs with one, two (opposite), three, or four sulfonates was rapid, while uptake of the disulfonated (on adjacent rings) porphyrin was slower. Products bearing one to four sulfonates localized equally well *in vivo*, but sites of localization varied considerably. Drugs with one sulfonate, or two sulfonates on adjacent rings partitioned into neoplastic cells, analogs with two (opposite), three, or four sulfonates partitioned to tumor stroma. Plasma binding studies show that drugs with one or two (adjacent) sulfonates bound to VLDL, LDL, and HDL components of plasma, while the tri- and tetrasulfonated analogs bound progressively more to albumin. These results suggest that tumor location can occur via two pathways: one mediated by lipoprotein binding leading to dye accumulation in neoplastic cells and another associated with albumin binding and leading to dye accumulation in stromal elements of neoplastic tissues [76].

F. TOXICITY OF PHTHALOCYANINES

The literature provides encouraging information regarding the lack of toxicity of Pc dyes. Cu-Pc was certified as a food color in Germany [27], and is permanently listed as an approved color additive for coloring polypropylene sutures for use in general and ophthalmic surgery, and for coloring contact lenses in the United States [107]. Phthalocyanine blue may be used as a colorant for polymeric coatings that are used in producing, manufacturing, packing, processing, preparing, treating, transporting, or holding food. The LD_{50} for Phthalocyanine Blue 15 and Phthalocyanine Green 7 is more than 10 g/kg body weight [102]. Doses of Cu-Pc sulfonate up to 100 mg/kg given to mice, guinea pigs, rabbits, cats, and dogs did not induce any toxic effects [164]. Two commercial Cu-Pc sulfonates at concentrations of 5000 ppm were nontoxic to protozoans, small crustaceans, roundworm, small fish, and, injected, to rabbit, rat, and gluinea pig [106]. Mice receiving five daily injections of uranyl tetrasulfonated Pc totalling 5000 mg/kg, showed no ill effects after 19 months, nor did their offspring [60]. An aqueous suspension of Cu-Pc (30 mg/kg) injected intravenously to rats showed no toxic effects after 1 month [71]. Finally, no toxic effects were noted in rats 24 h postinjection of [Tc]-Pc tetrasulfonate [128]. Mice injected with Ce-sulfonated Pc (11 mg/kg), ClGa-sulfonated Pc (15 mg/kg), or ClAl-sulfonated Pc (28 mg/kg) showed no apparent dark toxicity [33]. Mutagenicity tests of blue chalks colored with Cu-Pc indicated the industrial pigment contained mutagenic impurities of an undefined structure. Awareness of this fact led to a product free of mutagenic impurities [63]. In this context it is noted that Cu-Pc sulfonate binds nonpolar mutagens, a property which has been used for separation of polycyclic aromatics containing three or more fused rings in various materials using the dye attached to cotton through a covalent bond and elution with ammoniacal methanol [64].

Conversely, sulfonated Pcs tested for possible noxious effects on the developing chick embryo were found to induce a highly reproducible caudal malformative syndrome (trunk and tailessness, various anomalies of the limbs). The main effect observed in about 15% of the malformed specimens is associated with unilateral microphthalmy and, less frequently, with coelosomy. Developmental disturbances of the caudal axial organs, of the mesonephros, and of the limbs are observed microscopically. The initial pathological changes, at the microscopic level, are necrosis and hemorrhages in the caudal, axial, and paraxial area. The allantois is poorly developed or even absent. Skeletal changes involve anomalies of the ribs and of the vertebral column and total or partial absence of the pelvic girdle bones. The high mortality, mainly during the first week, is due, first of all, to the developmental disturbances including the poor development or absence of the allantois. Control experiments with $CuCl_2$ suggest the etiological role of copper [134]. Recently, cationic Cu-Pcs were found to produce DNA damage as expressed by an increase of the sister chromatid exchanges frequency in plant cell chromosomes [73].

ClAl-Pc sulfonate has a very low toxicity when administered intravenously in normal mice. A dose of 100 mg/kg had no effect on mice kept in complete darkness. A dose of 200 mg/kg caused 7% mortality in mice kept in complete darkness. The morbidity rate in the treated mice was followed for 14 days. No abnormal behavior among the treated mice was observed [162].

Appendix Compendium of *in Vivo* PDT Experiments with PC Dyes

Photosensitizer	Test	Dose (mg/kg)	Tested tissue	PDT conditions	Evaluation time	Treatment result	Reference
Al-Pc sulfonate	Rats	3 i.v.	Urethelial tumor	Quartz lamp + filter, $\lambda > 590$ nm, 360 J/cm²	4–24 h	Hemorrhage and necrosis	[136]
Al-Pc sulfonate	Rats	0.1–100 i.v.	Normal liver	Ar dye lasers, $\lambda = 675$ nm, 100 mW, 3 h postinjection	2–4 days	Necrosis	[31]
Al-Pc sulfonate	Rats	0.2–2.5 i.v.	Fibrosarcoma	Ar dye lasers, $\lambda = 675$ nm, 50 mW, 3–24 h postinjection	0.1–96 h	Necrosis	[153]
Zn-Pc in liposomes	Mice	2.5 i.p.	Fibrosarcoma	Halogen lamp + filter, $\lambda = 600$–690 nm, 300 J/cm², 24 h postinjection	15 h	Necrosis and vacuolization of endothelium	[97]
Al-Pc sulfonate	Mice	10–50 i.p.	Mammary carcinoma	Ar dye lasers, $\lambda = 610$ nm, 135 J/cm², 24 h postinjection	0–16 days	Reduction of tumor size	[52]
Al-Pc sulfonate	Mice	5 i.v.	Fibrosarcoma, lung carcinoma, cell sarcoma, colorectal carcinoma	Ar dye lasers, $\lambda = 675$ nm, 100 J, 24 h postinjection	5 days	Necrosis, reduction of tumor size	[41,42]
Al-Pc sulfonate	Rats	5 i.v.	Normal colon	Ar dye lasers, $\lambda = 675$ nm, 50 J, 1 h postinjection	72 h	Disappearance of mucosal crypt cells, necrosis	[7,8]
Al-Pc sulfonate	Mice	0.5–5 i.v.	Murine glioma	Ar dye lasers, $\lambda = 675$ nm, 1–200 J, 4–48 h, postinjection	24–48 h	Necrosis	[133]
H₂-, Al-, Ce-, Ga- Pc sulfonate	Mice	2.4–9	Mammary tumor	Halogen lamp + filter, $\lambda > 590$ nm, 300 J, 24 h postinjection	72 h–3 mo	Necrosis, tumor regression	[33]
Zn-Pc sulfonate	Mice	11.3 i.v.	Mammary tumor	Halogen lamp + filter, $\lambda > 590$ nm, 107 W/m², 24 h postinjection	3–13 days	Necrosis, tumor regression	[35]

REFERENCES

1. C. D. Abernathey, R. E. Anderson, K. L. Kooistra and E. R. Laws, Jr., *Neurosurgery*, 21 (1987) 468.

2. J. J. Abramson, J. R. Cronin and G. Salama, *Arch. Biochem. Biophys.*, 263 (1988) 245.

3. H. Ali, R. Langlois, J. R. Wagner, N. Brasseur, B. Paquette and J. E. van Lier, *Photochem. Photobiol.*, 47 (1988) 713.

4. Anon., *Aldrichim. Acta*, 20 (1987) 88.

5. H. Auler and G. Banzer, *Z. Krebsforsch.*, 53 (1942) 65.

6. J. Barltrop, B. B. Martin and D. F. Martin, *Microbios*, 37 (1983) 95.

7. H. Barr, C. J. Tralau, A. J. MacRobert, N. Krasner, P. B. Boulos, C. G. Clark and S. G. Bown, *Br. J. Cancer*, 56 (1987) 111.

8. H. Barr, C. J. Tralau, P. B. Boulos, A. J. MacRobert, R. Tilly and S. G. Bown, *Photochem. Photobiol.*, 46 (1987) 795.

9. E. Ben-Hur, R. Kol, Y. M. Heimer, Y. Shiloh, E. Tabor and Y. Becker, *Radiat. Environ. Biophys.*, 20 (1981) 21.

10. E. Ben-Hur, A. Carmichael, P. Riesz and I. Rosenthal, *Int. J. Radiat. Biol.*, (1985) 837.

11. E. Ben-Hur and I. Rosenthal, *Int. J. Radiat. Biol.*, 47 (1985) 145.

12. E. Ben-Hur and I. Rosenthal, *Photochem. Photobiol.*, 42 (1985) 129.

13. E. Ben-Hur and I. Rosenthal, *Radiat. Res.*, 103 (1985) 403.

14. E. Ben-Hur and I. Rosenthal, *Cancer Lett.*, 30 (1986) 321.

15. E. Ben-Hur and I. Rosenthal, *Photochem. Photobiol.*, 43 (1986) 615.

16. E. Ben-Hur and I. Rosenthal, *Lasers Life Sci.*, 1 (1986) 79.

17. E. Ben-Hur, *Photobiochem. Photobiophys. Suppl.* (1987) 407.

18. E. Ben-Hur, T. Fujihara, F. Suzuki and M. M. Elkind, *Photochem. Photobiol.*, 45 (1987) 227.

19. E. Ben-Hur, J. A. Siwecki, H. C. Newman, S. W. Crane and I. Rosenthal, *Cancer Lett.*, 38 (1987) 215.

20. E. Ben-Hur, M. Green, A. Prager, R. Kol and I. Rosenthal, *Photochem. Photobiol.*, 46 (1987) 651.

21. E. Ben-Hur, R. Kol, E. Riklis, R. Marko and I. Rosenthal, *Int. J. Radiat. Biol.*, 51 (1987) 467.

22. E. Ben-Hur, R. Kol, R. Marko, E. Riklis and I. Rosenthal, *Int. J. Radiat. Biol.*, 54 (1988) 21.

23. E. Ben-Hur, E. Heldman, S. W. Crane and I. Rosenthal, *FEBS Lett.*, 236 (1988) 105.

24. E. Ben-Hur, H. Miller, B. Miller and I. Rosenthal, *Ber. Bunsen-Gesellschaft Phys. Chem.*, 93 (1989) 284.

25. E. Ben-Hur, I. Rosenthal and C. C. Leznoff, *J. Photochem. Photobiol.; B: Biology*, 2 (1988) 243.

26. B. H. J. Bielski and E. Saito, *J. Phys. Chem.*, 75 (1971) 2263.

27. N. M. Bigelow and M. A. Perkins, *The Chemistry of Synthetic Dyes and Pigments*, H. A. Lubs, Ed., Reinhold, New York, 1955, pp. 577–605.

28. H. F. Blum, *Photodynamic Action and Diseases Caused by Light*, Reinhold, New York, 1941.

29. R. Bonnett and M. C. Berenbaum, in *Porphyrin Photosensitization*, in *Adv. Exp. Med. Biol.*, Vol. 160, Plenum, New York, 1982, 241–250.

30. S. G. Bown, J. Wieman, C. J. Tralau, C. Collins, P. R. Salmon and C. G. Clark, *Gut*, 26 (1985) A563.

31. S. G. Bown, C. J. Tralau, P. D. Coleridge Smith, D. Akdemir, and T. J. Wieman, *Br. J. Cancer*, 54 (1986) 43.

32. N. Brasseur, H. Ali, D. Autenrieth, R. Langlois and J. E. van Lier, *Photochem. Photobiol.*, 42 (1985) 515.

33. N. Brasseur, H. Ali, R. Langlois, J. R. Wagner, J. Rousseau and J. E. van Lier, *Photochem. Photobiol.*, 45 (1987) 581.

34. N. Brasseur, H. Ali, R. Langlois and J. E. van Lier, *Photochem. Photobiol.*, 46 (1987) 739.

35. N. Brasseur, H. Ali, R. Langlois and J. E. van Lier, *Photochem. Photobiol.*, 47 (1988) 705.

36. M. S. Brudnaia, K. M. Kirpichnikova, I. I. Komissarchuk and S. V. Levin, *Tsitologiya*, 26 (1984) 184.

37. P. A. Cadby, E. Dimitriadis, H. G. Grant, A. D. Ward, I. J. Forbes, *Porphyrin Photosensitization*, in *Adv. Exp. Med. Biol.*, Vol. 160, Plenum Press, New York, 1982, 251–263.

38. P. Cappelle, M. de Backer, O. de Witte, G. Feuillade and G. Lepoutre, *C. R. Acad. Sci. Paris*, 284 (1977) 597.

39. W. S. Chan, R. Svensen, D. Phillips and I. R. Hart, *Br. J. Cancer*, 53 (1986) 255.

40. W. S. Chan, J. F. Marshall, R. Svensen, D. Phillips and I. R. Hart, *Photochem. Photobiol.*, 45 (1987) 757.

41. W. S. Chan, J. F. Marshall and I. R. Hart, *Photochem. Photobiol.*, 46 (1987) 867.

42. W. S. Chan, J. F. Marshall, G. Y. F. Lam and I. R. Hart, *Cancer Res.*, 48 (1988) 3040.

43. R. J. Dam, C. A. Burke and O. H. Griffith, *Biophys. J.*, 14 (1974) 467.

44. J. R. Darwent, P. Douglas, A. Harriman, G. Porter and M. C. Richoux, *Coord. Chem. Rev.*, 44 (1982) 83.

45. J. R. Darwent, I. McCubbin and D. Phillips, *J.C.S., Faraday Trans. II*, 78 (1982) 347.

46. J. R. Darwent, I. McCubbin and G. Porter, *J.C.S., Faraday Trans. II*, 78 (1982) 903.

47. M. G. Debaker, O. Deleplanque, B. van Vlierberge and F. X. Sauvage, *Laser Chem.*, 8 (1988) 1.

48. T. J. Dougherty, K. R. Weishaupt and D. G. Boyle, *Cancer, Principles and Practice of Oncology*, V. T. DeVita Jr., S. Hellman and S. A. Rosenberg, Eds., J. B. Lippincot, Philadelphia, 1982, pp. 1836–1844.

49. T. J. Dougherty, *Photochem. Photobiol.*, 45 (1987) 879.

50. T. J. Dougherty, *Photochem. Photobiol.*, 46 (1987) 569.

51. L. Edwards and M. Gouterman, *J. Mol. Spectr.*, 33 (1970) 293.

52. J. F. Evensen and J. Moan, *Photochem. Photobiol.*, 46 (1987) 859.

53. G. Ferraudi, G. A. Arguello, H. Ali and J. E. van Lier, *Photochem. Photobiol.*, 47 (1988) 657.

54. F. H. G. Figge, G. S. Weiland, and L. O. J. Manganiello, *Proc. Soc. Exp. Biol. Med.*, 68 (1948) 640.

55. P. A. Firey and M. A. J. Rodgers, *Photochem. Photobiol.*, 45 (1987) 535.

56. P. A. Firey, T. W. Jones, G. Jori and M. A. J. Rodgers, *Photochem. Photobiol.*, 48 (1988) 357.

57. C. S. Foote, in *Free Radicals in Biology*, W. A. Pryor, Ed., Vol. II, Academic Press, New York, 1976, pp. 85–129.

58. C. S. Foote, in *Biochemical and Clinical Aspects of Oxygen*, W. S. Caughey, Ed., Academic Press, New York, 1979, pp. 603–626.

59. C. S. Foote, in *Singlet Oxygen*, H. H. Wasserman and R. W. Murray, Eds., Academic Press, New York, 1979, pp. 139–176.

60. N. A. Frigerio, Metal Phthalocyanines. U.S. Patent no. 3,027,391, 1962.

61. G. Geyer, U. Helmke and A. Christner, *Acta Histochem.*, 40 (1971) 80.

62. A. Harriman and M. C. Richoux, *J.C.S., Faraday Trans. II*, 76 (1980) 1618.

63. H. Hayatsu, Y. Ohara, T. Hayatsu and K. Togawa, *Mutat. Res.*, 124 (1983) 1.

64. H. Hayatsu, T. Oka, A. Wakata, Y. Ohara, T. Hayatsu, H. Kobayashi and S. Arimoto, *Mutat. Res.*, 119 (1983) 233.

65. W. M. Heller and S. L. Stoddard, *Stain Technol.*, 61 (1986) 71.

66. B. W. Henderson and A. C. Miller, *Radiat. Res.*, 108 (1986) 196.

67. M. Hercher, W. Chu and D. L. Stockman, *IEEE J. Quant. Electronics*, 4 (1968) 954.

68. K. G. Hofer, and B. Brizzard and M. G. Hofer, *Eur. J. Cancer*, 15 (1979) 1449.

69. D. J. Hunting, B. J. Gowans, N. Brasseur and J. E. van Lier, *Photochem. Photobiol.*, 45 (1987) 769.

70. A. Jesionek and V. H. Tappeiner, *Muench. Med. Wochenschr.*, 47 (1903) 2042.

71. I. Joris, U. DeGirolami, K. Wortham and G. Majno, *Stain Technol.*, 57 (1982) 177.

72. A. Juarranz, J. C. Stockert, M. Canete and A. Villanueva, *Cell. Mol. Biol.*, 31 (1985) 379.

73. A. Juarranz, M. J. Hanzen and J. C. Stockert, *J. Plant Physiol.*, 132 (1988) 557.

74. D. Kessel and T. J. Dougherty, in *Porphyrin Photsensitization*, in *Adv. Exp. Med. Biol.*, Vol. 160, Plenum, New York, 1982.

75. D. Kessel, P. Thomson, B. Muselman and C. K. Chang, *Cancer Res.*, 47 (1987) 4642.

76. D. Kessel, P. Thomson, K. Saatio and K. D. Nantwi, *Photochem. Photobiol.*, 45 (1987) 787.

77. K. M. Kirpichnikova and D. L. Rozental, *Tsitologiya*, 16 (1974) 826.

78. K. M. Kirpichnikova, S. V. Levin and E. V. Shuvalova, *Tsitologiya*, 23 (1981) 297.

79. K. M. Kirpichnikova and E. V. Shuvalova, *Tsitologiya*, 26 (1984) 914.

80. K. M. Kirpichnikova and E. V. Shuvalova, *Tsitologiya*, 30 (1988) 482.

81. K. Kojima, T. Oda, T. Nakahira and S. Iwabuchi, *Kagakubu Kenkyu Hokoku*, 33 (1981) 25.

82. R. Kol, E. Ben-Hur, E. Riklis, R. Marko and I. Rosenthal, *Lasers Med. Sci.*, 1 (1986) 187.

83. J. J. Komissarchik and S. V. Levin, *Nature (London)*, 204 (1964) 203.

84. S. E. Kornguth, T. Kalinke and W. Pietro, *Biochim. Biophys. Acta*, 924 (1987) 19.

85. R. Kugel, A. Svirmickas, J. J. Katz and J. C. Hindman, *Opt. Commun.*, 23 (1977) 189.

86. R. Langlois, H. Ali, N. Brasseur, J. R. Wagner and J. E. van Lier, *Photochem. Photobiol.*, 44 (1986) 117.

87. R. D. Lillie, in *H. J. Conn's Biological Stains*, 8th ed., Williams & Wilkins, Baltimore, 1969, pp. 323–324.

88. R. Lipson, E. Baldes and A. Olsen, *J. Natl. Cancer Inst.*, 26 (1961) 1.

89. P. Maillard, P. Krausz and C. Giannotti, *J. Organomet. Chem.*, 197 (1980) 285.

90. F. Mayer-Betz, *Deut. Arch. Klin. Med.*, 112 (1913) 476.

91. J. F. A. McManus and M. Bailie, *Fed. Proc.*, 22, no. 2, part I (1963) p. 190.

92. J. McVie, R. S. Sinclair and T. G. Truscott, *J.C.S. Faraday Trans. II*, 74 (1978) 1870.

93. K. M. Meek, E. J. Scott and C. Nave, *J. Microsc.*, 139 (1985) 205.

94. K. M. Meek, G. F. Elliott and C. Nave, *Coll. Relat. Res.*, 6 (1986) 203.

95. D. Mendelson, J. Tas, and J. James, *Histochem. J.*, 15 (1983) 1113.

96. P. B. Merkel, R. Nilson and D. R. Kearns, *J. Am. Chem. Soc.*, 94 (1972) 1029.

97. C. Milanesi, R. Biolo, E. Reddi and G. Jori, *Photochem. Photobiol.*, 46 (1987) 675.

98. J. Moan, *Photochem. Photobiol.*, 43 (1986) 681.

99. J. Moan, E. O. Pettersen and T. Christensen, *Br. J. Cancer*, 39 (1979) 398.

100. J. Moan, Q. Peng, J. F. Eversen, K. Berg, A. Western and C. Rimington, *Photochem. Photobiol.*, 46 (1987) 713.

101. A. R. Monahan, J. A. Brado and A. F. DeLuca, *J. Phys. Chem.*, 76 (1972) 1994.

102. F. H. Moser and W. H. Rhodes, in Kirk-Othmer *Encyclopedia of Chemical Technology*, 3rd ed., J. Wiley & Sons, New York, Vol. 17, 1982, pp. 777–787.

103. F. H. Moser and A. L. Thomas, *Phthalocyanine Compounds*, Reinhold, New York, 1963, p. 302.

104. F. H. Moser and A. L. Thomas, *The Phthalocyanines*, CRC Press, Boca Raton, FL, 1983.

105. S. H. Moss, and K. C. Smith, *Photochem. Photobiol.*, 33 (1981) 203.

106. E. Neuzil and J. Ballenger, *Compt. Rend. Soc. Biol.*, 146 (1952) 1108.

107. Office of the Federal Register, U.S. National Archives and Records Administration, Code of Federal Regulation, Food and Drugs, Vol. 21, p. 315. Washington D.C., 1985.

108. T. Ohno, S. Kato, A. Yamada and T. Tanno, *J. Phys. Chem.*, 87 (1983) 775.

109. H. Ohtani, T. Kobayashi, T. Ohno, S. Kato, T. Tanno and A. Yamada, *J. Phys. Chem.*, 88 (1984) 4431.

110. H. Ohtani, T. Kobayashi, T. Tanno, A. Yamada, D. Wohrle and T. Ohno, *Photochem. Photobiol.*, 44 (1986) 125.

111. C. R. Orford and D. L. Gardner, *Histochem. J.*, 17 (1985) 223.

112. B. Paquette, H. Ali, R. Langlois and J. E. van Lier, *Photochem. Photobiol.*, 47 (1988) 215.

113. B. Paquette, H. Ali and J. E. van Lier, *Photochem. Photobiol.*, 47 (1988) 11S.

114. J. A. Parrish, in *Porphyrin Photosensitization*, in *Adv. Exp. Med. Biol.*, Vol. 160, Plenum, New York, 1982, pp. 91–108.

115. Z. J. Petryka, J. C. Bommer and B. F. Burnham, *Photochem. Photobiol.*, 45 (1987) 80S.

116. R. Polany, G. Reinert, G. Hoelzle, A. Pugin and R. Vonderwahl, Ger. Offen. 2,812,261. cf. C.A., 90: 40159y, 1979.

117. H. B. Przywarska, L. Trynda and E. Antonini, *Eur. J. Biochem.*, 52 (1975) 567.

118. H. B. Przywarska and L. Trynda, *Eur. J. Biochem.*, 87 (1978) 569.

119. P. Qian, J. F. Evensen, C. Rimington and J. Moan, *Cancer Lett.*, 38 (1987) 1.

120. O. Raab, *Z. Biol.*, 39 (1900) 524.

121. N. Ramakrishnan, M. E. Clay, L. Xue, H. H. Evans, A. R. Antunez and N. L. Oleinick, *Photochem. Photobiol.*, 48 (1988) 297.

122. E. Reddi, G. L. Castro, R. Biolo and G. Jori, *Br. J. Cancer*, 56 (1987) 597.

123. R. S. Robinson, A. J. Roberts and I. D. Campbell, *Photochem. Photobiol.*, 45 (1987) 231.

124. I. Rosenthal, C. Murali Krishna, P. Riesz and E. Ben-Hur, *Radiat. Res.*, 107 (1986) 136.

125. I. Rosenthal, C. Murali Krishna, G. Yang, T. Kondo and P. Riesz, *FEBS Lett.*, 222 (1987) 75.

126. I. Rosenthal, E. Ben-Hur, S. Greenberg, S. Concepcion-Lam, D. M. Drew and C. C. Leznoff, *Photochem. Photobiol.*, 46 (1987) 959.

127. I. Rosenthal, E. Ben-Hur, S. Greenberg, S. Concepcion-Lam, D. M. Drew and C. C. Leznoff, *Oxygen Radicals in Biology and Medicine*, Eds. M. G. Simic, K. A. Taylor, J. F. Ward, and C. von Sonntag, Plenum, New York, 1988, pp. 467–472.

128. J. Rousseau, D. Autenrieth and J. E. Van Lier, *Int. J. Appl. Radiat. Isot.*, 34 (1983) 571.

129. J. Rousseau, H. Ali, G. Lamoureux, E. Lebel and J. E. Van Lier, *Int. J. Appl. Radiat. Isot.*, 36 (1985) 709.

130. D. L. Rozental and K. M. Kirpichnikova, *Tsitologiya*, 16 (1974) 734.

131. D. L. Rozental and S. V. Levin, *Tsitologiya*, 18 (1976) 1090.

132. T. N. Salthouse, *Nature (London)* 206 (1966) 1277.

133. D. R. Sanderman, R. Bradford, P. Buxton, S. G. Bown and D. G. T. Thomas, *Br. J. Cancer*, 55 (1987) 647.

134. S. Sandor, O. Prelipceanu and I. Checiu, *Morph. Embry.*, 31 (1985) 173.

135. J. E. Scott, *J. Microsc.*, 119 (1980) 373.
136. S. H. Selman, M. Kreimer-Birnbaum, K. Chaudhuri, G. M. Garbo, D. A. Seaman, R. W. Keck, E. Ben-Hur and I. Rosenthal, *J. Urol.*, 136 (1986) 141.
137. M. Shimoni, I. Rosenthal and E. Ben-Hur, *J. Labelled Cpds. Radiopharm.*, 22 (1985) 863.
138. E. V. Shuvalova and K. M. Kirpichnikova, *Tsitologiya*, 30 (1988) 442.
139. B. Simic-Glavaski, *Cell Biophys.*, 7 (1985) 205.
140. C. R. J. Singer, S. G. Bown, D. C. Linch, E. R. Huehns and A. H. Goldstone, *Photochem. Photobiol.*, 46 (1987) 745.
141. C. R. J. Singer, D. C. Linch, S. G. Bown, E. R. Huehns and A. H. Goldstone, *Br. J. Haematol.*, 68 (1988) 417.
142. M. Sonoda, C. Murali Krishna and P. Riesz, *Photochem. Photobiol.*, 46 (1987) 625.
143. P. P. Sorokin and J. R. Lankard, *IBM J. Res. Dev.*, 10 (1966) 162.
144. J. D. Spikes and R. Livingston, *Adv. Radiat. Biol.*, 3 (1969) 29.
145. J. D. Spikes, *Photochem. Photobiol.*, 43 (1986) 691.
146. J. D. Spikes and J. C. Bommer, *Int. J. Radiat. Biol.*, 50 (1986) 41.
147. J. D. Spikes and J. C. Bommer, *Photochem. Photobiol.*, 45 (1987) 79S.
148. R. C. Straight, J. A. Dixon and J. D. Spikes, *Lasers Surg. Med.*, 5 (1985) 139.
149. T. Tanno, D. Wohrle, M. Kaneko and A. Yamada, *Ber. Buns. Phys. Chem.*, 84 (1980) 1032.
150. H. V. Tappeiner and A. Jodlbauer, *Deut. Arch. Klin. Med.*, 80 (1904) 427.
151. J. Tas, D. Mendelson and C. J. Noorden, *Histochem. J.*, 15 (1983) 801.
152. L. W. Tice and R. J. Barrnett, *J. Cell Biol.*, 25 (1965) 23.
153. C. J. Tralau, A. J. MacRobert, P. D. Coleridge-Smith, H. Barr and S. G. Bown, *Br. J. Cancer*, 55 (1987) 389.
154. C. J. Tralau, H. Barr, D. R. Sanderman, T. Barton, M. R. Lewin and S. G. Bown, *Photochem. Photobiol.*, 46 (1987) 777.
155. L. Trynda, H. P. Boniecka and T. Kosciukiewicz, *Inorg. Chim. Acta*, 135 (1987) 55.
156. K. Uchida, M. Soma, S. Naito, T. Onishi and K. Tamaru, *Chem. Lett.*, 5 (1978) 471.
157. G. Valduga, S. Nonell, E. Reddi, G. Jori and S. E. Braslavsky, *Photochem. Photobiol.*, 48 (1988) 1.
158. T. H. Van Kuppevelt, F. P. Cremers, J. G. Domen and C. M. Kuyper, *Histochem. J.*, 16 (1984) 671.
159. J. E. Van Lier, H. Ali and J. Rousseau, in *Porphyrin Localization and Treatment of Tumors*, D. R. Doiron and C. J. Gomer, Eds., Liss, New York, 1984, pp. 315–319.
160. J. R. Wagner, H. Ali, R. Langlois, N. Brasseur and J. E. van Lier, *Photochem. Photobiol.*, 45 (1987) 587.
161. F. S. Waldrop and H. Puchtle, *Arch. Pathol.*, 99 (1975) 529.
162. H. Weintraub, A. Abramovici, A. Altman, E. Ben-Hur and I. Rosenthal, *Lasers Life Sci.*, 2 (1988) 185.
163. B. C. Wilson and W. P. Jeeves, in *Photomedicine*, E. Ben-Hur and I. Rosenthal, Eds., Vol. II, CRC, Boca Raton, FL, 1987, pp. 127–177.
164. F. R. Wrenn, Jr., M. L. Good and P. Handler, *Science*, 113 (1951) 525.
165. S. K. Wu, H. C. Zhang, G. Z. Cui, D. N. Xu, and H. J. Xu, *Acta Chim. Sinica*, 43 (1985) 10.
166. M. M. Zuk, K. Tyczkowska, E. Ben-Hur, H. C. Newman, I. Rosenthal and S. W. Crane, *J. Chrom., Biomed. Appl.*, 433 (1988) 367.

* Contribution from the Agricultural Research Organization, The Volcani Center, Bet Dagan, Israel. No. 2257-E, 1987 series.

Index